Polarization Engineering for LCD Projection

Wiley-SID Series in Display Technology

Series Editor:
Anthony C. Lowe

Display Systems:
Design and Applications
Lindsay W. Macdonald and **Anthony C. Lowe (Eds)**

Electronic Display Measurement:
Concepts, Techniques and Instrumentation
Peter A. Keller

Projection Displays
Edward H. Stupp and **Matthew S. Brennesholz**

Liquid Crystal Displays:
Addressing Schemes and Electro-Optical Effects
Ernst Lueder

Reflective Liquid Crystal Displays
Shin-Tson Wu and **Deng-Ke Yang**

Colour Engineering:
Achieving Device Independent Colour
Phil Green and **Lindsay MacDonald (Eds)**

Display Interfaces:
Fundamentals and Standards
Robert L. Myers

Digital Image Display:
Algorithms and Implementation
Gheorghe Berbecel

Flexible Flat Panel Displays
Gregory Crawford (Ed.)

Polarization Engineering for LCD Projection
Michael G. Robinson, Jianmin Chen, and **Gary D. Sharp**

Polarization Engineering for LCD Projection

Michael G. Robinson, Jianmin Chen, and Gary D. Sharp
Colorlink Inc., USA

John Wiley & Sons, Ltd

Copyright © 2005   John Wiley & Sons Ltd, The Atrium, Southern Gate,
Chichester, West Sussex PO19 8SQ, England

Telephone   (+44) 1243 779777

Email (for orders and customer service enquiries): cs-books@wiley.co.uk
Visit our Home Page on www.wiley.com

All Rights Reserved. No part of this publication may be reproduced, stored in a retrieval system or transmitted in any form or by any means, electronic, mechanical, photocopying, recording, scanning or otherwise, except under the terms of the Copyright, Designs and Patents Act 1988 or under the terms of a licence issued by the Copyright Licensing Agency Ltd, 90 Tottenham Court Road, London W1T 4LP, UK, without the permission in writing of the Publisher. Requests to the Publisher should be addressed to the Permissions Department, John Wiley & Sons Ltd, The Atrium, Southern Gate, Chichester, West Sussex PO19 8SQ, England, or emailed to permreq@wiley.co.uk, or faxed to (+44) 1243 770620.

This publication is designed to provide accurate and authoritative information in regard to the subject matter covered. It is sold on the understanding that the Publisher is not engaged in rendering professional services. If professional advice or other expert assistance is required, the services of a competent professional should be sought.

Other Wiley Editorial Offices

John Wiley & Sons Inc., 111, River Street, Hoboken, NJ 07030, USA

Jossey-Bass, 989 Market Street, San Francisco, CA 94103-1741, USA

Wiley-VCH Verlag GmbH, Boschstr. 12, D-69469 Weinheim, Germany

John Wiley & Sons Australia Ltd, 33 Park Road, Milton, Queensland 4064, Australia

John Wiley & Sons (Asia) Pte Ltd, 2 Clementi Loop #02-01, Jin Xing Distripark, Singapore 129809

John Wiley & Sons Canada Ltd, 22 Worcester Road, Etobicoke, Ontario, Canada M9W 1L1

Wiley also publishes its books in a variety of electronic formats. Some content that appears in print may not be available in electronic books.

Library of Congress Cataloging-in-Publication Data

Robinson, Michael G.
  Polarization engineering for LCD projection / Michael G. Robinson, Jianmin Chen & Gary D. Sharp.
    p.   cm.
  Includes bibliographical references and index.
  ISBN-13 978-0-470-87105-8 (cloth : alk. paper)
  ISBN-10 0-470-87105-9 (cloth : alk. paper)
  1. Liquid crystal displays. 2. Polarization (Light) 3. Projectors.
  I. Chen, Jianmin, 1963– II. Sharp, Gary D. III. Title.
  TK7872.L56R76 2005
  621.3815′ 422—dc22
                                                                                2005004148

British Library Cataloguing in Publication Data

A catalogue record for this book is available from the British Library

ISBN-13 978-0-470-87105-8 (HB)
ISBN-10 0-470-87105-9 (HB)

Typeset in 10/12pt Times by Integra Software Services Pvt. Ltd, Pondicherry, India
Printed and bound in Great Britain by Antony Rowe Ltd, Chippenham, Wiltshire
This book is printed on acid-free paper responsibly manufactured from sustainable forestry
in which at least two trees are planted for each one used for paper production.

We dedicate this book to our wives Eleri, Jinjin, and Patty,
and to our children Nia, Hannah, Megan, Catherine, Benjamin,
Ian, Kai, and Malia

Contents

Series Editor's Foreword	XIII
Preface	XV

1 Introduction — 1
- 1.1 The Case for Projection — 1
- 1.2 History and Projection Technology Overview — 2
  - 1.2.1 Cinema Film — 2
  - 1.2.2 CRT-based Projection Systems — 3
  - 1.2.3 Schlieren Optics-based Projector — 5
  - 1.2.4 Microdisplay-based Projection Systems — 7
  - 1.2.5 Other Projection Technologies — 16
- 1.3 Scope of the Book — 17

2 Liquid Crystal Projection System Basics — 21
- 2.1 Introduction — 21
- 2.2 Brightness and Color Sensitivity of the Human Eye — 22
  - 2.2.1 Brightness — 22
  - 2.2.2 Brightness Uniformity — 24
  - 2.2.3 Color — 24
  - 2.2.4 White — 26
  - 2.2.5 Color Distinction and Just Noticeable Differences (JNDs) — 28
  - 2.2.6 Contrast — 28
  - 2.2.7 Size, Resolution, Registration, and Distortion — 29
  - 2.2.8 Electronic and Panel-related Metrics — 30
- 2.3 Photometric Measurement — 30
- 2.4 Summary of What Constitutes a "Good" RPTV Display in the Current Marketplace — 30
- 2.5 System Engineering — 30
  - 2.5.1 Rear-projection Screens — 31
  - 2.5.2 Folding Mirrors — 34

		2.5.3	Projection Optics	35
		2.5.4	Color Management and Modulation Subsystem	37
		2.5.5	Illumination System	37
		2.5.6	Light Source	40
	2.6	Étendue Considerations		41

3 Polarization Basics 47

- 3.1 Introduction 47
- 3.2 Electromagnetic Wave Propagation 47
  - 3.2.1 Polarization of Monochromatic Waves 48
  - 3.2.2 Complex Number Representation 50
  - 3.2.3 Jones' Vector Representation 51
  - 3.2.4 Stokes' Parameters 53
  - 3.2.5 Poincaré Sphere 54
- 3.3 Interaction with Media 57
  - 3.3.1 Reflection and Refraction of Plane Waves 57
  - 3.3.2 Matrix Formulation for Isotropic Layered Media 60
  - 3.3.3 Matrix Formulation for Anisotropic Layered Media 61
- 3.4 Index Ellipsoid Visualization 70
- 3.5 Modeling Techniques 72

4 System Components 77

- 4.1 Introduction 77
- 4.2 Retarders 77
- 4.3 Polarizers 83
  - 4.3.1 Absorptive Polarizer 83
  - 4.3.2 Reflective Polarizers 85
- 4.4 Interference Filters 88
  - 4.4.1 Anti-reflection Coatings 88
  - 4.4.2 Quarter-wave Stack 89
  - 4.4.3 Normal Incidence Dichroic Filters 90
  - 4.4.4 Dichroic Beam Splitters 92
- 4.5 Polarizing Beam Splitters (PBSs) 94
  - 4.5.1 Dichroic Cube PBS 94
  - 4.5.2 Multilayer Birefringent Cube PBS (MBC PBS) 97
  - 4.5.3 Wire Grid Plate PBS 98
- 4.6 Other Components 100
  - 4.6.1 Mirrors 100
  - 4.6.2 Light-pipe 100
  - 4.6.3 Substrates 101

5 Liquid Crystal Displays (LCDs) 105

- 5.1 Description and Brief History 105
- 5.2 Anisotropic Properties of Liquid Crystals 109
- 5.3 Frank Free Energy and Electromagnetic Field Contribution to Free Energy 110
- 5.4 Alignment Layer and LC Pretilt Angle 111
- 5.5 Rotational Viscosity 113
- 5.6 Electro-optical Effect of LCs 113
- 5.7 LC Modes for Projection 114
  - 5.7.1 Electrically Controlled Birefringence (ECB) Mode 114
  - 5.7.2 90° TN and VA 90° TN Mode 117

		5.7.3	45° Reflective TN Mode	120
	5.8	5.7.4	63.6° Mixed TN (MTN) Mode	121
		5.7.5	90° MTN Mode	123
	5.8	FOV of LCDs		124

6 Retarder Stack Filters — 129

- 6.1 Introduction — 129
- 6.2 Principle and Background of RSFs — 130
  - 6.2.1 Single Stage Polarization Interference — 130
  - 6.2.2 Multilayer Polarization Interference — 132
- 6.3 RSFs in LC Projection Systems — 134
  - 6.3.1 Optical Filters — 134
  - 6.3.2 Color Splitters/Combiners — 136
- 6.4 Design of RSFs — 137
  - 6.4.1 Impulse Response of a Birefringent Network — 137
  - 6.4.2 Design Methodology — 140
  - 6.4.3 Impulse Response to RSF Angular Profile Mapping — 140
- 6.5 Properties of Retarder Stacks — 143
  - 6.5.1 Unitary Jones' Matrix Representation — 143
  - 6.5.2 Properties of Symmetric RSF Designs — 143
  - 6.5.3 General Properties of Symmetric RSF Designs — 145

7 System Contrast — 153

- 7.1 Introduction — 153
- 7.2 On-axis Contrast — 154
  - 7.2.1 Head-on Contrast of LC Mode — 154
  - 7.2.2 Normal Incidence Pre- and Post-polarizers — 157
- 7.3 Off-axis Effects — 159
  - 7.3.1 Homeotropic Liquid Crystals — 159
  - 7.3.2 Off-axis Property of Sheet Polarizers — 160
  - 7.3.3 Geometrical PBS Compensation — 166
- 7.4 PBS/LCOS Compensation — 175
  - 7.4.1 VA LCOS Mode — 176
  - 7.4.2 General LCOS Mode — 178
  - 7.4.3 Influence of the Reflections from Interfaces on System Contrast — 182
- 7.5 ANSI Contrast Enhancement — 186
- 7.6 Skew Ray Compensated Retarder Stack Filters — 187
- 7.7 Alternative Projection Systems — 191
  - 7.7.1 Off-telecentric Wire Grid PBS System — 191
  - 7.7.2 Off-axis System — 192
- 7.8 Overall System Contrast — 194

8 Color Management — 197

- 8.1 Introduction — 197
- 8.2 System Color Band Determination — 197
- 8.3 Color Management in Projection Systems — 201
  - 8.3.1 Spatial Color Separation and Recombination — 202
  - 8.3.2 Temporal Color Separation — 209

9 Transmissive Three-panel Projection System — 217

- 9.1 Introduction — 217
- 9.2 Brief System Description — 217
- 9.3 System Throughput — 219
  - 9.3.1 Lamp Flux Output, Φ — 219
  - 9.3.2 Illumination Efficiency, η_{ill} — 219
  - 9.3.3 Color Management System Efficiency, η_{cm} — 220
  - 9.3.4 Color Correction Efficiency, η_{cc} — 222
  - 9.3.5 Modulation System Efficiency, η_m — 223
  - 9.3.6 Imaging System Efficiency, η_{im} — 224
  - 9.3.7 Total System Lumen Output, Φ_{out} — 224
- 9.4 Contrast — 225
  - 9.4.1 Negative c-plate Compensation — 227
  - 9.4.2 Splayed Negative Birefringent Film Compensation Scheme — 227
  - 9.4.3 Negative o-plate Compensation — 230
  - 9.4.4 Positive o-plate Compensation Scheme — 230
  - 9.4.5 Liquid Crystal Polymer (LCP) Compensation Scheme — 233

10 Three-panel Reflective Systems — 237

- 10.1 Introduction — 237
- 10.2 3 × PBS/X-cube System — 238
  - 10.2.1 Description of Basic Operation — 238
  - 10.2.2 Comparison to Transmissive System — 239
  - 10.2.3 Brightness — 240
  - 10.2.4 Contrast — 240
  - 10.2.5 Systems Upgrades — 242
  - 10.2.6 Alternative PBS Solutions — 242
- 10.3 Polarization Color Filter Systems — 247
  - 10.3.1 The CQ3 Three-PBS Architecture — 248
  - 10.3.2 System Analysis — 250
- 10.4 Three-panel LCOS System Comparison — 255

11 Single and Dual Panel LC Projection Systems — 257

- 11.1 Introduction — 257
- 11.2 Generic Color Sequential Single Panel Reflective LC System — 257
  - 11.2.1 System Description — 257
  - 11.2.2 Single Panel LCOS System Throughput — 258
  - 11.2.3 System Contrast — 261
  - 11.2.4 Temporal System Issues — 262
- 11.3 Example Single Panel Color Sequential Systems — 267
  - 11.3.1 Scrolling Color System — 267
  - 11.3.2 Field Sequential Single Panel System — 267
- 11.4 Two-panel Systems — 268
  - 11.4.1 White Color Balance — 269
  - 11.4.2 Color Break-up — 269
  - 11.4.3 Two-panel Architectures — 270

11.5	Commercialized Single Panel Projection Systems Based on Spatial Color Separation	273
	11.5.1 Angular Color Beam Separation with Panel-based Microlens Arrays	273
	11.5.2 Holographic Micro-optic Color Separation	273
	11.5.3 Flat-panel LCD Projection	274

Appendix A 277

Index 281

Series Editor's Foreword

Projection TVs and data displays are producing increasingly beautiful images. Precisely how these remarkable improvements in performance have been achieved is a mystery to many. I would imagine that this statement holds true within the technical community and even to some extent within that part of it which has a working knowledge of liquid crystal display effects, TFTs and single crystal silicon transistor backplane technology.

This latest offering from the Wiley-SID Series in Display Technology demystifies the subject. It explains polarization engineering, an extremely complex topic, with elegance and clarity from basic theory and fundamentals to worked examples of commercially available systems.

All of the bulk optical materials and, in systems with three LCDs, of the order of 21 optical surfaces per colour must be optimised for performance in terms of polarisation and reflectivity. The challenge of designing projection systems which simultaneously deliver high luminous efficiency, high contrast and low volume at acceptable cost are formidable. Nothing but a comprehensive and holistic treatment of every element of a projection system is sufficient to achieve the levels of performance now being realised and that is exactly what this book provides. An entire chapter is devoted to an important optical element only recently developed. This is the essentially lossless retarder stack filter, which can rotate the plane of polarisation of certain wavelength bands and leave others unchanged. A detailed discussion is also presented of on- and off-normal compensation of LCD cells, which will be of value to engineers involved in the development of projection and direct view displays alike.

To return to my first point, beauty is apparent not just in LCD projection displays themselves and in the images they produce but also in the manner of their optical engineering. This book will be of great interest to display engineers in general and of great value to those actively involved in the field.

Anthony Lowe
Series Editor
Braishfield, England

Preface

Large-screen TVs are today common in many households, with a vast choice of moderately priced (greater than 42″ diagonal) TVs on display at most retail outlets. The transformation has been fuelled by increasing demand for digital HDTV and high-resolution Internet content. In response many display technologies have been introduced onto the market such as direct view LCD, plasma, and projection systems based on digital light processing (DLP) and liquid crystal microdisplays. The technologies compete in both price and performance with the once dominant cathode ray tube (CRT). However, these technologies do not have the same screen size limitation as that of the CRT, which is restricted to screen sizes less than ∼40″ diagonal. Above 40″ a CRT TV requires more than three people to lift it due primarily to the thickness of the glass. Alternative large-area TVs have been available in the form of CRT projection systems, though their modest performance (e.g., brightness, resolution, color) was never a serious threat below 30″. The new technologies offer superior performance to the CRT in almost all categories.

Of the emerging large-area display technologies, projection offers the fewest technical hurdles to increasing screen size, as it relies on the projection of postage stamp-sized panels with magnifications limited solely by source brightness. The dominant microdisplay panel technology has been transmissive high-temperature polysilicon (HTPS), such as that used for example in the Sony Grand WEGA RPTV. More recently, projectors based on DLP technology have increased market share. These systems rely on time-sequential color and gray scale, using a single panel containing an array of micro-electromechanical mirrors. Even more recently, JVC introduced RPTVs using liquid crystal on silicon (LCOS) panels. Using conventional silicon VLSI fabrication techniques, this technology is seen as the best at coping with demands of increasing resolution. It remains to be seen whether projection systems will dominate the big-screen display market, but there is little doubt that it will populate significantly the higher end, where the largest and highest resolution displays reside.

To date, display contrast has been the toughest performance metric for all projection technologies. Contrast is the ratio of white to black projected intensity. Unlike a CRT display, where contrast is relatively easy to control, a projection display relies on modulating intense illumination. In DLP systems, contrast was initially made acceptable by deflecting the tilting mirrors about the mirror diagonal, thus significantly reducing edge scattering effects. Recently, increased tilting angles and more effective light shielding beneath the mirrors have allowed systems to achieve >1500:1. At this level, DLP systems improve significantly upon their CRT predecessors. This throws down the gauntlet for LC projection systems, requiring a more thorough control of polarization to increase contrast. Unfortunately, preserving polarization is no easy task when dealing with complex birefringent materials such as LCs, and is further complicated by the effects of seemingly innocuous components such as mirrors, windows, lenses, AR coatings, and polarizing beam splitters (PBSs). It is this key aspect that primarily concerns this book, since the subtle polarization effects in LC projection systems must be clearly understood and controlled in order to achieve acceptable performance.

The central pillar of this book is thus the system contrast chapter (Chapter 7). Within this chapter all major contributions to three-dimensional polarization mixing are addressed within the context of LC projection systems. In order to understand the phenomena and their compensation, the book attempts in the preceding chapters to familiarize the reader with general projection system issues and polarization effects of key components. In general, polarization requires some understanding of its mathematical treatment, the essentials of which are covered in Chapter 3. Its first sections introduce the various analytical descriptions of polarization and derive solutions that describe basic polarization manipulation. Later sections are geared toward understanding the basis of the sophisticated mathematical simulation tools used to derive many of the results presented in later chapters. The polarization properties of passive components and active LC films are then considered in separate chapters, before devoting a complete chapter to the relatively new retarder stack filter (RSF) component. RSFs have become key components in many commercially successful LCOS projection systems although their working principle and design are relatively new to most optical engineers. This chapter gives insight into the manipulation of polarization by RSFs, covering their manufacture, design, and subtle physical and performance symmetries. Such symmetries can enhance the performance of projection systems since they can be made compatible with geometrical symmetries between, for example, PBSs. The subsequent chapters introduce projection systems that use polarization engineering techniques. Since color management and input/output beam separation are related, a separate chapter is devoted to the different approaches taken to date. Although not demanding of polarization engineering in most cases, the workable solutions require knowledge of component polarization performance. The last three chapters then consider specific projection engine architectures, covering first the transmissive three-panel system, and completing the book with two chapters devoted to three-, two-, and single-panel systems.

The book is geared primarily towards engineers actively working in the area of LC projection displays. However, its general treatment of projection systems should allow those interested in the field to become acquainted with the current state of the art. It covers, for example, what constitutes a "good" display and describes aspects of projection systems relating to color and throughput that only have slight relevance to polarization engineering. This material, although accessible in other texts, is presented in a concise form. Despite its projection system emphasis, the book's content should appeal to those interested in more general polarization engineering issues. It covers aspects of components and their integration that span any system based on polarization control.

Throughout the book there is a considerable amount of material that has either not been published before or would not be familiar to most of the target audience. It represents a detailed description of the concentrated research and development that ColorLink Inc.

has undertaken in the field of projection systems over the last 10 years. It is rare that three engineers with expertise in the complementary areas of projection systems, LCs, and polarization components have had the opportunity to work closely together on a single research topic for such a long period of time. Although a startup, ColorLink has participated in many high-profile projection programs (most no longer in existence) and weathered many ups and downs in this industry. Throughout, we have been constantly challenged by our customers and the ever changing demands of the industry. It was our goal to share the results of this work with those interested in display and polarization. We would particularly like to thank ColorLink's president, Leo Bannon, for all his support.

The work is of course never finished, and we continue our effort to understand the subtle polarization effects that limit the performance of projection systems. However, the current success of commercial projection products, based on the concepts covered in this book, make it opportune to publish at this time. We hope you enjoy it!

MGR, JC, GDS
January 2005

1

Introduction

1.1 The Case for Projection

Daily life increasingly relies on electronic displays. Indeed, the information age is unimaginable without them. An electronic display is a device or system that converts an electronic signal representing video, graphic, or text information to a viewable image of this information. A display can be virtual, direct view, or projection. With a virtual display, there is no real image in space and the image information is brought to a focus only on the retina. Such displays are limited to one observer only. Direct-view displays are most familiar to the average person. The most common direct-view displays are cathode-ray tubes (CRTs) in televisions (TVs) and computer monitors. Other direct-view technologies, such as plasma displays, organic light emitting diode displays (OLEDs), and liquid crystal displays (LCDs), are starting to challenge the dominant position of the CRT in display applications. Active matrix LCD (AMLCD) computer monitors outshipped CRTs for the first time in 2003. These displays are all capable of high resolution and satisfactory luminance. However, it is difficult and expensive to make a direct-view display large enough to accommodate several viewers simultaneously.

The human eye has an angular resolution of approximately 1 minute of arc. Assuming an image is displayed at a distance of 2 meters from the viewer, the size of the display must be as large as $\sim 70''$ to fully resolve the high-definition television (HDTV) content, which is shown in 1920×1080, ~ 0.6 mm, full-color pixels (see Figure 1.1). It is certainly challenging, and expensive, to make a direct-view display of this size at present.

Projection displays utilize an optical imaging system to magnify a small picture created either by conventional direct-view technologies, such as CRTs, or by modulating the light from an illumination system with a device called a light valve or panel. A projection display can be operated either in front-projection mode, where the viewer and projector are on the same side of the screen, or in rear-projection (RP) mode, where the viewer and projector are

---

Polarization Engineering for LCD Projection M. G. Robinson, J. Chen and G. D. Sharp
© 2005 John Wiley & Sons, Ltd

2 INTRODUCTION

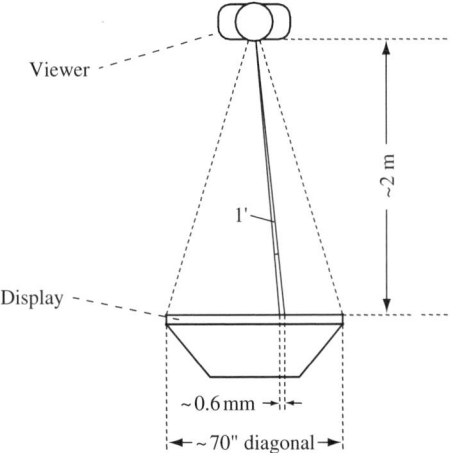

Figure 1.1 Viewing geometry for direct-view HDTV

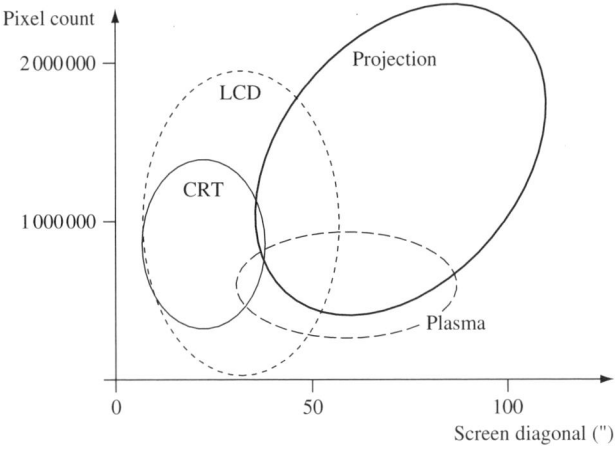

Figure 1.2 Display technologies as a function of screen size and resolution

on opposite sides of the screen. At the present time, projection systems offer the only economical solution to large, high-resolution displays. Figure 1.2 shows where projection displays figure in the display market with regard to resolution and screen size [Stupp E. H., 1999, p. 4].

1.2 History and Projection Technology Overview

1.2.1 Cinema Film

The history of projection systems begins with cinema movie projectors, which are the earliest and most familiar projection systems to the public. This type of projector is able to deliver a large, high-resolution image viewable by a large audience. The first machine patented in

HISTORY AND PROJECTION TECHNOLOGY OVERVIEW 3

the United States, that showed animated pictures or movies, was a device called the "wheel of life" patented by William Lincoln in 1867 [http://inventors.about.com/library/inventors/blmotionpictures.htm]. However, the Frenchman Louis Lumiere and his brother Auguste are often credited with inventing the first motion picture camera and projector in 1895. They presented the first projected moving photographic pictures to a paying audience. The first commercially successful projector was invented by Thomas Edison in 1896. The advantage of the film projector is that it displays very high-resolution images, which no modern projection technologies have surpassed as yet. Other types of film projectors include slide projectors and overhead projectors commonly used in classrooms.

The system layout of a typical cinema projection system is shown in Figure 1.3. It consists of an illuminator (lamp), film rotation drums, a sync shutter, and a projection lens. The film frame rate is 24 frames/sec, but is illuminated through a sync shutter operated at double the frequency to avoid flicker. A 16 mm diagonal format is the typical film size used for motion pictures. Although the projection system is relatively simple and cheap, the film is not in digital format and must be physically copied for individual media content. It is therefore expensive to distribute the media and is clearly incompatible with the modern digital information age.

1.2.2 CRT-based Projection Systems

The most common projection systems are CRT based, as they dominate the middle and low-end rear-projection system market [Wolf M., 1937]. Three monochrome tubes, each optimized for luminance and beam width of a specific primary color, are imaged onto the screen. Since the path of the electron beam is relatively short, the beam spot size can be better controlled, minimizing any smearing effects. These features are required in projection systems to produce good resolution and chromaticity with high brightness. There are two configurations for CRT projectors, using either three lenses or a single lens as shown in

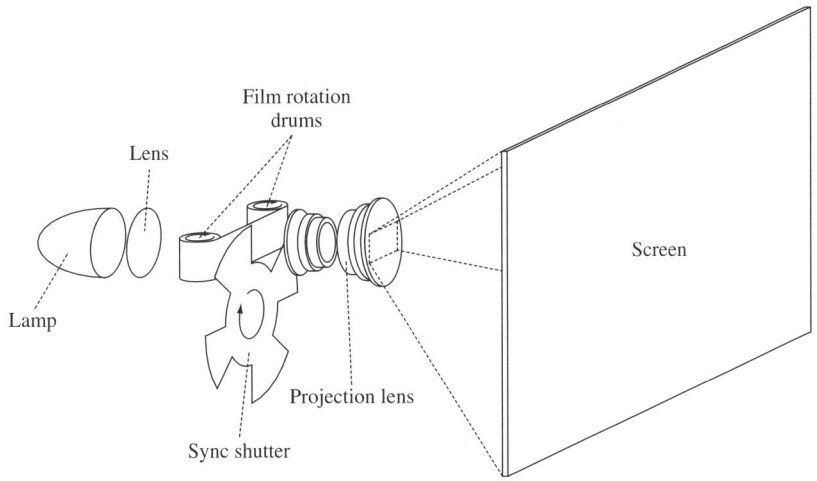

Figure 1.3 System layout of a typical cinema projection system

4 INTRODUCTION

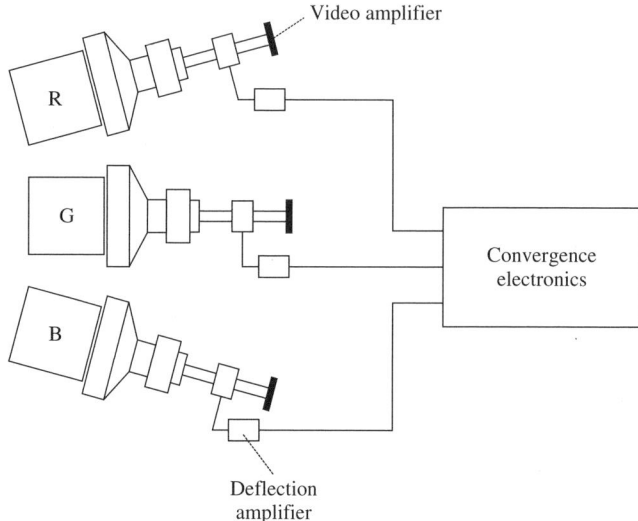

Figure 1.4 Three-lens CRT light engine. The red and blue lenses are tilted to partially correct geometrical errors

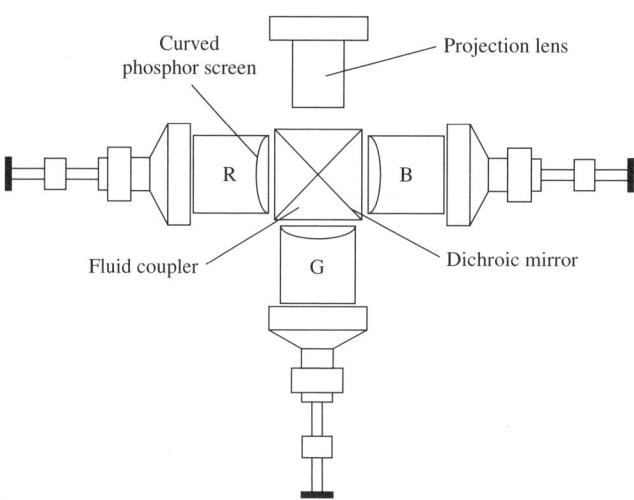

Figure 1.5 Single lens CRT light engine

Figure 1.4 and Figure 1.5 respectively. The optical coupling between the tube and projection lens is enhanced by the cooling fluid placed between the tube face front and the first optical surface of the lens. Furthermore, the tube faceplate is usually curved to improve the light collection by the lens [Stupp E. H., 1999, p. 202; Malang A. W., 1989].

Convergence of CRT projection systems is a major challenge. For good image quality, it is desirable to converge the images from three tubes to within about a half pixel. Since red and blue channels are in an off-axis arrangement in the three-lens CRT projection system, the off-axis tubes will generate a trapezoidal image (keystone distortion) [Hockenbrock R., 1982]. The angular dependence of the Fresnel reflection coefficients can cause color non-uniformity, which can be reduced by tilting the red and blue lenses. Suitable deflection circuits must also be implemented to correct for these errors.

Even though single lens CRT projectors are free from convergence errors arising from trapezoidal distortion, there are many other sources degrading convergence of CRT projectors due to optical, electrical, and magnetic issues. [George J. G., 1995]. Issues specific to single lens systems include the long back focal length (bfl) due to the dichroic combiner, and the relatively high $f_{/\#}$ required to avoid color non-uniformity stemming from the angular sensitive dichroic filter. High $f_{/\#}$ systems are typically low in brightness.

CRT projection cabinets are usually bulky. There is a trade-off between the cabinet size and image quality. A shorter focal length lens decreases the optical throw distance and allows a thin cabinet. However, it increases offset angles between tubes, which results in poor image quality due to increased electron beam deflection.

1.2.3 Schlieren Optics-based Projector

Among the earliest optical configurations employed in electronic projection systems was the schlieren optics-based projector. It was originally developed for the study of defects in lenses using dark-field optics [Fischer, F., 1940; Glenn W. E., 1958; Glenn W. E., 1979; Johannes H., 1979]. Diffracted beams can be either stopped or projected onto a screen depending on whether dark-field or bright-field optics are used. The higher contrast dark-field system is shown in Figure 1.6.

The projection panels in this system are diffractive light valves specifically based on phase gratings, which produce angular separation between the modulated and unmodulated beams. Systems can operate in either reflective or transmissive mode. The phase profile for the light valve is shown in Figure 1.7. The phase profile is flat in its non-diffracting state while imparting a spatially varying phase profile in its diffracting state. The maximum diffraction efficiency of a typical square phase profile can be achieved with $(\pi, 0)$ phase modulation.

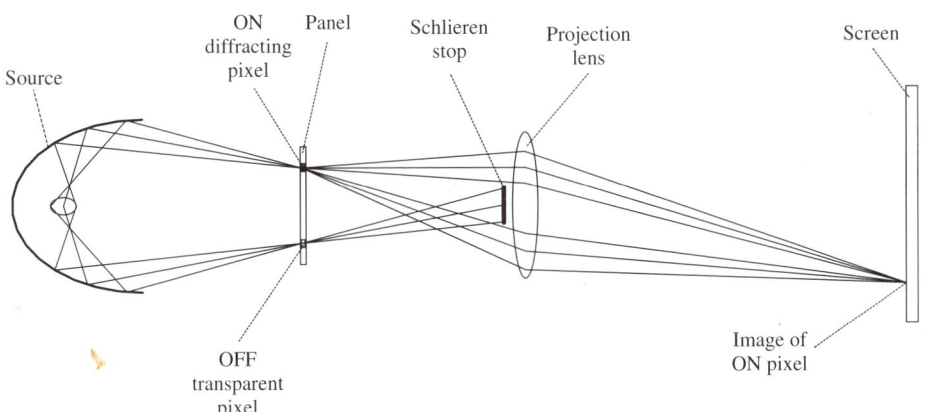

Figure 1.6 The principle of a schlieren projection system based on dark-field optics

6 INTRODUCTION

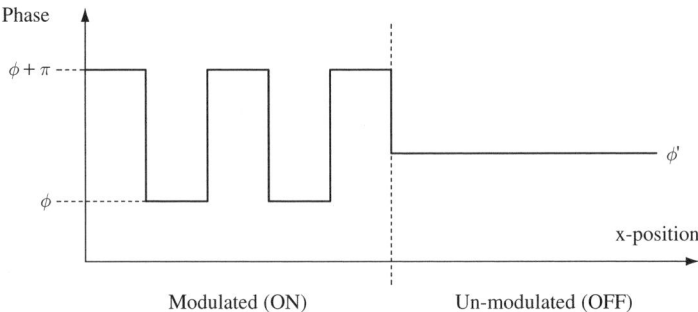

Figure 1.7 Phase profile of a diffraction grating light valve used in schlieren projection systems

Two high-output systems have been made using dark-field schlieren optics: the Ediophor® [Johannes H., 1979; William S. A., 1997] and the Talaria® [Glenn W. E., 1958] systems. Both are based on electron beam-written diffractive gratings in oil films. The Eidophor® is operated in reflective mode, while the Talaria® is operated in transmission. Eidophor® projectors use three separate modulators for the three primary colors. They are among the highest luminance projectors ever made (~9500 lumens). Talaria® is no longer in production and can be operated with one, two, or three light valves.

A Schlieren projection system based on LC diffractive light valves was proposed by Bos *et al.* [Bos P. J., 1995]. It consists of a periodic structure with alternating left- and right-handed twisted nematic (TN) LC strips (Figure 1.8). Such a structure is realized by patterning LC alignment. When the retardance (Δnd) of the LC cell satisfies the first minimum condition (see Chapter 5), the output beams from adjacent strips have the desired π phase difference. Furthermore, an advantage of this light valve is polarization insensitivity, as no polarizer is required. The gray scale can be well controlled by the applied voltage. Other diffraction structures based on LC light valves have been subsequently proposed [Yang K. H., 1998; Wang B., 2002], many of which can be operated in reflective mode.

In principle, dark-field schlieren systems can deliver high-contrast images. However, the demanding requirement of defect-free optical components results in expensive optical systems that are difficult to manufacture. Disclination lines at the boundary between two adjacent strips in LC diffractive light valves also degrade system contrast and reduce light throughput.

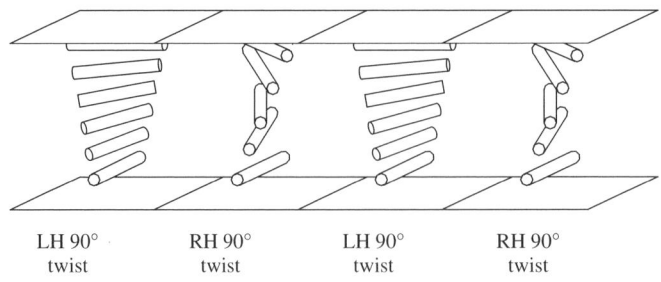

Figure 1.8 LC diffractive light valve based on periodic alternative right/left-handed TN stripes

1.2.4 Microdisplay-based Projection Systems

Microdisplay-based projection is quickly overtaking CRT-based projection in the large-screen projection TV market. In the near future, microdisplay projection will displace CRT projection due to its superb image resolution, and brightness. There are three major microdisplay technologies, based on digital micromirror devices (DMDs), high-temperature polysilicon (HTPS), and liquid crystal on silicon (LCOS) technologies. Each technology has unique properties that influence the quality of the image.

1.2.4.1 Digital Micromirror Device (DMD)

The DMD was developed by Texas Instruments Inc. (TI) [Hornbeck L. J., 1983; Sampsell J. B., 1994; Hornbeck L. J. 1996] and is based on micro-electromechanical systems (MEMS) technology. Its fabrication is compatible with integrated circuit (IC) manufacturing. It consists of an array of aluminum mirrors (one per pixel), which are suspended above individual electrically addressed SRAM (Static Random Access Memory) cells by two thin metal torsion hinges attached to posts. A small tilting yoke, address electrodes, torsion hinges, and landing electrodes are created by successive photolithographic mask steps. A square mirror is fabricated that is integral to the post formed by each via. The sacrificial layers are then removed simultaneously. Figure 1.9(a) shows a photomicrograph of a DMD mirror array and its detailed structure is illustrated in Figure 1.9(b).

The working principle of the DMD is shown in Figure 1.10. Electrostatic forces are created between the mirrors and address electrodes connected to the SRAM nodes, at which positive and negative voltages (representing 1 and 0) are applied. These forces twist the mirrors one way or the other about an axis through the torsion hinges until the yoke hits a

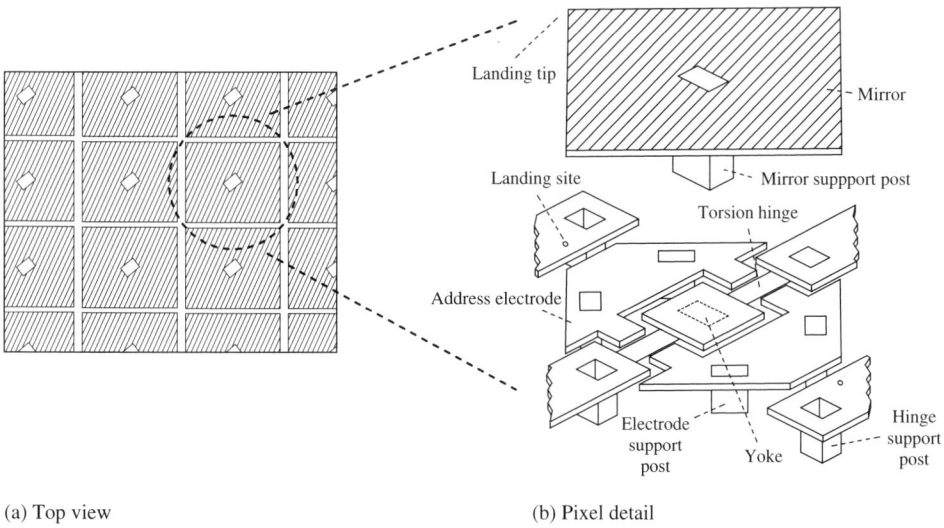

(a) Top view   (b) Pixel detail

Figure 1.9 (a) Top view of a portion of a DMD mirror array, (b) a schematic drawing of the construction of a DMD mirror element

8   INTRODUCTION

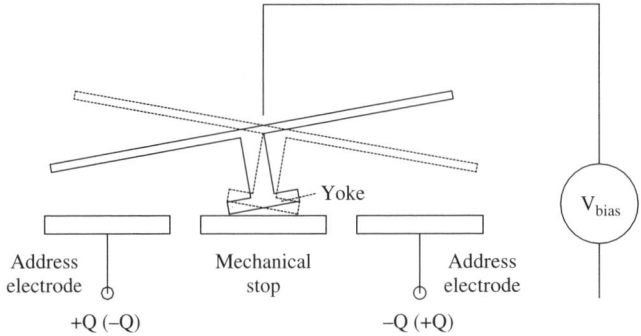

Figure 1.10 Principle of DMD operation

mechanical stop. The mirror rotation angle is typically ~10°, which determines the system $f_{/\#}$ and ultimately system brightness. TI recently developed DMD chips that operate with a tilt angle of ~12°.

The idea of a metal mirror suspended by deformable metal hinges was initially proposed by Van Raalte in 1970 [Van Raalte J. R., 1970]. However, the original device was operated in an analog mode, where the deflection was controlled by the voltage levels on the address electrodes. In practice, it is very difficult to use analog driving schemes to produce a uniform gray scale for an entire mirror array. The DMD developed by TI is bistable and its associated circuitry is entirely digital.

The layout of a projection system based on a DMD, also called digital light processing (DLP), is shown in Figure 1.11. A total internal reflection prism is used. When the DMD

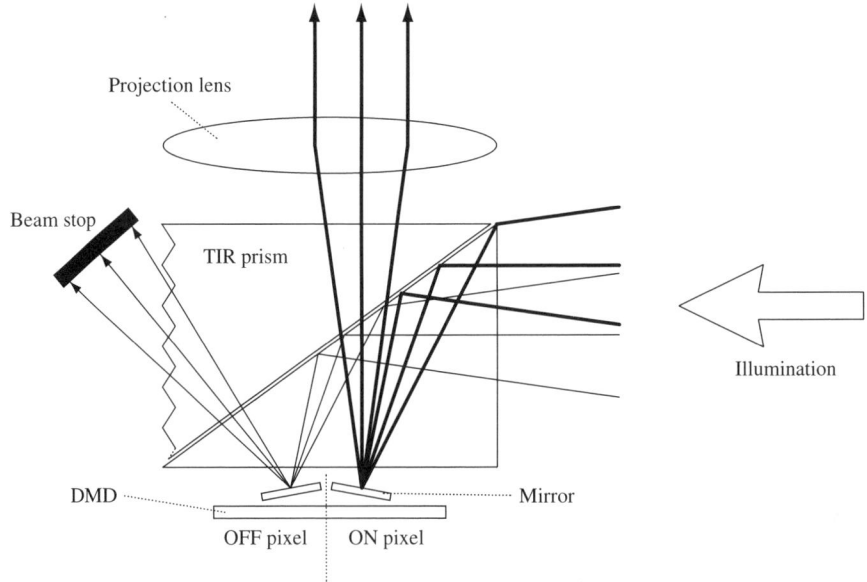

Figure 1.11 DLP projection system based on a DMD

is in the $+\theta$ state, light incident onto the mirror will be reflected into the projection lens, producing a bright state. Conversely, when the DMD is in the $-\theta$ state, the light is reflected away and is internally absorbed. The projection system is operated in a binary mode. In very expensive DLP projectors, there are three separate DMD chips, one each for each primary color. However, in DLP projectors under $10 000, there is only one chip where full color is created by sequential R, G, and B illumination by a color wheel. Typical color wheels consist of red, green, and blue segments, although white segments can be introduced to boost brightness.

The gray scale in DLP projection systems is generated by time multiplexing enabled by the fast ($\sim 15\mu s$) switching between the mirror ON and OFF positions. For example, a single panel system operated at a 60 Hz field rate (shown in Figure 1.12) has a color field duration of 5.56 ms. For 256 gray levels, the shortest address interval required is about $22\mu s$, which is comparable to the DMD's switching speed. Signal correction is needed to avoid errors in the low gray-scale levels due to the finite switching speed. In a real DMD, the driving scheme uses so-called bit plane weighting, which dramatically improves manufacturability [Sampsell J. B. 1994; Tew C. 1994].

There are several unique advantages of DLP projection systems. They include:

- Small package size, a feature most important in the mobile presentation market. Since DLP light engines consist of a single chip rather than three LCD panels, DLP projectors tend to be compact. All of the current 3 lb (1.4 kg) mini-projectors on the market are DLP based.

- High contrast ratio. TI has developed a new generation of DMD, which increases the mirror tilt angle from 10° to 12° and features an absorbing coating to the substrate under the mirrors. These improvements significantly improve the DLP system contrast. Over 1000:1 system contrast is quite common for DLP systems.

- High aperture ratio. DMD operates by deflecting suspended mirrors allowing the driving circuits to reside underneath. The gap between adjacent mirrors is usually less than $1\mu m$. Aperture ratios can therefore often exceed 90%. Visible pixel boundaries leading to the so-called 'screen door effect' are barely seen in DLP systems.

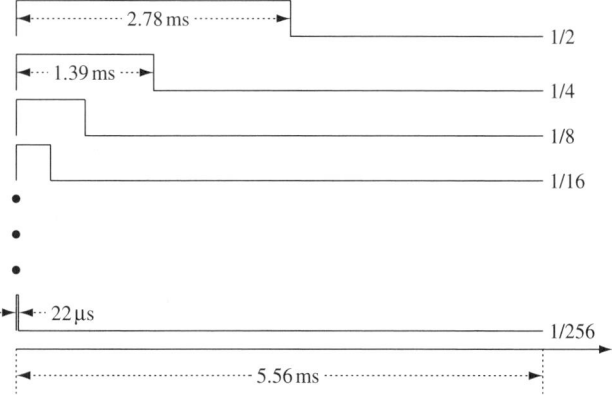

Figure 1.12 Simple time-multiplexing scheme to generate gray scale with a binary DMD

10 INTRODUCTION

- Good reliability. Tests indicate that current DMD performance is not degraded after thousands of hours of operation under harsh environmental conditions [Gouglass M. R., 1995; 1996].

- Polarization independence. No loss is associated with polarizing the source.

Each technology has its own weaknesses, and DLP systems are no exception. These include:

- Manufacturing complexity. The CMOS electronics in the underlying silicon substrate consists of a six-transistor SRAM circuit per pixel, and additional auxiliary addressing electronics. This drives the price of high-resolution DMD chips.

- Color break-up—rainbow effect. DLP systems based on a single panel DMD require a spinning color wheel to achieve full color, resulting in a visible artifact known as color break-up, or the "rainbow effect." At any given instant in time, the image on the screen is red, green, or blue, and the technology relies upon the relatively slow response of the human visual system for the perception of full color. Unfortunately, a proportion of people in the population are sensitive to color break-up, resulting in eyestrain or even headaches. TI and DLP vendors have made progress to address this issue, such as increasing field rates from 180 Hz (1×) to 360 Hz (2×). Today, many DLP projectors being built for the home theater market incorporate a six-segment color wheel, which has two sequences of red, green, and blue, and spins at 120 Hz. Since R, G, and B are refreshed twice in every rotation, the industry refers to this as a 4× rotation speed. This further doubling of the refresh rate has again reduced the number of people who can detect color artifacts, but nevertheless it remains a problem for a number of viewers even today.

- Temporal artifacts. Even in static images the binary nature of the DMD creates the sensation of temporal modulation. In fast-moving video images that cause the eye to move rapidly, object edges can become temporally unstable and appear fuzzy. Improvements to the addressing algorithms have reduced this effect, but it can still be perceived under certain viewing conditions. Low gray-scale contouring also results from the binary addressing.

- Poor color saturation. In most single panel DLP projectors, color wheels often contain a clear (white) segment to boost brightness. Though the image appears brighter, this reduces color saturation.

1.2.4.2 High-temperature Polycrystalline Silicon (HTPS)

The structure of a polycrystalline silicon LC panel is very similar to that of an amorphous silicon active matrix LCD (AMLCD) commonly used for laptops and monitor screens. The electron/hole mobility of polycrystalline silicon is, however, much higher than that of amorphous silicon, allowing the size of a thin-film transistor (TFT) to be made much smaller. It is also possible to make on-panel driver ICs. Due to the good aperture ratio (>60% is possible), high-brightness polysilicon LC projection systems are feasible. The Sony Grand WEGA RPTV based on HTPS is currently the highest volume selling microdisplay RPTV [Shirochi Y., 2003].

There are two methods of fabricating polycrystalline silicon. The more common method used for projection light valves is HTPS [Yamamoto Y., 1995]. Low-temperature (<600°C) polycrystalline silicon (LTPS) on glass is possible through either furnace or laser annealing.

However, the leakage currents are higher in LTPS than in HTPS. HTPS is fabricated with processes requiring temperatures in excess of 1000°C [Morozumi S., 1984]. Therefore, fused silica substrates must be used. HTPS panels offer extremely high performance in terms of degree of miniaturization, high definition, response speed, and reliability.

Almost all HTPS active matrix light valves are transmissive and are based on the 90° TN mode (see Chapters 5 and 9 for further details). The cross-section of such a device is shown in Figure 1.13. The device has a self-aligned gate to minimize parasitic capacitance, which causes DC offset of the pixel voltage relative to the signal voltage, resulting in image sticking. The active matrix structures, such as the row and column metal lines, TFT, and storage capacitor, must be covered by an opaque material (the black matrix) to avoid undesirable optical effects (ITO = Indium Tin Oxide). For instance, direct light exposure of the TFT will cause current leakage. The black mask also hides the low-contrast regions created by the disclination lines from edge fringe fields. Since the total area of the hidden structures behind the black matrix is nearly independent of pixel pitch, the aperture ratio is dramatically reduced as the display resolution increases (see Chapter 9).

All commercially available HTPS projection systems are operated in three-panel mode, due to the slow LC response time of the transmissive TN mode. Architecturally, like the DLP system, they have become standardized. The incident white light is split into three primary colors by dichroic filters. Each primary color passes through an HTPS panel sandwiched between two sheet polarizers. The output light is spatially modulated by the voltage applied to the pixels (see Chapter 5). A dichroic X-cube is used to combine the three colors immediately before the projection lens (Figure 1.14). Details of the system operation are covered in Chapter 9.

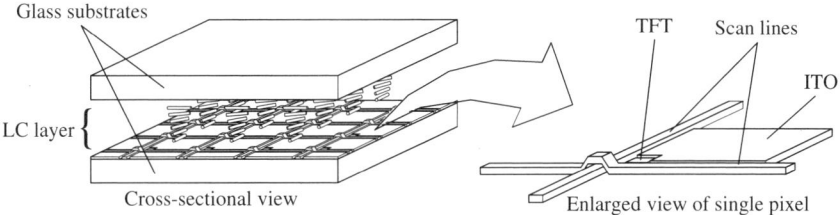

Figure 1.13 Cross-section of an HTPS light valve

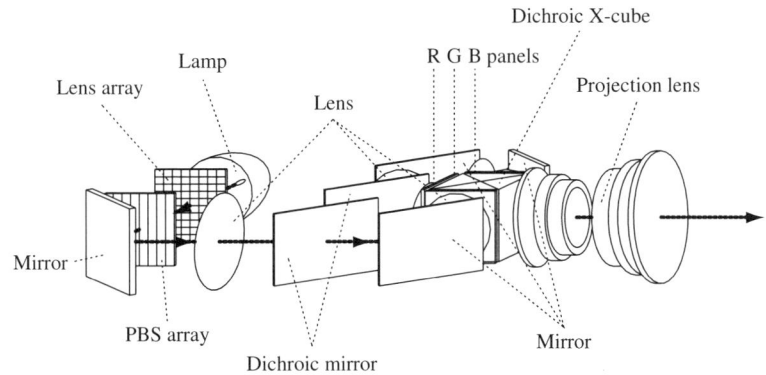

Figure 1.14 HTPS projection system

12 **INTRODUCTION**

The advantages of HTPS projectors include:

- Better color saturation. Since HTPS projectors use three separate RGB panels, colors tends to be rich and vibrant.
- Sharper image. LCDs deliver a somewhat sharper image than equivalent resolution DLP systems, whose tilted pixels can appear blurred at the corners. The difference is more noticeable when viewing high-resolution text than with video content.
- High ANSI lumen output. Three-panel HTPS systems deliver significantly higher output than single panel DLP projectors with the same wattage lamp [Itoh Y., 1997].
- Manufacturability. HTPS is the most mature microdisplay technology. Epson alone has manufactured more than 18 millions HTPS panels to date.

The disadvantages of HTPS projectors include:

- Screen door effect. The black matrix TFT element shielding structures necessary in HTPS panels creates visible pixelation. It looks as if the image is being viewed through a screen door. Several measures have been taken to alleviate this effect. The first one is through increased pixel count. A WXGA projector has over 1 million pixels while a VGA one has only 300 000 pixels. Second, the inter-pixel gaps, independent of resolution, are continually being reduced in width through technological advances. A third development is the use of microlens arrays (MLAs) (Figure 1.15), developed primarily to boost efficiency [http://www.espon.com]. Fortuitously, the concentration of light through the pixel aperture acts to de-emphasize the sharp pixel edges.
- Relatively low contrast. System contrast >1000:1 with HTPS projectors is difficult, as a result of the field of view (FOV) of the TN LCD mode. Retarder-based compensation is one very effective way to improve the FOV of LCs as described in detail in Chapter 9.
- Potential lifetime issues. LC alignment layers used in HTPS light valves are organic polyimides (PIs). PI is susceptible to UV and deep blue light photochemical damage, which reduces operation lifetimes. UV filters with a long-wavelength cut-off are helpful,

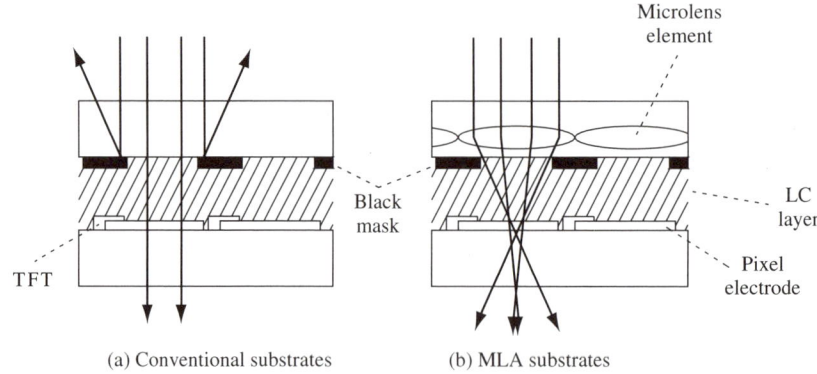

Figure 1.15 Increased throughput by using MLAs in HTPS projection systems

but tend to reduce the blue content of the final color imagery. The industry trend is toward UV filters with 50% transmission points between 430 and 435 nm.

1.2.4.3 Liquid Crystal on Silicon (LCOS)

In the previous sections, we have described two projection systems based on different microdisplay technologies: DLP and HTPS LCD. A new emerging technology is LCOS, which potentially has the advantages of both LCD and DLP and could surpass both technologies in final image quality.

The HTPS LCD is a transmissive technology using LC modulators, while the DLP panel is inherently a reflective display. LCOS is therefore a hybrid technology, using LC modulators on a passive mirror. Pixel brightness depends upon the polarization state from a double pass of the LC layer, which is controlled by an electric field. The field is induced by a voltage applied between the pixel mirror and a transparent conductor (see Chapters 7, 10, and 11). A cross-section of an LCOS light valve is shown in Figure 1.16. Traditional LCOS back planes provide analog addressing, but more recently full digital LCOS panels have been developed. The latter have unique advantages, such as stable and uniform gray levels, low fabrication costs, and high reliability [Shimizu S., 2004].

In three-panel LCOS projection systems, each panel separately modulates red, green, and blue light. The first three-panel projection system based on LCOS was developed by IBM [IBM, 1998] and used a Philips color prism to separate and recombine colored beams (Figure 1.17). While improvements were made to the performance of the Philips color prism [Greenberg M. R., 2000], it is difficult to maintain the state of polarization adequately in a dichroic prism, resulting in poor system contrast. As an alternative, off-axis systems were developed [Bone M. F., 1998; 2000], in which the incident and reflective beams do not counter-propagate (Figure 1.18). Sheet polarizers can be used instead of polarizing beam splitters (PBSs) to pre-polarize the incident beam and separately analyze the reflected

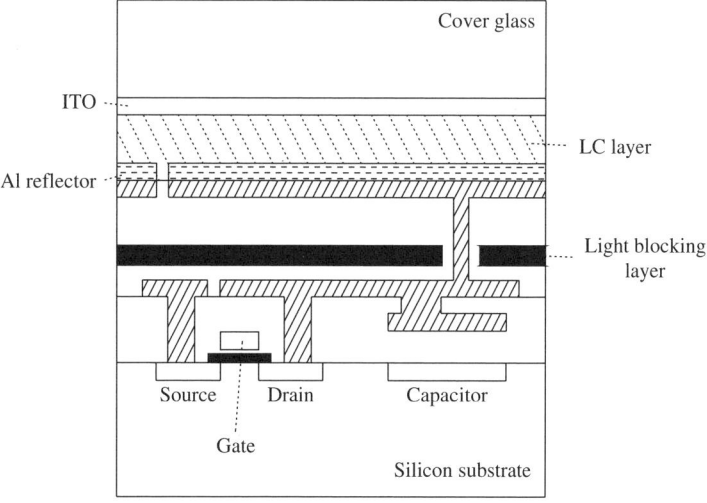

Figure 1.16 Cross-section of LCOS light valve

14 *INTRODUCTION*

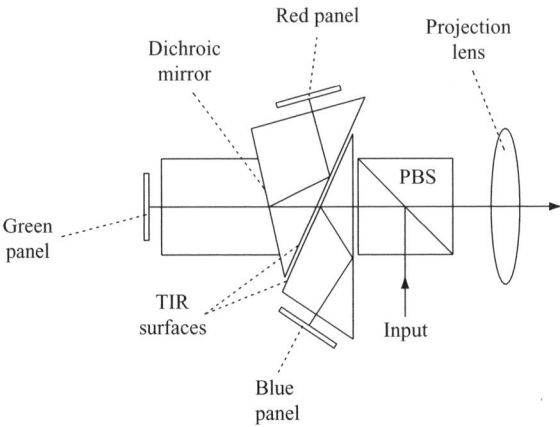

Figure 1.17 LCOS projection system based on the Philips color prism

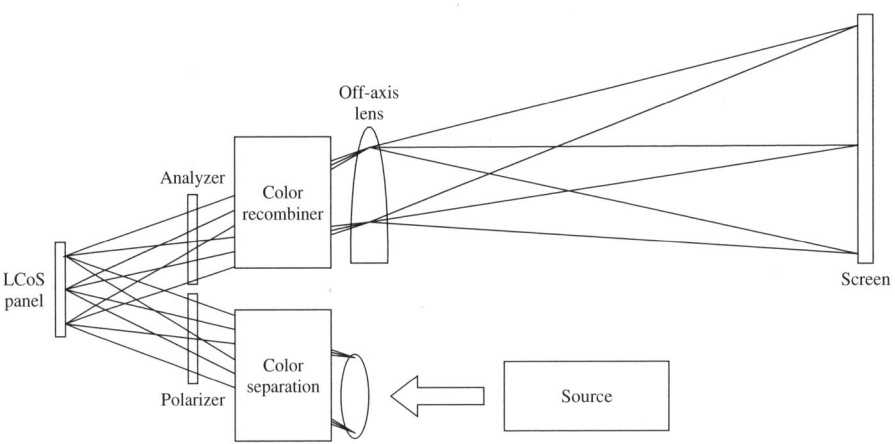

Figure 1.18 Off-axis LCOS projection system

light. With off-axis designs, however, the panel convergence and projection lens design are difficult.

After many years of development, the surviving color management architectures fall into two categories. The first is based on three PBSs with a combining X-cube (3 × PBS/X-cube), which is an extension of the HTPS LCD projection system. The PBSs can be of the conventional MacNeille type [Melcher R. L., 1998], the 3M reflective type, [Bruzzone C. L., 2003; 2004], or the wire grid type [Gardner E., 2003; Pentico C., 2003; Kurtz A. K., 2004] (Figure 1.19). The second group of LCOS projectors use retarder stacks with MacNeille PBSs, which combine the polarizing/analyzing functions with the splitting/recombining of color [Robinson M., 2000; Sharp G., 2002]. This technology is capable of very compact color management systems. The ColorQuad™ is one of many LCOS projection architectures based on this approach (Figure 1.20).

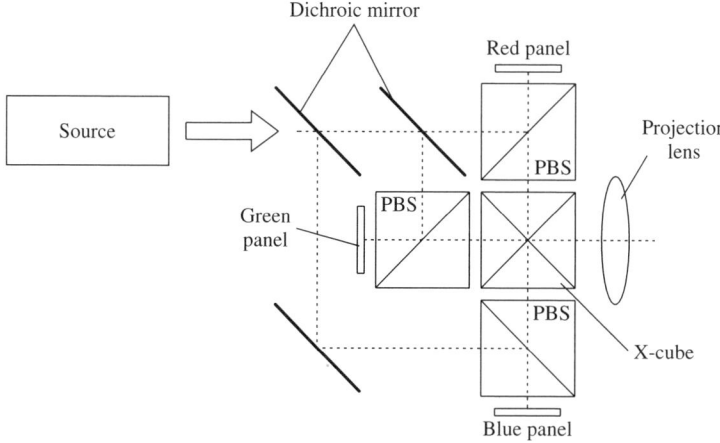

Figure 1.19 The 3× PBS/X-cube LCOS projection system

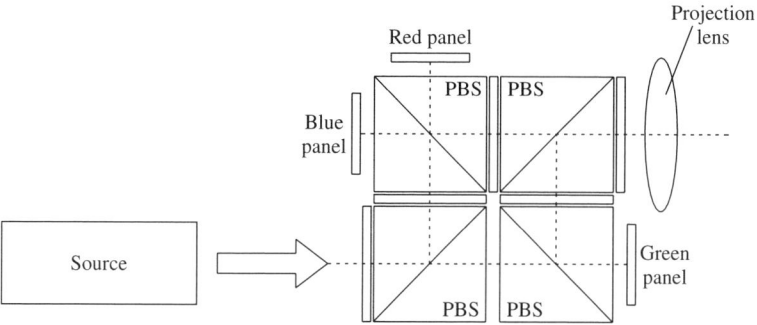

Figure 1.20 LCOS projection system based on the ColorQuad™

The merits of LCOS projection systems include:

- IC compatibility. The LCOS electronic substrate (back plane) is compatible with the standard silicon technology, allowing additional driving circuitry to be integrated into the back-plane design.

- Cost effectiveness for high resolution. LCOS is much more amenable to higher resolutions than HTPS and DLP. Due to standard silicon processes, it is relatively easy to scale up device resolution without suffering loss in manufacturing yield. LCOS can achieve HD resolution (1920 × 1080) in a 0.7″ panel and 1280 × 720 in a 0.5″ panel, which no other technology to date has accomplished.

- No screen door effect. This is due to high resolution and high fill factor (i.e., minimal space between pixels). Pixelation from an LCOS projection system is barely visible.

- Smooth picture. The pixel edges in LCOS tend to be smoother compared to the sharp edges of the micromirrors with DLP. Video images produced with this inter-pixel smoothing are more natural looking.

16 INTRODUCTION

- High contrast. System contrasts over 2000:1 have been demonstrated with vertical-aligned LCOS panels. This contrast is comparable to the contrast of DLP systems, and better than that of HTPS systems.

- High response speed. Compared to transmissive LCD panels, the response time of LCOS panels is much shorter. Since LCOS is operated in reflective mode, the cell gap is about half that of transmissive panels, which results in a 4 × faster response time. ON/OFF periods of less than 1 ms have been achieved using submicron cell gaps. With a suitable choice of a fast LCOS mode, sequential single and two-panel LCOS projection systems are feasible (see Chapters 5, 11) [Janssen P. 1993; Shimizu J. A., 2001].

Demerits include:

- Lifetime. Long-term reliability of LCOS systems, as with HTPS, is still a concern. However, the situation is much better if inorganic SiO_x LC alignment is used. D-ILA LCOS panels have recently been quoted as having a 300 000 hour lifetime under typical operating conditions [Bleha W. P., 2003; Schimizu S., 2004].

- Color break-up. In sequential color LCOS systems a rainbow effect is observed as in single panel DLP systems.

- Complexity. An LCOS optical system is more complex than either DLP or HTPS systems. Accurate control of the state of polarization is key to making LCOS products commercially competitive, which will be the main focus of this book.

1.2.5 Other Projection Technologies

In addition to the projection systems mentioned in previous sections, there are several other technologies under development. We will list them here without going into detail. For those seeking further information about specific technologies the following references should be consulted:

- Polymer-dispersed LC (PDLC) devices. Without applied voltage, PDLC devices scatter light but appear transparent when a high voltage is applied. Systems can deliver high brightness because they are operated under unpolarized light. However, it is difficult to deliver high contrast [Fergason J. L., 1992].

- Surface-stabilized ferroelectric LCs (SSFLCs) [Clark N. A., 1980]. This technology is based on a chiral smectic C* phase. Due to its ferroelectricity, it exhibits two states depending upon the polarity of applied voltage. The response is very fast, though switching is bistable in nature. Therefore, gray scale is generated by time multiplexing as for DLP systems. Due to issues with brightness loss from DC balancing requirements, SSFLCs are rarely used in projection systems. Currently, FLCs are primarily used in digital camera viewfinders and in head-mounted displays.

- Actuated mirror array (AMA) [Um G., 1992; 1995]. This technology was developed by Aurora Systems and Daewoo and is conceptually similar to the DMD. It is based on piezoelectric or electrostrictive-mechanical angular deflection of individual mirrors within an array. The mirror tilt angle can be continuously adjusted. However, the maximum mirror tilt angle is much smaller than that of the DMD, and systems can only work with large $f_{/\#}$s. Systems tend to have low brightness.

- Micromechanical diffractive grating light valves. Developed by Silicon Light Machines [Apte R. B., 1993], there are based on the diffraction of light by a physical grating formed on the surface of the device. Each pixel of this reflective device consists of two or more parallel reflective ribbons. Alternate ribbons can be pulled down electrostatically by approximately a quarter wavelength to create a diffractive grating, which can be used in schlieren projection systems.

- Light amplifiers. A low-intensity input optical image from a CRT is amplified for high-intensity projection on a screen [Beard T., 1973; Ledebuhr A. G., 1986]. It consists of a photoconductor layer, a light-blocking layer, a dielectric mirror, and a nematic LC cell. A bias voltage is applied to the outside electrodes. In regions where the optical writing signal is absent, the photoconductor layer has high impedance and the state of the LC is not altered. Where a writing signal is present, the impedance of the photoconductor decreases and a switching voltage appears across the LC layer. The image from the CRT is therefore reproduced as a spatial modulation pattern in the LC, which can be projected with high-intensity illumination.

1.3 Scope of the Book

This book consists of 11 chapters. Following this introductory chapter, the basics of LC projection systems are addressed in Chapter 2. The concepts of color, brightness, balanced white point, visual artifacts, and the requirements of contrast and uniformity are described in this chapter. LC projectors are based on polarized light. Controlling the state of polarization is therefore the key to making a good LC projection system. Required mathematical representations of polarized light, state of polarization calculations, and modeling techniques are presented in Chapter 3. In Chapter 4, the key projection system components, such as PBSs, retardation elements, various types of transmissive and reflective polarizers, dichroic filters, and anti-reflection (AR) coatings, are illustrated. The basic LC property and its electro-optical (EO) effect are presented in Chapter 5. Here, LC modes used in LC projectors are summarized. Retarder stacks have been widely used in LCOS color management systems and more recently in HTPS systems (RPTV) for color uniformity improvement. Chapter 6 is designated to describing their basic properties and design. Chapter 7 presents methods to optimize system contrast in LCOS projection systems and illustrates general compensation schemes to enhance head-on and off-axis contrast. Color management is a key part of projector system design, and options in color management are summarized in Chapter 8. The three-panel transmissive system is presented in Chapter 9, with emphasis initially on throughput. One weakness of HTPS projectors is system contrast. The principle of operation and methodologies for system contrast improvement complete the chapter. Three-panel LCOS architectures are not standardized as yet. Mainstream three-panel configurations and their performance comparison are covered in Chapter 10. Finally, sequential LCOS systems are gaining popularity due to the cost pressure from DLP and HTPS. Two-panel and one-panel LCOS systems are presented in Chapter 11.

References

[Apte R. B., 1993] R. B. Apte, F. S. A. Sandejas, W. C. Banyai, and D. M. Bloom, Grating light valves for high resolution displays, SID'03 Digest, pp.807–808, 1993.

[Beard T., 1973] T. Beard, W. F. Bleha, and S. Y. Wong, AC liquid crystal light valve, *Appl. Phys. Lett.*, 22, pp.90–92, 1973.

[Bleha W. P., 2003] W. P. Bleha, Robustness of D-ILATM projectors, Projection Summit Proceedings, 2003.

[Bone M. F., 1998] M. F. Bone, M. Francis, P. Menard, M. E. Stefanov, and Y. Ji, Novel optical system design for reflective CMOS technology, Proceedings of 5th Annual Flat Panel Display Strategic and Technical Symposium, p.81, 1998.

[Bone M. F., 2000] M. F. Bone, Front projection optical system design for reflective LCOS technology, Proceeding of Microdisplay Conference, 2000.

[Bos P. J., 1995] P. J. Bos, J. Chen, J. W. Doane, B. Simth, C. Holton, and W. E. Glenn, An optically active diffractive (OAD) device for a high efficiency diffractive light valve, *J. SID*, 3/4 p.195, 1995, and SID'95 Digest, pp.601–604, 1995.

[Bruzzone C. L., 2003] C. L. Bruzzone, J. J. Ma, D. J. W. Aastuen, and S. K. Eckhardt, High-performance LCOS optical engine using Cartesian polarizer technology, SID'03, Digest, pp.126–129, 2003.

[Bruzzone C. L., 2004] C. L. Bruzzone, J. J. Schneider, and S. K. Eckhardt, Photostability of polymeric Cartesian polarizing beam splitters, SID'04, Digest, pp.60–63, 2004.

[Clark N. A., 1980] N. A. Clark and S. T. Lagerwall, Submicrosecond bistable electro-optic switching in liquid crystals, *Appl. Phys. Lett.*, 36, pp.899–901, 1980.

[Fergason J. L., 1992] J. L. Fergason, Polymer encapsulated nematic liquid crystals for display and light control applications, SID'92 Digest, pp.571–574, 1992.

[Fischer F., 1940] F. Fischer, Auf dem Wege zur Fernseh-Grossprojektion, *Schweiz. Arch. Angew. Wiss. Tech.*, 6, pp.89–106, 1940.

[Gardner E., 2003] E. Gardner, and D. Hansen, An image quality wire-grid polarizing beam splitter, SID'03, Digest, pp.62–63, 2003.

[George J. G., 1995] J. G. George, Minimum adjustment analog convergence system for curved faceplate projection tubes, *IEEE Trans. Consumer Electron.*, 44, pp.536–539, 1995.

[Glenn W. E., 1958] W. E. Glenn, New color projection system, *J. Opt. Soc. Am.*, 48, pp.841–843, 1958.

[Glenn W. E., 1979] W. E. Glenn, Principle of simultaneous color projection television using fluid deformation, *SMPTE J.*, 79, pp.788–794, 1979.

[Gouglass M. R., 1995] M. R. Gouglass and D. M. Kozuch, DMD realiability assessment for large-area display, SID'95, Digest (Application Session), p.49, 1995.

[Gouglass M. R., 1996] M. R. Gouglass and C. G. Malemes, Reliability of displays using light processing, SID'96, Digest, pp.774–777, 1996.

[Greenberg M. R., 2000] M. R. Greenberg and B. J. Bryars, Skew ray compensated color-separation prism for projection display applications, SID'00, Digest, pp.88–91, 2000.

[Hockenbrock R., 1982] R. Hockenbrock and W. Rowe, Self-convergent, 3-CRT projetion TV system, SID'82, Digest, pp.108–109, 1982.

[Hornbeck L. J., 1983] L. J. Hornbeck, 128 × 128 deformable mirror device, *IEEE Trans. Electron Dev.*, ED-30, pp.539–545, 1983.

[Hornbeck L. J., 1996] L. J. Hornbeck, Active yoke hidden hinge digital micromirror device, US Patent 5,535,047, 1996.

[IBM, 1998] IBM, Special session for high-resolution displays, *IBM J. Res. Dev.*, 42, no. 3/4, 1998.

[Itoh Y., 1997] Y. Itoh, J. I. Nakamura, K. Yoneno, H. Kamakura, and N. Okamoto, Ultra-high-efficiency LC projector using a polarized light illumination system, SID'97, Digest pp.993–996, 1997.

REFERENCES

[Janssen P., 1993] P. Janssen, A novel high brightness single light valve HD color projector, Proceedings IDRC'93, Society for Information Displays, pp.249–252.

[Johannes H., 1979] H. Johannes, The history of the EIDOPHOR large screen television projector, Gretag, Rehensdorf/Zürich, Switzerland, 1989.

[Kurtz A. K., 2004] A. K. Kurtz, B. D. Silverstein, and J. M. Cobb, Digital cinema projection with R-LCOS display, SID'04, Digest, pp.166–169, 2004.

[Ledebuhr A. G., 1986] A. G. Ledebuhr, Full-color single-projection-lens liquid crystal light-valve projector, SID'86 Digest, pp.379–381,1986.

[Malang A. W., 1989] A. W. Malang, High brightness projection video display with concave phosphor surface, *Proc. SPIE*, 1081, pp.101–106, 1989.

[Melcher R. L., 1998] R. L. Melcher, M. Ohhata, and K. Enami, High-information-content projection display based on reflective LC on silicon light valves, SID'98, Digest, pp.25–28, 1998.

[Morozumi S., 1984] S. Morozumi, 4.25-in and 1.51-in b/w and full-color LC video displays addressed by poly-Si TFTs, SID'04, Digest, pp.316–317, 1984.

[Pentico C., 2003] C. Pentico, M. Newell, and M. Greenberg, Ultra high contrast color management systems for projection displays, SID'03, Digest, pp.130–133, 2003.

[Robinson M., 2000] M. Robinson, J. Korah, G. Sharp, and J. Birge, SID'00, Digest, pp.92–95, 2000.

[Sampsell J. B., 1994] J. B. Sampsell, An overview of the performance envelope of digital-micromirror-device-based projection display systems, SID'94 Digest, pp.669–672, 1994.

[Sharp G., 2002] G. Sharp, M. Robinson, J. Chen, and J. Birge, LCOS projection color management using retarder stack technology, *Displays*, 23, pp.139–144, 2002.

[Shimizu J. A., 2001] J. A. Schimizu, Scrolling color LCOS for HDTV rear projection, SID'01, Digest, pp.1072–1075, 2001.

[Shimizu S., 2004] S. Shimizu, Y. Ochi, A. Nakano, and M. Bone, Fully digital D-ILATM device for consumer applications, SID'04 Digest, pp.72–75, 2004.

[Shirochi Y., 2003] Y. Shirochi, K. Murakami, H. Endo, S. Arakawa, H. Kitagawa, and K. Uchino, 50/60 V Hivision LCD rear-projection TV (Grand WEGA) with excellent picture quality, SID'03, Digest, pp.114–117, 2003.

[Stupp E. H., 1999] Edward H. Stupp and Mathew S. Brennesholtz, *Projection Displays*, John Wiley & Sons, Ltd, Chichester, p.4, 1999.

[Tew C., 1994] Claude Tew, Electronic control of a digital micromirror device for projection displays, *IEEE Solid-State Circuits Digest of Technical Papers*, vol. 37, p.130, 1994.

[Um G., 1992] G. Um, A new display projection system, SID'92 Digest, pp.455–459, 1992.

[Um G., 1995] G. Um, D. Foley, A. Szilagyi, J. B. Ji, Y. B. Jeon, and Y. K. Kim, Recent advances in actuated mirror array (AMA) projector development, Proceedings of the 15th International Display Research Conference (Asia Display 95), pp.95–98, 1995.

[Van Raalte J. R., 1970] J. R. Van Raalte, A new schlieren light valve for television projection, *Appl. Opt.*, 9, pp.2225–2230, 1970.

[Wang B., 2002] Bin Wang, Philip J. Bos, and David B. Chung, A new type of liquid crystal diffractive light valve with very small pixel size, SID'02 Digest, p.962, 2002.

[William S. A., 1997] S. A. William, A history of Eidophor projection in North America, *Proc. SPIE*, 3013, pp.7–13, 1997.

[Wolf M., 1937] M. Wolf, The enlarged projection of television pictures, *Philips Tech. Rev.*, 2, pp.249–253, 1937.

[Yamamoto Y., 1995] Y. Yamamoto, T. Morita, Y. Yamana, F. Funada, and K. Awane, High-performance low temperature poly-Si TFT with self-aligned offset gate structure by anodic oxidation of Al for a driver monolithic LCDs, Proceedings of the 15th International Research Conference (Asia Display 95), pp.941–942, 1995.

[Yang K. H., 1998] K. H. Yang and M. Lu, Nematic LC modes and LC phase gratings for reflective spatial light modulators, *IBM J. Res. Dev.*, 42, pp.401–410, 1998.

2

Liquid Crystal Projection System Basics

2.1 Introduction

It is most effective to introduce the basics of projection display by starting with the end goal: that is, the generation of imagery that is pleasing to the eye in all relevant respects. Such a display has the following qualitative attributes:

1. High brightness.
2. Large color gamut, with saturated red, green, and blue.
3. Balanced white point.
4. High contrast.
5. Uniformity in brightness, color, and dark state.
6. Large screen size.
7. High resolution.
8. Lack of spatial artifacts (with a photographic-like quality).
9. Lack of temporal artifacts, such as flicker, electronic noise, motion blurring, and color break-up.
10. Attractive packaging, such as cabinet depth and weight.

Polarization Engineering for LCD Projection   M. G. Robinson, J. Chen and G. D. Sharp
© 2005 John Wiley & Sons, Ltd

To quantify several of these attributes, it is necessary to understand the human visual system in terms of color and brightness sensitivity, as well as resolving power at various contrast levels. The chapter will begin therefore by summarizing color science, as it is currently understood, which relates the eye's perception of imagery to display metrics. This establishes display requirements in terms of brightness, color, uniformity, size, resolution, etc., together with means of quantifying them. Subsequently, the requisite projection component technologies are introduced, beginning with the screen and working backward through the projection lens toward the light source. The functional requirements of the subsystem responsible for color management and light modulation will be discussed briefly, with a more rigorous treatment left to later chapters. It should be noted that basic projection technology has been discussed in detail elsewhere [Stupp E. H., 1999], and, where applicable, suitable references have been given should the reader need greater explanation.

2.2 Brightness and Color Sensitivity of the Human Eye

Extensive study has established that there is a dense array of three types of sensors, or cones, in the fovea region of the eye's retina, the region responsible for observing detailed imagery [Wyszecki G., 1982; Williamson S. J., 1983; Bahadur B., 1993]. The three cone types detect electromagnetic radiation within overlapping wavelength bands associated with the sensation of red, green, and blue: the additive primary colors. The extent to which these spatially scattered sensors are excited gives the sensation of color and brightness. Image generation is thus an exercise in controlling the ratio of three primary colors (spectra that relate to red, green, and blue) at a spatial resolution close to that sampled by the eye. A display need only produce three colors, with a full-color palette perceived through the appropriate admixtures of RGB. The eye is tricked into seeing full-color imagery using only these primaries, with absolutely no sensation of synthetic color.

In the strictest sense, a 3D representation is necessary to fully describe color, since its perception depends upon the retinal irradiance of each of the three primary color spectra. For example, the white that is perceived from a backlit display under room lighting conditions appears gray when viewed in sunlight. In order to quantify brightness-dependent color variation, a standard white relating to the viewing environment is required. Since projection displays are viewed under user-defined ambient conditions, the 3D mapping of color is unsuitable as a display engineering metric. Color is therefore decoupled from brightness. Displays are made to map video projection signals with the correct RGB ratios, but with absolute intensity values (i.e., brightness) dependent upon the user-defined display setting.

2.2.1 Brightness

Perceived brightness of an object is dependent upon three factors: (a) the amount of light captured by the eye; (b) the degree to which the light is focused onto the retina; and (c) the spectral content of the light. Figure 2.1 illustrates the imaging of light from an object onto the retina.

The amount of light from a given point on the object that is collected by the eye is a function of the distance D between the object and eye, the diameter d of the pupil, and the amount of light contained within the capture solid angle $\Omega = \pi d^2/D^2$. The amount of light

BRIGHTNESS AND COLOR SENSITIVITY OF THE HUMAN EYE

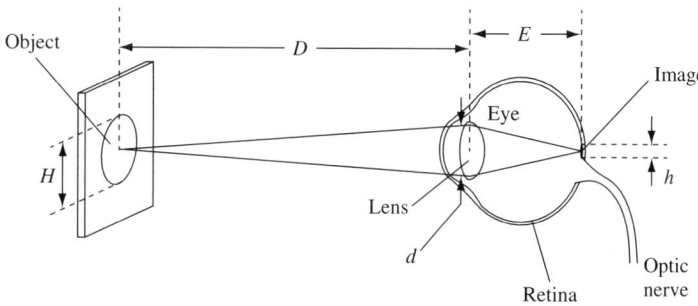

Figure 2.1 Human visual system schematic

radiated from an object per solid angle per unit area is defined as its radiance. Assuming an object has a radiance at a given wavelength λ of $R(\lambda)$ W/m²/sr, then the irradiance, $I(\lambda)$, of the retina is given by $R(\lambda)\Omega/\Delta^2$ W/m², where $\Delta(=E/D)$ is the demagnification of the image. Expanding this simple expression yields $I(\lambda) = \pi d^2 R(\lambda)/E^2$. This irradiance gives the sensation of brightness for any given wavelength. For any given viewing conditions, which determine the pupil diameter d, the irradiance is proportional to the object radiance and is independent of the viewing distance, D. This is clearly the case in practice, where the size of an object depends upon proximity, but the perceived brightness does not.

The sensitivity of the eye is a function of wavelength. The eye's sensitivity curve is shown in Figure 2.2 [Williamson S. J., 1983] and is labeled as $\bar{y}(\lambda)$.

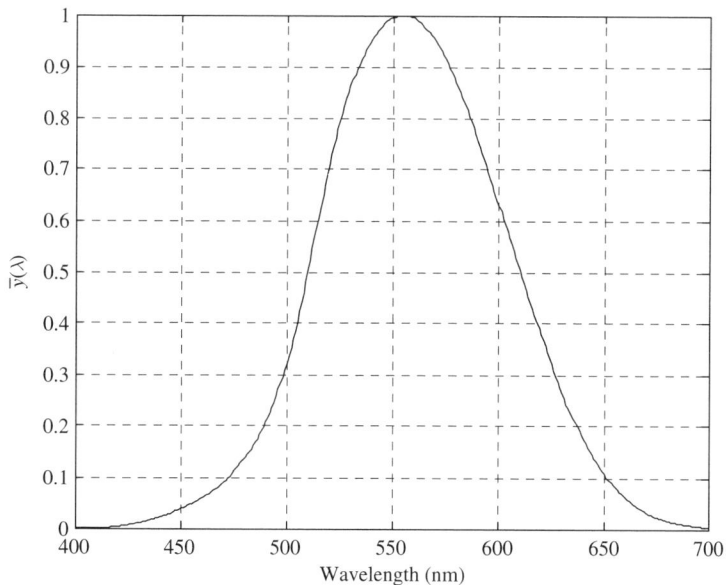

Figure 2.2 Eye sensitivity weighting function

24 LC PROJECTION SYSTEM BASICS

The definition of the color-independent brightness B of an object, or a display, with spectral radiance $R(\lambda)$ is:

$$B = 683 \int_{-\infty}^{\infty} R(\lambda) \cdot \overline{y}(\lambda) d\lambda \qquad (2.1)$$

and has the units of nits or cd/m² (lm/sr/m²). The 683 factor comes from the conversion of watts to lumens at the peak of the normalized photopic weighting curve of Figure 2.2. The brightness of an RPTV display, showing a fully on white frame, should exceed 600 nits.

2.2.2 Brightness Uniformity

The sensitivity of the eye to brightness uniformity depends upon spatial frequency. Slowly varying brightness levels within the image are often not noticed, whereas more localized intensity variation of the same magnitude stand out. Work by Robson *et al.* [Robson J. G., 1966] has attempted to quantify this effect. Figure 2.3 is a plot of sinusoidal intensity variation with varying period (*x*-axis), with varying contrast along the *y*-axis. It shows that at a periodicity of 2–3 cycles/degree, a brightness variation of a few percent is discernible. However over the area of a large-screen TV viewed at typical viewing distance of 2 m, 10% intensity variation is barely noticeable.

2.2.3 Color

Two common 2D coordinate systems are used to describe color quantitatively: the CIE 1931 (x, y), and the modified CIE 1976 (u', v') coordinates [Judd D. B., 1975;

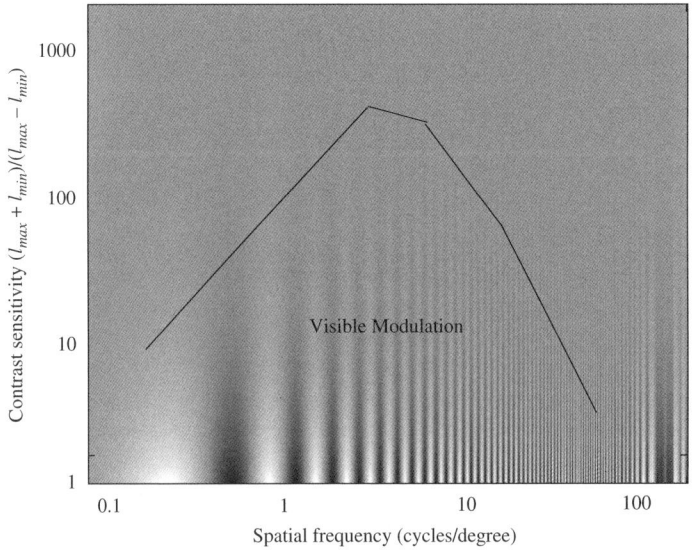

Figure 2.3 Contrast sensitivity after Campbell and Robson showing the sensitivity of the eye to intensity variation as a function of spatial frequency [Campbell F. W., 1968]

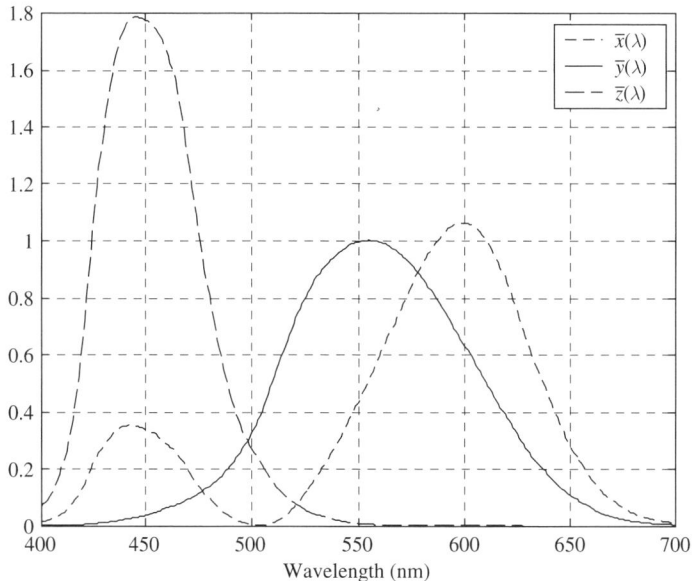

Figure 2.4 Tristimulus curves describing the spectral sensitivity of the three retinal sensors

Hunt R. W., G. 1987]. They both derive from the experimentally determined color matching functions or tristimulus $\bar{x}(\lambda), \bar{y}(\lambda), \bar{z}(\lambda)$ curves shown in Figure 2.4 [Wright W. D., 1928; Guild J., 1931].

Each curve represents the sensitivity of a specific retinal sensor to a particular monochromatic irradiance. Note that $y(\lambda)$ is the same photopic weighting curve shown in Figure 2.2. Mathematically we obtain the theoretical sensor stimulus (X, Y, Z) for a spectrum $S(\lambda)$ using the following integral expressions:

$$X = 683 \int_{-\infty}^{\infty} S(\lambda) \cdot \bar{x}(\lambda) \, d\lambda$$
$$Y = 683 \int_{-\infty}^{\infty} S(\lambda) \cdot \bar{y}(\lambda) \, d\lambda \quad (2.2)$$
$$Z = 683 \int_{-\infty}^{\infty} S(\lambda) \cdot \bar{z}(\lambda) \, d\lambda$$

from which the (x, y) color coordinates are calculated:

$$x = \frac{X}{X+Y+Z} \quad y = \frac{Y}{X+Y+Z} \quad (2.3)$$

The color coordinates for monochromatic visible light (between approximately 400 and 700 nm) form a locus representing colors perceived as fully saturated, outside of which colors are not defined. Since there are three sensors in the eye, effectively detecting red, green, and blue, it is not surprising that this boundary forms a near triangle with red, green, and blue wavelengths at the corners (see Figure 2.5).

26 LC PROJECTION SYSTEM BASICS

The (x, y) coordinate system is linear. If the eye experiences irradiance from two separate spectral sources $(S_1 = (x_1, y_1)$ and $S_2 = (x_2, y_2))$ the resultant color point $S_3 (= (x_3, y_3))$ lies on a line between S_1 and S_2 as described by equation (2.4):

$$(x_3, y_3) = \frac{1}{(W_1 + W_2)} (W_1 \ W_2) \begin{pmatrix} x_1 & y_1 \\ x_2 & y_2 \end{pmatrix} \qquad (2.4)$$

where:

$$W_1 = X_1 + Y_1 + Z_1, \text{ etc.}$$

Further manipulation yields:

$$\frac{S_1 S_3}{S_2 S_3} = \frac{Y_1}{Y_2} \qquad (2.5)$$

which implies that any color point S_3 on the line $(S_1 S_2)$ between S_1 and S_2 (see Figure 2.5) can be reached by adjusting the intensity ratio of the spectra S_1 and S_2. With three-color mixing, a triangular gamut of accessible colors is defined, the vertices being the individual (RGB) color points. To access all colors demanded of a TV signal, three primary colors must be chosen that encompass standard triangular color gamuts. These standards include for example the SMPTE and EBU standards shown in Figure 2.5.

Projection systems designed to display TV imagery should endeavor to create primary color spectra close to those of standard TV gamuts to avoid distortion in color reproduction and any overall brightness penalty.

2.2.4 White

White corresponds to a mixed spectrum that stimulates all three retinal sensors nearly equally. Traditionally, white has been associated with solar illumination, which closely resembles the spectrum of a black body radiating at 5800 K. More recently, with the advent of artificial fluorescent lighting, white is now more associated with radiation containing a higher blue content than the sun. Color science defines white as that which is emitted from black bodies of varying temperatures. Physical calculations accurately determine the emitted spectra from heated black bodies, and their associated color coordinates form a continuous line called the Planckian locus (see Figure 2.5).

Displays that independently modulate three primary colors can achieve any desired white color. The extent to which each color is modulated can be calculated. Assuming the red, green, and blue primaries have spectra $R(\lambda)$, $G(\lambda)$, and $B(\lambda)$ respectively, first a matrix $\mathbf{M}$ representing the individual retinal stimuli is calculated for each of these colors as outlined above:

$$\mathbf{M} = \begin{pmatrix} X_R & X_G & X_B \\ Y_R & Y_G & Y_B \\ Z_R & Z_R & Z_R \end{pmatrix} \qquad (2.6)$$

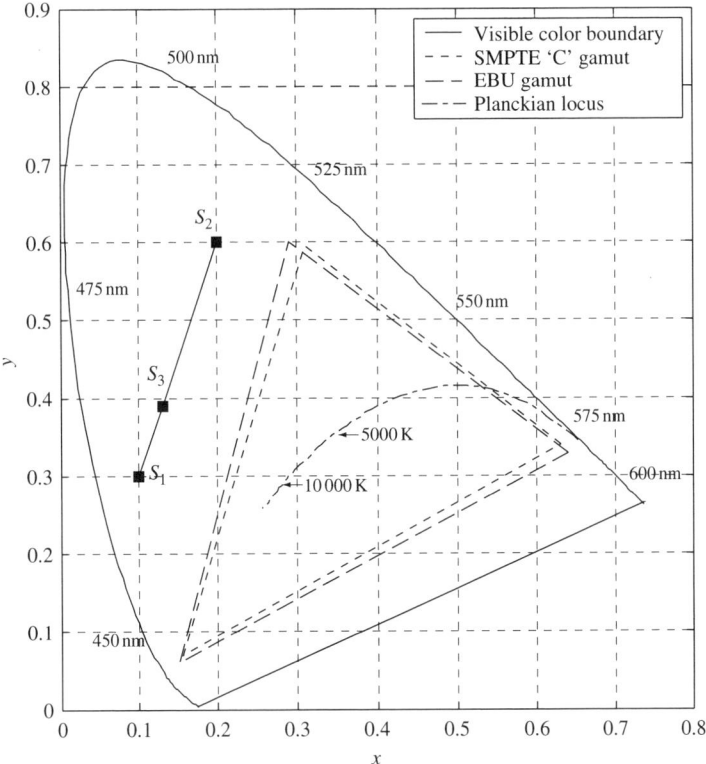

Figure 2.5 The x, y color coordinate space showing the boundary of valid colors achieved solely by monochromatic light, two standard TV gamuts, and the Planckian locus associated with black-body radiation

The color coordinates resulting from the combination of primaries with relative intensities $r:g:b$ is then given by:

$$\begin{pmatrix} x \\ y \\ z \end{pmatrix} = k \cdot \mathbf{M} \cdot \begin{pmatrix} r \\ g \\ b \end{pmatrix} \qquad (2.7)$$

where k is a proportionality constant, and $z = 1 - x - y$.

Replacing the values of (x, y) with coordinates of a desired white color (x_d, y_d) allows a coefficient vector Ω representing the relative modulations of each spectrum to be calculated by the expression:

$$\Omega = \mathbf{M}^{-1} \begin{pmatrix} x_d \\ y_d \\ 1 - x_d - y_d \end{pmatrix} \qquad (2.8)$$

28 LC PROJECTION SYSTEM BASICS

Typically in a three-panel system, the coefficient vector Ω is normalized to its largest component to yield actual color modulation values.

Typical video displays demand a white corrected color temperature of ~6500 K to show what is considered natural-looking video content, while computer displays have adopted a bluer 10 000 K white standard. In practice, it is expected that displays will be required to adjust between these two limits. Furthermore, although a particular color temperature has a specific color coordinate, there is an allowable region above and below the Planckian locus where the eye cannot clearly distinguish a color difference. These so-called corrected color temperatures are an example of the insensitivity that the eye has to hue. In projection, this means that higher brightness can be obtained by allowing imperceptible surplus green to be displayed at full white.

2.2.5 Color Distinction and Just Noticeable Differences (JNDs)

The resolution limit of the eye in distinguishing between colors that have marginally different color coordinates is termed a *just noticeable difference* or JND. Experimentally, a JND is found to be ~0.01 in (x, y) coordinate space for red and white, ~0.02 in the green, but significantly lower <0.005 for blue light [MacAdam D. L., 1937]. The discrepancy between the different regions of color space led to the topological modification of the (x, y) coordinate system, resulting in a color space (u', v') with nearly spatially invariant JND. The mapping between the (x, y) and the (u', v') coordinates is given by:

$$u' = \frac{4x}{-2x + 12y + 3} \quad v' = \frac{9y}{-2x + 12y + 3} \tag{2.9}$$

The equivalent (u', v') color space, analogous to Figure 2.5, is shown in Figure 2.6. A JND in this coordinate system is almost spatially invariant, with a value of ~0.004 [Jones A. H., 1968].

An excellent display has color non-uniformities below a JND. In practice, it is difficult to achieve this, particularly when displaying low gray-level desaturated colors. Color non-uniformities of $\Delta u'v' \sim 0.01$ are currently acceptable in good-quality RPTV systems.

2.2.6 Contrast

There are two distinct types of contrast specifications that are used to define display performance. The first is sequential contrast, defined as the ratio between the brightness of a fully white to a fully dark screen. The other is ANSI contrast [ANSI], defined as the ratio between white and dark areas when a 4×4 checkerboard pattern is displayed. The former is the more stringent, as it relates to the color of low-brightness images, whereas the latter is more forgiving due to spatial contrast enhancement of the eye. Indeed low ANSI contrast seems not to be noticed with white on black images unless there is a glow associated with boundaries. However, low ANSI contrast does act to degrade color variations, causing images to be less vibrant and more "flat".

In a typical 8-bit color encoding scheme, object brightness is sampled, and thus can only be displayed, with a coarseness of ~1/512 of the maximum signal. This means that finite leakage close to this level is not clearly resolvable from signal sampling error. Hence ~1000:1

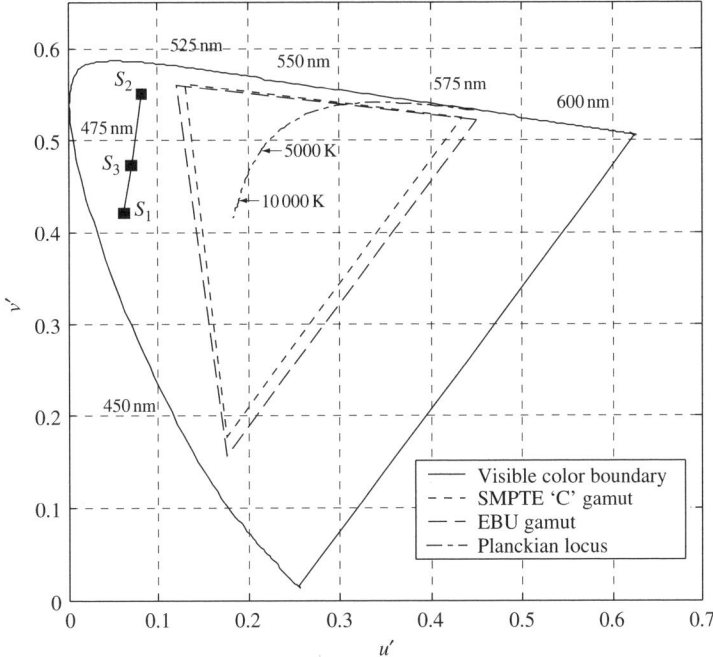

Figure 2.6 The modified color space (u', v') with spatially invariant JND

sequential contrast in a typical 8-bit color scheme could be considered adequate, but higher contrasts can act to hide OFF-state color issues (uniform non-neutral or non-uniform neutral OFF-states).

ANSI contrast represents the extent to which light from bright areas of a displayed image contaminates dark areas. Though a problem in all projection display technologies, ANSI contrast is particularly difficult to control in reflective LC microdisplay projectors (Section 7.5). Good systems are currently approaching 200:1, but the 500:1 mark is desired for the same reason argued for sequential contrast.

2.2.7 Size, Resolution, Registration, and Distortion

Experiments indicate that the eye can resolve $>1/60°$ [Robson J. G., 1966], which corresponds to resolving ~0.5 mm at a distance of ~2 m. This is consistent with a 70″ diagonal screen displaying a full HD format image of 1920 × 1080 full-color pixels. At this resolution, registration of superimposed RGB pixels should be within ~0.5 pixel throughout the image. Distortion in video images is not easily discernible. However, high-resolution projectors will be used to display generated images containing straight lines, which, if curved, can easily be detected. Deviation of a line close to the screen boundary in an RPTV by ~1% of the image size is noticeable. A realistic specification with either pin-cushion, sag, or barrel distortion bulge, is <0.5%.

30 LC PROJECTION SYSTEM BASICS

2.2.8 Electronic and Panel-related Metrics

Flicker, color depth, cross-talk, signal-dependent non-uniformities, etc., are all important for an attractive display image. Because these relate to panel properties, and are not affected by the optical engineering of the projector, they will not be discussed here. Color break-up as a consequence of color sequential illumination is also a separate issue that will be covered in detail in Chapter 11.

2.3 Photometric Measurement

Measurements of brightness and color as perceived by the eye are termed photometric. It is not in the scope of this book to describe in detail the methods of photometric measurement, which are covered in detail elsewhere [Bahadur B., 1993]. It is important, however, to briefly describe the relationship between luminance output and brightness, since luminance output is most relevant to projector design. Brightness is often difficult to quantify from the perspective of projector output, being heavily dependent on screen properties and viewing position. However, since projectors illuminate diffusing screens with near-uniform irradiance, image brightness is directly proportional to output luminance for any given screen and viewing position. Illuminance is the photopically weighted irradiance incident on the screen and is measured in lux (lm/m^2). Total illuminance relates to brightness for a given size image and is measured in lumens ($=$ lux.A, where A is the total image area).

Measurements of illuminance and color are inherently spatially dependent. To determine any single system metric, averages and standard deviations are taken over specific screen locations. ANSI lumens, a measurement of total illuminance, is the average lux incident on the screen measured over nine grid points of the projected image multiplied by the total image area [ANSI]. Brightness uniformity is then associated with illuminance uniformity, which is related directly to the standard deviation of the nine ANSI lux measurements. Color and color uniformity are measured in a similar manner. Sequential contrast is the ratio of the ON-state to OFF-state ANSI lumens, while ANSI contrast is measured as the ratio of the average center white-square to dark-square illuminance using a 4×4 checkerboard [ANSI].

2.4 Summary of What Constitutes a "Good" RPTV Display in the Current Marketplace

What constitutes a "good" display is summarized in Table 2.1.

2.5 System Engineering

A typical rear-projection system is shown in Figure 2.7.

When relating the display qualities described above to system components and their design, it is best to consider system elements in sequence from viewer to source. Front projectors constitute the subsystem without screen and folding mirrors.

Table 2.1 RPTV display properties and targets for achieving good performance

Display property	Target
Brightness	>600 nits
Viewing angle (>50% center brightness)	60° horizontal, 10° vertical
Brightness uniformity	>90%
Color R,G,B (u', v')	e.g., EBU standard (0.45,0.52), (0.12,0.56), (0.18,0.16)
Balanced white color temperature	6500–10 000 K
Color uniformity $\Delta u'v'$	<0.01
Contrast (sequential)	>1000:1
Contrast (ANSI)	>150:1
Screen size diagonal	>50″
Image resolution	>1280 × 720 (1920 × 1080)
Pixel resolution	<0.5 pixel
Pixel registration (lateral color)	<0.5 pixel

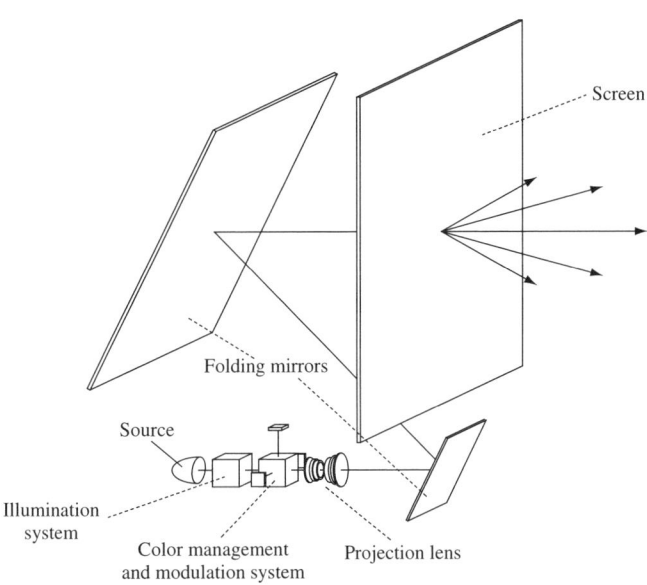

Figure 2.7 Schematic of an RPTV system

2.5.1 Rear-projection Screens

A rear-projection screen ideally performs the following functions:

(a) Directs light efficiently toward viewers to control overall brightness levels and uniformity.
(b) Discriminates from ambient light to maintain high contrast.
(c) Preserves image resolution.

(a) Directing Light

Light emanating from the projection lens of a rear-projection system typically has a large divergence angle (about ±40°) to minimize cabinet depth. Each point on the screen receives near-collimated light from markedly different angles. A viewer placed in front of a projector with the screen removed would see a bright but very small portion of the total image. To create a viewable image a screen must redirect light equally from all parts of the image toward the viewer, and preferably away from unwanted viewing regions (such as the ceiling) to increase efficiency. Commercial screens accomplish this in three steps. First, light is redirected toward a central viewing area using a Fresnel lens whose focus is equivalent to the throw distance between projection lens and screen. Second, light is spread asymmetrically over all expected viewing positions on either side of the central position (see Figure 2.8). Finally, light is diffused to avoid the viewer experiencing any sharp intensity variations. The latter two functions are best carried out with arrays of micro-optical elements for asymmetric spreading, followed by more conventional scattering diffusers. These three functions are shown in Figure 2.8.

Typical screens have the asymmetric angular spread shown in Figure 2.9 [Shimizu Y., 2003]. This concentration of light in the direction of the viewer, determined by screen gain, makes the image brighter.

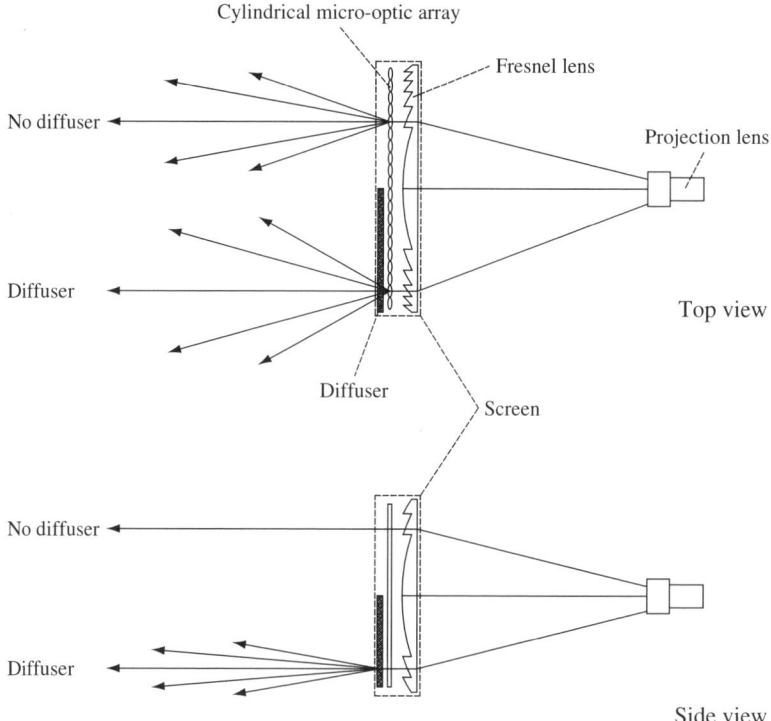

Figure 2.8 Diagram showing the effect of the screen on projected light

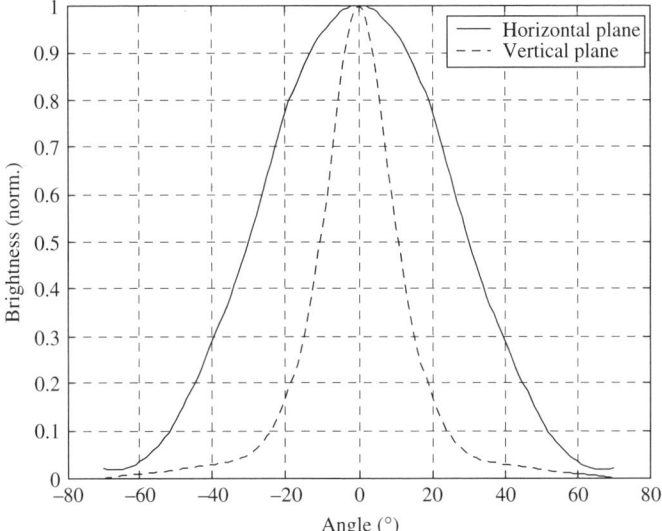

Figure 2.9 Typical screen angular intensity profile

Gain is the ratio of screen brightness (in nits) to the brightness of an ideal Lambertian screen (i.e., one that is uniformly bright for any forward viewing direction). Mathematically:

$$L = \frac{G\Phi}{\pi} \qquad (2.10)$$

where L is the luminance of the screen in nits (brightness of the image), G is the screen gain, and Φ is the screen illuminance in lux. Typical gain of an LC RPTV screen is ∼5.0. Higher gain screens, though brighter on-axis, have a very restricted field of view (FOV). To achieve the 600 nit target with a 5.0 gain screen 350 lumens is required.

(b) Ambient Discrimination
The reflection of ambient light from the front surface of a screen acts to reduce contrast and affect performance, particularly the color depth. A good RPTV screen discriminates between reflected ambient light and that emanating from the displayed image by using an absorbing polarizing layer and/or angular discrimination techniques. Polarizing screens are, however, rarely used since they often result in a significant lowering of brightness.

Angular discrimination works on the principle that near-collimated incident light from the back of the rear-projection screen can be squeezed through small apertures without significant optical loss. These same apertures can then act to absorb any incident ambient light, avoiding unwanted reflections and maximizing contrast. This is possible since it is compatible with the micro-optical elements that are used to carry out the angular spreading. Figure 2.10 shows this generic approach [Shimizu Y., 2003].

(c) Maintaining Image Resolution
To maintain pixel resolution (∼0.6 mm for a 70″ diagonal screen displaying 1920 × 1080 images), the pitch of the micro-optical array structure must be less than one-third of a pixel,

34 LC PROJECTION SYSTEM BASICS

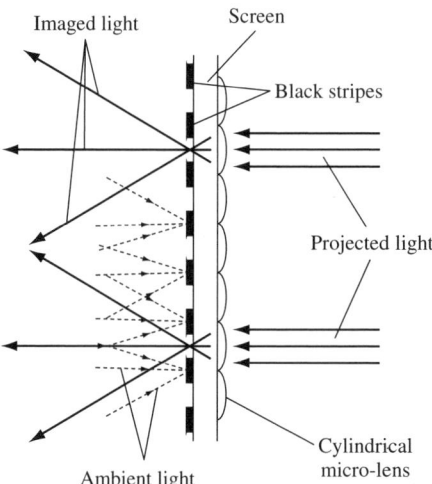

Figure 2.10 Ambient suppression through spatial filtering

or <0.2 mm, to avoid spatial artifacts in the form of Moiré fringes [Stupp E. H., 1999, chapter 14].

The most common screen used in commercial RPTV systems is supplied by Toppan and uses both angular divergence and ambient suppression techniques outlined above [Shimizu Y., 2003]. More recently screen technologies have progressed using advanced polarizing films [Wolfe C. R., 2004], Cholesteric–LC films [Umeya M., 2004], and crossed cylindrical lens arrays [Nagata Y., 2004].

2.5.2 Folding Mirrors

The function of the folding mirrors is to allow an optical system to be contained within a TV cabinet. Typically there are two folding mirrors in an RPTV system as shown in Figure 2.7, although often the first mirror is situated within the projection lens. Geometrical constraints limit cabinet thickness to ∼450 mm for typical lens divergence and screen size. Recently, however [Peterson M., 2004], an off-axis imaging system was developed that allows ∼50″ diagonal displays within much thinner cabinets (∼150 mm thick).

Folding mirrors can affect brightness through low reflectivity, but cost limits the choice of reflecting coatings to aluminum (∼92%) or in some cases more expensive coatings containing silver (e.g., Balzers silflex ∼98%). Folding mirrors can also affect color uniformity in systems that project light with color-dependent polarization. Such systems typically use a polarization-sensitive dichroic X-cube to combine primary colored light prior to projection. The angular dependence of mirror reflectivity of orthogonal polarization components (discussed later in Chapter 4) in these cases can lead to green or magenta vertical edges appearing in the image. Fortunately these effects can be corrected with a suitable *GM* output polarization filter (see Chapter 6).

2.5.3 Projection Optics

Projection optics describe the projection lens assembly, which consists of many (~10) lens elements acting to image panel pixels onto the screen. To achieve a good projection display, the lens is required to yield the following [Lee J., 2002]:

- **High resolution.** A typical minimum requirement in order to image individual pixels on a high-definition, 70″ display is 50% MTF (Modulation Transfer Function) at a screen resolution of ~0.75 line pairs per millimeter (lp/mm) equivalent to 40 lp/mm at the panel. MTF is the extent to which contrast is lost when imaging black and white stripes and is calculated by $(I_{max} - I_{min})/(I_{max} + I_{min})$, where I_{max} and I_{min} are the intensities at the center of the black and white stripes, respectively [Hecht E., 1980].

- **High magnification.** Magnification between 50 and 100 times is required since panel sizes are typically <1″ diagonal whereas desired image sizes are >50″.

- **Large field.** The field is effectively the area of the panel, which at >2000 times the required resolution is deemed large.

- **Low level of lateral color.** Lateral color is the extent to which the separate primary colored images are superimposed. Less than half a pixel variation in the relative positions of the individual primary colored pixels is acceptable.

- **Low distortion.** Distortion is the extent to which rectangular panels are imaged without bowed edges. Typically distortions of less than 1% [= (width − distorted width) × 100/width] are desired for RPTV systems to allow computer imagery to be displayed with high fidelity.

- **Short throw (large divergence).** Throw is defined as the distance between the exiting lens element and the image. Short-throw lenses (<1 m) result in large divergence and thin cabinets.

- **Long back focal length (bfl).** The bfl is the physical distance in air between the object and the first element of the projection lens. To accommodate color management components, a large bfl is required (>30 mm for a typical X-cube element). The imaging condition, however, requires the object to be at the focal length f of the lens, which for short-throw scenarios is often ~10 mm. Lenses that have bfl $<f$ are called inverse telephoto consisting of a positive lens group followed by a negative set as shown in Figure 2.11.

- **Telecentric.** Telecentricity is the condition where the principal ray from any point in the object is parallel to the optic axis of the lens. This requirement is necessary in most LC projection systems as it avoids any non-uniformity resulting from angular dependent component performance. For example, X-cubes, PBSs, and LCs all have highly angular dependent optical properties. To enforce telecentricity, a stop is placed at the focal plane of the projection lens on the image side. In the case of inverse telephoto lenses the stop is typically situated at the focal plane of the first positive element group. Figure 2.11 shows a schematic ray trace through such a lens.

- **High-angle capture (low $f_{/\#}$).** The diameter of the stop and the focal length of the first positive group define the cone of light that is collected. This cone is the same for all

36 LC PROJECTION SYSTEM BASICS

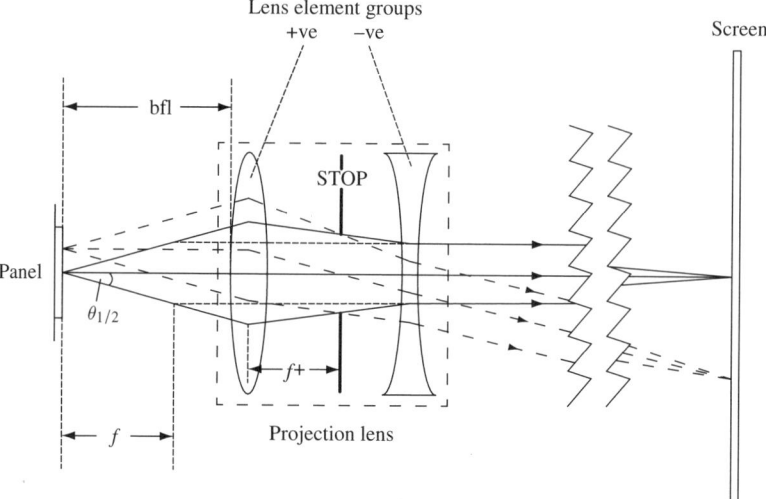

Figure 2.11 Concept of an LC imaging optical system

object points and determines how much light is captured. This relates directly to the brightness of the image, and is defined by an $f_{/\#}$, where $f_{/\#} = 1/2 \sin(\theta_{1/2})$, $\theta_{1/2}$ being the half-cone angle. The lower the $f_{/\#}$, the higher the extreme ray angles seen by the lens, in general requiring extra elements to deliver the required optical performance. Commercial projection lenses therefore operate at $<f_{/2.5}$ and are typically $f_{/2.8}$.

- **Low convergent back reflection.** Lens elements have anti-reflection coatings to maximize transmission, but the number of elements and their curvature can, particularly in LCOS systems, cause dark regions of the image to become brighter through convergent reflection. To avoid a reduction of ANSI contrast or ghosting arising from this with non-ideal lenses, polarization isolation techniques can be used (Section 7.5).

A typical lens design is shown in Figure 2.12.

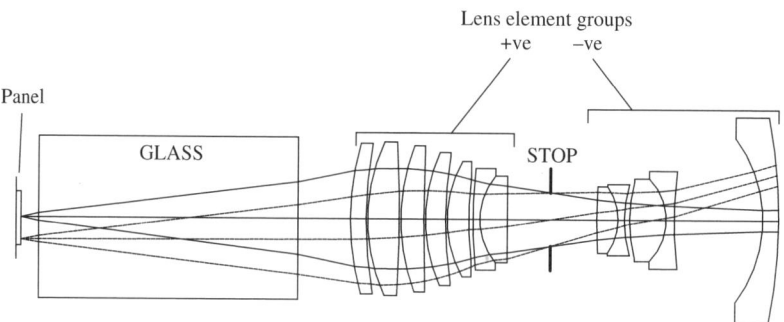

Figure 2.12 Typical projection lens [e.g., Moskovich J., 2003] used in LC video projection displays

2.5.4 Color Management and Modulation Subsystem

This subsystem is only briefly introduced here. It can be considered a series of optical elements situated before and after the panels that allow projection of a full-color image from monochrome panels. Two color management issues discussed here concern the impact of component cosmetic defects and aberrations on system performance.

High-resolution imaging requires minimal optical distortion between the imaging lens and object. Since components are typically required in this space they must have a tight defect specification. Defects can be localized, in the form of scratches and digs, or be more dispersed as phase distortion or gradual non-uniformity. These two types can be considered separately: (a) scratch and dig specifications and (b) phase deformation.

The main effect of a scratch or dig is to create color non-uniformity in the image. In the case where polarization is maintained, light is scattered by these defects out of the capture of the lens, reducing the intensity of one color. A 1% reduction in any one color can cause a localized color variation of close to a JND in the image. To avoid this, digs of diameter $<150\,\mu m$ and a scratch of width $<15\,\mu m$ are acceptable at distances >3 mm from the panel. In the case where polarization is altered by scattering from defects, more stringent system-specific limits are necessary.

Phase deformation is due to local variation in the refraction of rays by components that are not optically flat. To have $<2.5\,\mu m$ equivalent ray shift at the panel (i.e., a quarter of a pixel), a ray must not be refracted by $>0.000\,25/d$ radians, where d is the distance in mm between the phase distortion and the panel. This implies that phase distortions are more problematic at greater distances from the object, with a maximum allowable distortion of $\sim 13/d$ waves/inch.

2.5.5 Illumination System

The purpose of the illumination system is:

(a) To create homogenized, telecentric illumination at the panel.

(b) To efficiently polarize the incoming beam.

(a) Homogenization
Current projection systems use two approaches: pared lens arrays [Wang S., 1988; van den Brant A. H. J., 1992]; and an integrator rod or light-pipe [Moskovich J., 1997]. The paired lens array approach is shown in Figure 2.13 in which two identical arrays of rectangular lenses, matched in aperture ratio to the panel, are placed one focal distance apart. An array of images is created by the first lens array in the plane of the second. Each source image is incident on a single element of the second array. Superimposed images of the rectangular elements of the first array are then formed at infinity. A combining lens placed directly after the second lens array acts to map this rectangular light patch to the location of the panel. To enforce telecentricity, a second combining lens is introduced at its focal length from the first.

In a light-pipe system, the source is focused onto the entrance of a rectangular glass rod (or pipe), which forces the light to be multiply reflected prior to exiting (see Figure 2.14).

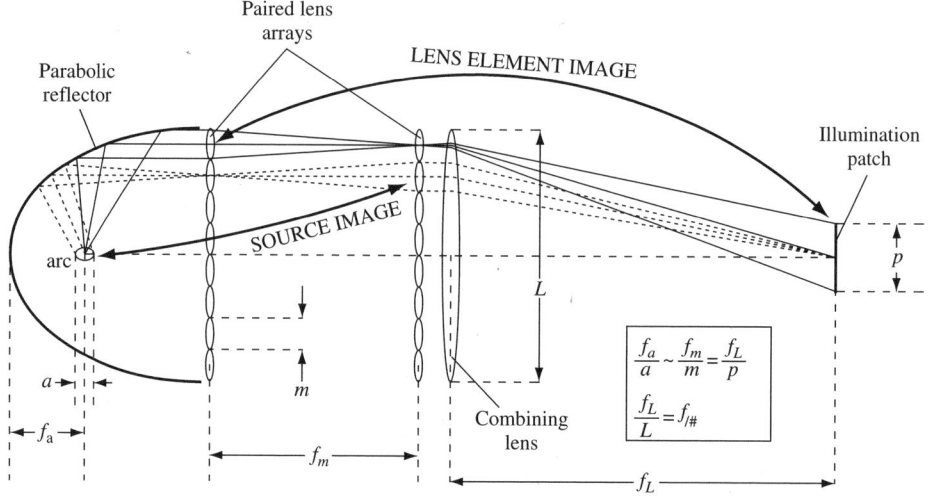

Figure 2.13 Paired lens array system for creating uniform panel illumination

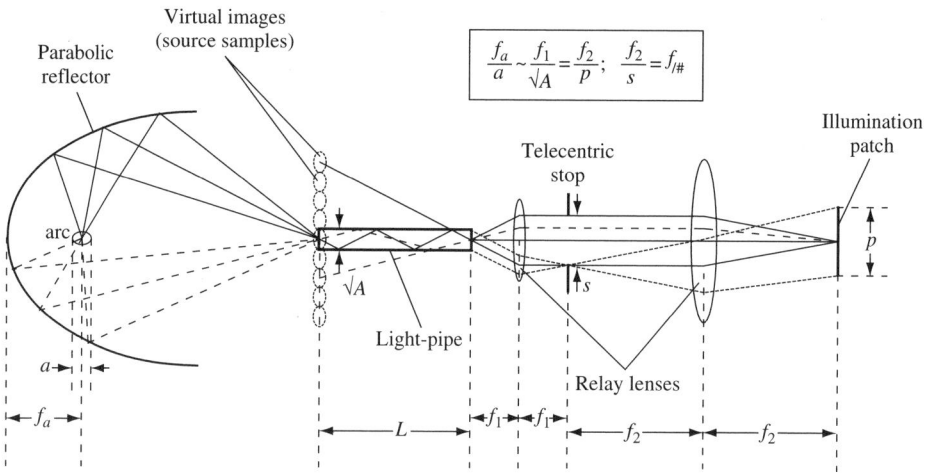

Figure 2.14 Schematic of the light-pipe homogenizing system

The multiple reflections effectively split the source into an array of sources similar to that seen in the lens array approach. The uniform light exiting the rectangular pipe is then relayed onto the panel with a telecentric relay system.

A homogenizing system must provide uniform illumination with high efficiency. The degree of uniformity of the above cases is determined either by the number of elements in the paired lens arrays, or by the length of the light-pipe. In either case, uniformity relates directly to the number of samples of the source. To ensure >90% uniformity with universally accepted ultra-high-pressure (UHP) arc sources, at least an 8×8 sampling is required, which in the case of lens arrays is the minimum array size. For the light-pipe, the number of samplings is related to the length of the pipe, which is typically chosen to have four bounces for the extreme angles of the focused source. Mathematically $L \approx 4n\sqrt{A}/NA$ where n is the rod refractive index, A is the area of the pipe, and NA is the numerical aperture of the focused source [Stupp E. H., 1999, pp. 121–123]. Increasing the number of samplings improves uniformity but at the expense of throughput stemming from edge effects.

From a system standpoint, efficient design comes through matching the étendue of the homogenizer to the remainder of the optical projection system (Section 2.6) [McGettigan T., 2004]. In LC projectors, this results in the early removal of unusable light, avoiding unwanted light leakage that reduces contrast. High efficiency is also achieved through high-quality imaging of the light patch onto the panel to avoid soft edges. With soft edges, light is wasted because the panel must be overfilled to avoid edge non-uniformities. Typical overfill for a two-lens illumination optic system is between 5% and 10%.

(b) Polarization Conversion

Polarization conversion techniques are typically used to efficiently create a polarized illumination beam. The most common is the system first commercialized by Seiko Epson [Itoh Y., 1997], in which a linear array of PBSs is incorporated into the paired lens array homogenizer. Shown in Figure 2.15, it uses lens arrays whose elements have focal lengths half that of an unpolarized system, producing a sparse array of source images. The PBS array splits the beams into orthogonal polarizations that fill the aperture, allowing conversion to a single polarized beam with a half-wave retarder film.

With a light-pipe, polarization conversion is more difficult because polarization of reflected light within the pipe is not well preserved. Many approaches have been proposed [Stupp E. H., 1999, pp. 137–140] with the most recent being based on wire grid polarizers [Duelli M., 2002] (see Figure 2.16). Commercially available polarization conversion light-pipes remain rare.

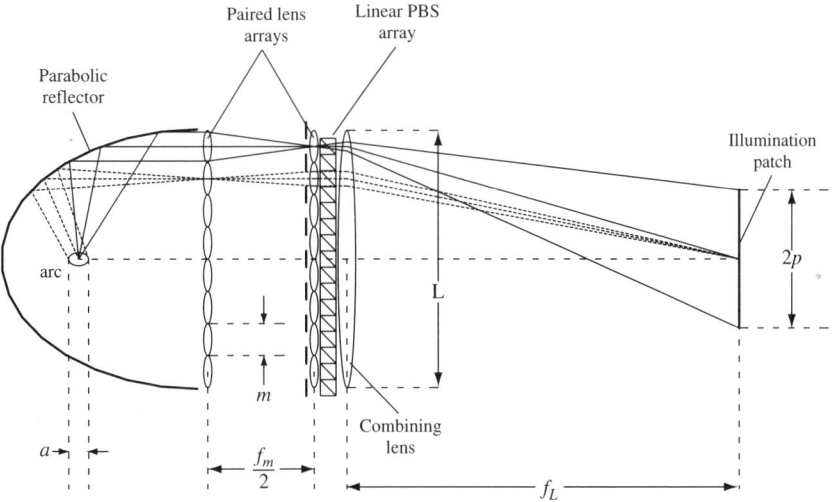

Figure 2.15 Polarization conversion with paired lens arrays. The net effect is to increase the illumination light patch for a given convergence angle, i.e., it doubles the optical extent of the illumination or output étendue

40 LC PROJECTION SYSTEM BASICS

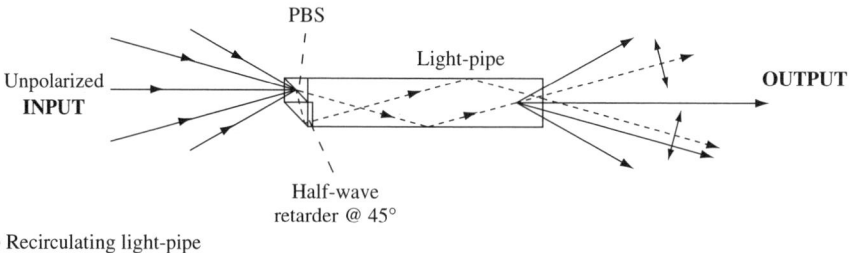

(a) Recirculating light-pipe

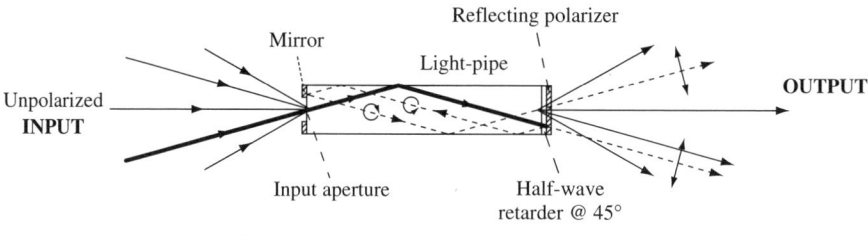

(b) PCS light-pipe

Figure 2.16 Schematics of light-pipe polarization conversion systems

Though in principle polarization conversion can be accomplished with 100% efficiency, passive optical losses and étendue considerations result in significantly lower efficiency (see Section 2.6).

2.5.6 Light Source

All commercially available LC projectors use UHP mercury lamps as a light source. Developed by Philips [Fisher E., 1992; 1998], the lamp produces the brightest available source of light, which is necessary to illuminate small microdisplays. Brightness of a UHP lamp is determined by input electrical power and arc gap. Currently 1 mm is a typical arc gap with powers up to 250 W. Although progress is being made in reducing stable arc sizes [Moench H., 2000], lifetime issues favor lower wattage lamps, with the 100 W, 1 mm lamp quoting >10 000 h lifetime.

Common to all UHP lamps is the characteristic spectrum shown in Figure 2.17. Though essentially white, it is somewhat red deficient and green rich. The high-pressure requirement exists to boost the red output. The peaked nature of this spectrum demands precise color management particularly since the yellow 580 nm spike must be almost entirely removed. Light from these lamps is typically collected via parabolic or elliptical reflectors for use with the paired lens or light-pipe homogenizer.

In order to facilitate high sequential contrast and low gray-level bit depth, the source is ideally attenuated in accordance with the video signal. For example, when displaying a fully

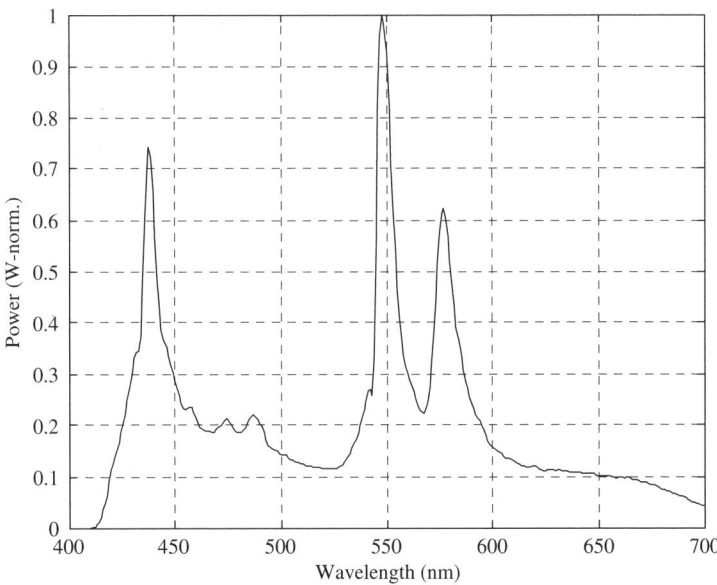

Figure 2.17 Spectrum of the standard UHP illumination source (includes UV filter)

black screen between video segments, ideal infinite sequential contrast could be obtained by simply turning off the source. Likewise in dark scenes with brightness levels only a small fraction of maximum, the source could be attenuated, allowing the full bit depth of the panel to be available over the scaled luminance level. Unfortunately, the UHP source is not easily actively attenuated without unacceptable lifetime and arc instability issues. To achieve these advantages without altering the arc output, recent systems employ an auto-iris as a dynamic aperture stop. This electromechanical device could also be LC based, though throughput issues dictate that this approach is most suitable for polarization-based projectors.

2.6 Étendue Considerations

Étendue is often confused with brightness, and for many it is a new concept. Étendue is the extent to which an optical system allows the unimpeded propagation of light. It is determined purely by geometrical considerations and is derived directly from the Lagrangian invariant [Doany F. E., 1998; Stupp E. H., 1999, pp. 241–254; McGettigan T., 2004]. For any optical system, the Lagrangian invariant is the product of the acceptance solid angle and the size of an aperture at any given plane (see Figure 2.18). If all optical elements comply with this condition, then the system has a common invariant, which is then the étendue of the system. For unmatched systems, the invariant can differ and the étendue is then the value of the smallest invariant, or the weakest link in the chain.

Mathematically, the limiting optical invariant and therefore the étendue of the system in Figure 2.18 is:

$$n^2 \int \Omega(r) \cdot d^2 r$$

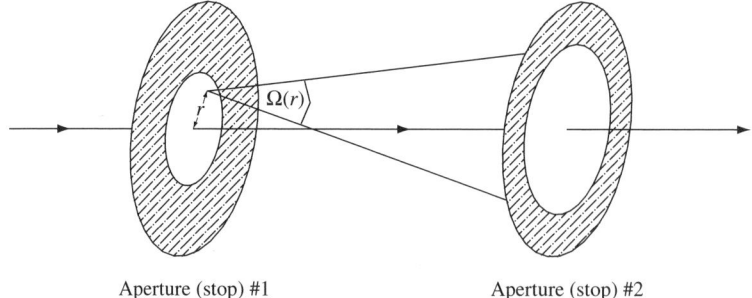

Figure 2.18 Optical two-stop subsystem

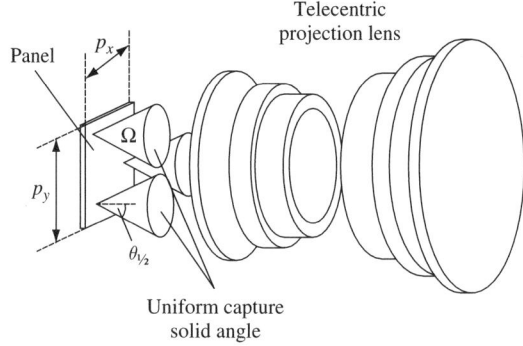

Figure 2.19 Étendue as defined by the panel and projection lens

where $\Omega(r)$ is the capture solid angle defined by the second aperture of a point at r within the first aperture. The integration is over the entire first aperture and n is the refractive index of the medium. In the more practical case of an LC projection system, the étendue is almost always determined by the panel size and the capture of the telecentric projection lens as shown in Figure 2.19.

Since the capture solid angle Ω is the same for all panel positions in an LC projection system, the étendue is simply $p_y p_x \Omega$, or $2\pi[1-\cos(\theta_{1/2})].A$ where $\theta_{1/2}$ is the half-cone angle and A is the area of the panel. Taking the small-angle approximation, which is applicable for most projection systems, and expressing $\theta_{1/2}$ in terms of the lens $f_{/\#}$, the étendue is $\pi A/4(f_{/\#})^2$. The imaging system étendue for a typical 0.7" diagonal panel with an $f_{/2.8}$ lens is $\sim 14\,\text{mm}^2$ sr.

The importance of étendue is in determining optical system throughput. Any given source emits light into an optical system with efficiency dependent on the system étendue. If the étendue of a system is equal to the product of the source area and the total emission solid angle, then all light from the source can pass through the system assuming no other losses. LC projection systems have their étendue determined by the imaging system as calculated above, which is typically less than that required to capture all light from standard UHP arc sources. The extent to which light can be coupled into a system of a given étendue is primarily determined by the arc size. Figure 2.20 shows the coupling efficiency versus

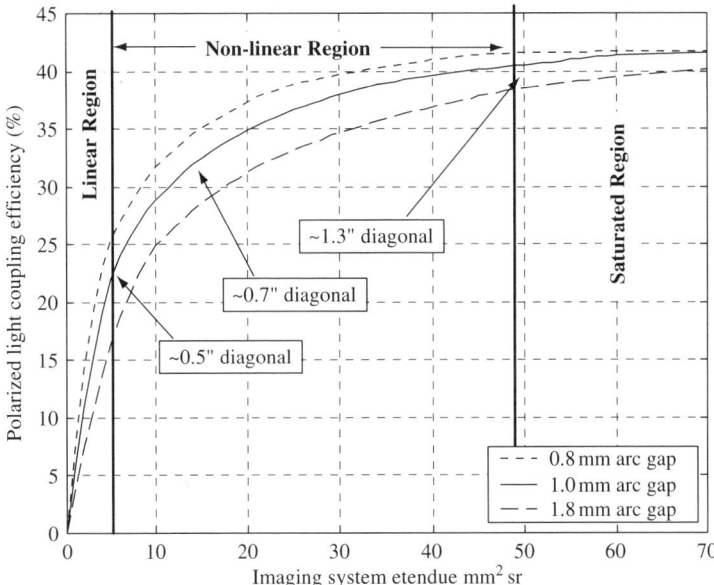

Figure 2.20 Light coupling efficiency between UHP sources of various arc gap sizes and polarized LC projection systems as a function of étendue

system étendue for UHP sources with different arc gaps and with a standard parabolic or elliptical reflector arrangement.

There are three distinct regions of the curves shown in Figure 2.20:

1. A linear region, where output lumens are directly proportional to the area of the panel for a given imaging $f_{/\#}$. Panels below 0.5″ diagonal typically fall into this region.

2. A non-linear region, where most projection systems operate. Panel sizes are 0.5″ to 1.3″ in diagonal (7–50 mm² sr).

3. A saturated region where the system's étendue exceeds that required to effectively capture all light.

For any LC microdisplay projection system where the étendue is in the linear or non-linear regions of the coupling curve, higher coupling efficiency is achieved with a smaller arc gap. For a system using a 1 mm arc gap and an $f_{/2.8}$ imaging system, going from a 0.7″ to a 0.5″ diagonal panel reduces the display brightness by ∼30%.

Optical conversion systems can have a significant impact on étendue since they create a mismatch between input and output étendue by spatially separating otherwise superimposed light beams. Examples of optical conversion systems include the polarization conversion systems (PCSs) described above and scrolling color illumination systems using either spiral color wheels [Dewald D. S., 2001] or prisms [Shimizu J. A., 2001]. Taking the most common example of a conversion system, i.e., the paired lens array PCS, the effect of étendue mismatch can be illustrated. In this case, the input étendue is the product of the total area of the entrance apertures adjacent to the second lens array and the solid angle subtended

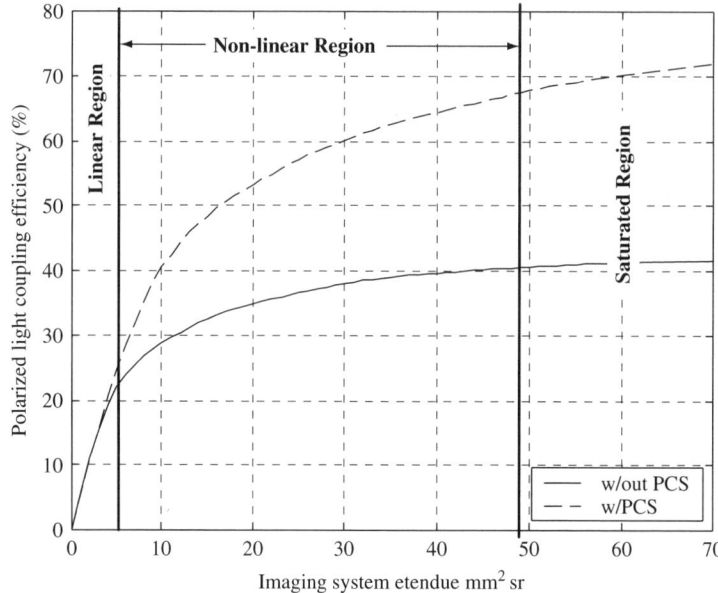

Figure 2.21 Source–system coupling efficiency as a function of system limiting étendue for systems with and without a PCS

by the light from elements of the first array. In the system where no PBS array is present, the area of the entrance apertures at the second array is equal to the total area of the array. Introducing a PBS array without changing solid angles forces the system to have a striped aperture that halves the entrance aperture at the second lens array. The étendue of the input is thus half that of the output. Matching the output étendue to the limiting imaging system therefore halves the input étendue relative to a system with no PCS. Since the source is clipped by even the original input étendue, light is further lost when a PCS is introduced. However, as the maximum throughput in a polarized LC system is 100% with a PCS and only 50% in the original no PCS case, there can be a net gain in throughput. Figure 2.21 shows the coupling efficiency with and without a PCS as a function of system limiting étendue. For panel diagonal sizes >0.5″ (~5 mm² sr at $f_{/2.8}$) there is a net gain, with ~30% throughput increase possible for 0.7″ panels at $f_{/2.8}$.

In most cases the modest 50% increase in throughput is sufficient to warrant its general implementation in both front-projection and RPTV systems.

References

[ANSI] ANSI IT7.215–1992, Data projection equipment and large screen displays – test methods and performance characteristics.

[Bahadur B., 1993] B. Bahadur, *Liquid Crystals: Applications and Uses*, World Scientific, Singapore, vol. 2, chapter 8, 1993.

[van den Brant A. H. J., 1992] A. H. J. van den Brant and W. A. G. Timmers, Optical illumination system and projection apparatus comprising such a system, US Patent 5,098,184, 1992.

[Campbell F. W., 1968] F. W. Campbell and J. G. Robson, Application of Fourier analysis to the visibility of gratings, *J. Physiol. (London)*, 197, pp.551–566, 1968.

[Dewald D. S., 2001] D. Scott Dewald, S. M. Penn, and M. Davis, Sequential color recapture and dynamic filtering, SID'01, Digest, p.1076, 2001.

[Doany F. E., 1998] F. E. Doany, R. N. Singh, A. E. Rosenbluth, and G. L.–T. Chiu, Projection display throughput efficiency of optical transmission and light-source collection, *IBM J. Res. Dev*, 42, pp.387–399, 1998.

[Duelli M., 2002] M. Duelli, T. McGettigan, and C. Pentico, Integrator rod with polarization recycling functionality, SID'02, Digest, p.1078, 2002.

[Fisher E., 1992] E. Fisher and H. Hoerster, High pressure mercury vapor discharge lamp, US Patent 5,109,181, 1992.

[Fisher E., 1998] E. Fisher, Ultra high performance discharge lamps for projection TV systems, 8th International Symposium on Light Sources, Greifswald, Germany, 1998.

[Guild J., 1931] J. Guild, The colorimetric properties of the spectrum, *Philos. Trans. R. Soc.*, A230, pp.149–187, 1931.

[Hecht E., 1980] E. Hecht and A. Zajac, *Optics*, Addison-Wesley World Student Series Edition, 5th printing, p.420, Addison–Wesley, Reading, MA, 1980 .

[Hunt R. W. G., 1987] R. W. G. Hunt, *Measuring Color*, Ellis Horwood, Chichester, 1987.

[Itoh Y., 1997] Y. Itoh, J. I. Nakamura, K. Yoneno, H. Kamakura, and N. Okamoto, Ultra-high-efficiency LC projector using a polarized light illumination system, SID'97, Digest, p.993, 1997.

[Jones A. H., 1968] A. H. Jones, Optimum color analysis characteristics and matrices for color television cameras with three receptors, *J. SMPTE*, 77, p.108, 1968.

[Judd D. B., 1975] D. B. Judd and G. Wyszecki, *Color in Business, Science and Industry*, John Wiley & Sons, Inc., New York, 1975.

[Lee J., 2002] J. Lee, The quality analysis of long back-focal projection lens, SID'02, Digest, p.642, 2002.

[MacAdam D. L., 1937] D. L. MacAdam, Transformations of I.C.I. color specifications, *J. Opt. Soc. Am.*, 17, p.294, 1937.

[McGettigan T., 2004] T. McGettigan, Etendue and its application in light engine lumens budgeting, SID'04, Applications Tutorials, p.1, 2004.

[Moench H., 2000] H. Moench, G. Derra, E. Fischer, and X. Riederer, ARC stabilisation for short arc projection lamps, SID'00, Digest, p.84, 2000.

[Moskovich J., 1997] J. Moskovich, Telecentric lens systems for forming an image of an object composed of pixels, US Patent 5,625,495, 1997.

[Moskovich J., 2003] J. Moskovich, Compact, telecentric projection lenses for use with pixelized panels, US Patent 6,563,650, 2003.

[Nagata Y., 2004] Y. Nagata, A. Kagotani, and K. Ebina, An advanced projection screen with a wide vertical view angle, SID'04 Digest, p.846, 2004.

[Peterson M., 2004] M. Peterson, D. Slobodin, J. Gohman, and S. Bierhuizen, Rear projection display system, US Patent 6,728,032, 2004.

[Robson J. G., 1966] J. G. Robson, Spatial and temporal contrast sensitivity functions of the visual system, *J. Opt. Soc. Am.*, 56, pp.1141–1142, 1966.

[Shimizu J. A., 2001] J. A. Shimizu, Scrolling color LCOS for HDTV rear projection, SID'01, Digest, p.1072, 2001.

[Shimizu Y., 2003] Y. Shimizu, S. Iwata, T. Yoshida, S. Takahashi, and T. Abe, A fine-pitch screen for rear projection TV, SID'03 Digest, p.886, 2003.

[Stupp E. H., 1999] E. H. Stupp and M. S. Brennesholtz, *Projection Display*, John Wiley & Sons, Ltd, Chichester, 1999.

[Umeya M., 2004] M. Umeya, M. Hatano, and N. Egashira, New front-projection screen comprised of cholesteric-LC films, SID'04, Digest, p.842, 2004.

[Wang S., 1988] S. Wang and L. Ronchi, Principles and design of optical array, in E. Wolf (ed.), *Progress in Optics*, 25, pp.279–347, North-Holland, Amsterdam, 1988.

[Williamson S. J., 1983] S. J. Williamson and H. Z. Cummins, *Light and color in nature and art*, John Wiley & Sons, Inc., New York, 1983.

[Wolfe C. R., 2004] C. R. Wolfe, M. Paukshto, and p.Smith, Design and performance of high contrast polarized light front-projection screens, SID'04, Digest, p.838, 2004.

[Wright W. D., 1928] W. D. Wright, A re-determination of the trichromatic coefficients of the spectral colours, *Trans. Opt. Soc.*, 30, pp.141–164, 1928.

[Wyszecki G., 1982] G. Wyszecki and W. S. Stiles, *Color Science*, John Wiley & Sons, Inc., New York, 1982.

3

Polarization Basics

3.1 Introduction

Polarization control is fundamental to LCD projection. The understanding of the basic properties and mathematical representations of polarization is therefore essential. This chapter introduces the various mathematical constructs that describe the polarization of light and presents the methods by which optical materials manipulate the state of polarization. It covers the basic electromagnetic theory of plane waves, polarization representations, and the relationship between them. Interaction with isotropic materials will be briefly introduced as a precursor to propagation through anisotropic materials. The concepts of birefringence and retardance will be reviewed. The last sections will be devoted to the mathematical modeling of polarization for light propagating through multiplayer anisotropic media.

3.2 Electromagnetic Wave Propagation

Electromagnetic theory describes light as a propagating electromagnetic wave [Born M., 1980, chapter 1]. It generally requires four basic field vectors for its complete description: the electric field strength E, the electric displacement D, the magnetic field H, and the magnetic flux density B. Of these four vectors, E, which vibrates in time and space as the beam propagates, is chosen to define the state of polarization of light waves. Once the polarization of E has been determined, the three remaining field vectors D, H, and B can be found through Maxwell's field equations and associated material relations. In an isotropic medium, the direction of vibration is always orthogonal to the direction of propagation. For this transverse mode, there are two independent directions of vibration, which we can choose arbitrarily. Fourier analysis of the temporal (t) and spatial (r) variation of a light wave's electric field $E(r, t)$ yields spectral components with frequencies from ultraviolet (UV) to

---

Polarization Engineering for LCD Projection M. G. Robinson, J. Chen and G. D. Sharp
© 2005 John Wiley & Sons, Ltd

infrared (IR). Visible light extends from 4×10^{14} to 8×10^{14} hertz (Hz). For projection applications, we can assume that the system is linear relative to light intensity. This means that we can treat the transmission or reflectance of light independently for each wavelength. In addition, we can also calculate the optical response independently for each angle of incidence. Complete solutions then constitute a linear sum of the individual monochromatic plane wave solutions. In this manner, we need only consider in this book the mathematical manipulation of monochromatic plane optical waves.

3.2.1 Polarization of Monochromatic Waves [Born M., 1980; Yeh P., 1988; Azzam R. M. A., 1989; Yeh P., 1999]

A monochromatic plane wave propagating in an anisotropic and homogeneous medium can be written:

$$E = A \cos(\omega t - k \cdot r) \tag{3.1}$$

where ω is the angular frequency, k is the wavevector, and A is a constant vector representing the amplitude. The magnitude of the wavevector k is related to the frequency by the equation:

$$k = n\omega/c = 2\pi n/\lambda \tag{3.2}$$

where n is the refractive index of the medium, c is the speed of light in vacuum, and λ is the wavelength of light in vacuum. The refractive index in general depends upon wavelength, which is known as dispersion. It is a real number for transparent materials, but complex for materials that absorb. Since electromagnetic waves are transverse in nature, the electric field vector is always perpendicular to the propagation vector; that is:

$$k \cdot E = 0 \tag{3.3}$$

From Maxwell's equations, it can be proven that the time evolution of the electric field vector for monochromatic light is sinusoidal; that is, the electric field oscillates with a single frequency. Assuming a light wave propagating along the z-axis, the electric field vector must lie in the xy plane. In this way, two mutually independent components of the electric field vector can be expressed as:

$$\begin{aligned} E_x &= A_x \cos(\omega t - kz + \delta_x) \\ E_y &= A_y \cos(\omega t - kz + \delta_y) \end{aligned} \tag{3.4}$$

We have used two independent positive amplitudes A_x, A_y and two independent phases δ_x, δ_y to reflect the mutual independence of the two components. Since amplitude is positive by definition, the phase angles are defined within the range $-\pi < \delta \leq \pi$. The locus of two independent orthogonal oscillations at the same frequency is an ellipse, which is analogous to the classic motion of a two-dimensional harmonic oscillator. After several steps of elementary algebra that eliminates $(\omega t - kz)$ from equation (3.4), we obtain:

$$\left(\frac{E_x}{A_x}\right)^2 + \left(\frac{E_y}{A_y}\right)^2 - \frac{2 E_x E_y \cos \delta}{A_x A_y} = \sin^2 \delta \tag{3.5}$$

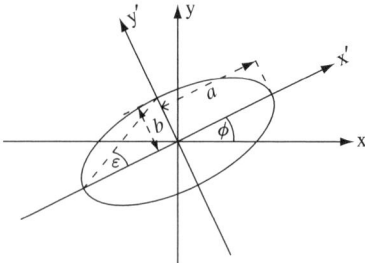

Figure 3.1 Polarization ellipse of monochromatic light

where $\delta = (\delta_y - \delta_x)$ and $-\pi < \delta \leq \pi$. By rotating coordinates, we are able to simplify equation (3.5). Let x' and y' be the new set of axes along the principal axes of the ellipse as shown in Figure 3.1. The equation of the ellipse in this new coordinate system becomes:

$$\left(\frac{E'_x}{a}\right)^2 + \left(\frac{E'_y}{b}\right)^2 = 1 \tag{3.6}$$

where:

$$a^2 = A_x^2 \cos^2 \phi + A_y^2 \sin^2 \phi + 2A_x A_y \cos \delta \cos \phi \sin \phi$$
$$b^2 = A_x^2 \sin^2 \phi + A_y^2 \cos^2 \phi - 2A_x A_y \cos \delta \cos \phi \sin \phi \tag{3.7}$$
$$\tan(2\phi) = \frac{2A_x A_y}{A_x^2 - A_y^2} \cos \delta$$

Complete specification of the elliptical polarization requires:

1. The azimuthal angle ϕ ($-\pi/2$ to $\pi/2$). The angle between the major axis of the ellipse and the x-axis.

2. The ellipticity e. The ratio of the length of the semi-minor axis of the ellipse b to the length of its semi-major axis a; that is:

$$e = \frac{b}{a} \tag{3.8}$$

3. The handedness of the polarization ellipse. The polarization is right-handed if the ellipse rotates in a clockwise sense, and left-handed if the ellipse rotates in an anti-clockwise sense, when looking along the direction of propagation at any given position in space. It is convenient to incorporate the handedness in the definition of the ellipticity e by allowing the ellipticity to assume positive and negative values, corresponding to right-handed and left-handed polarizations, respectively. The ellipticity angle is defined as:

$$\varepsilon = \tan^{-1}(e) \tag{3.9}$$

50 POLARIZATION BASICS

where ε is limited to $\pm\pi/4$. Since e relates directly to δ, its sign also defines the handedness of the polarization state, such that $\delta > 0$ corresponds to left-handed, while $\delta < 0$ is right-handed.

4. The amplitude. A measure of the strength of vibration. Its square is proportional to the energy density of the wave at any fixed point of observation:

$$A = (a^2 + b^2)^{1/2} \tag{3.10}$$

5. The absolute phase. The angle between the initial position of the electric field vector at time $t = 0$ and the major axis of the ellipse. Since all detectors respond in a time average manner, only relative phase is of consequence to the problems faced in this book.

Figure 3.2 shows polarization ellipses for various phase angles $\delta(A_x = A_y)$. As we can see, the circular and linear states of polarization (SOPs) are the special cases of the more general state of elliptical polarization, and are generated when the ellipticity, e, assumes the special values of ± 1 and 0, respectively. It happens that circular and linear SOPs are the most important states in an LCD projection system. For example, in order to achieve good contrast and maximum brightness in a reflective LCOS projection system, the black SOP at the mirror must be linear while the white SOP must be circular. Since the LC is sandwiched between crossed polarizers in a transmissive HTPS projection system, the SOP should be linear and matched to the front polarizer to ensure a good contrast [Chen J., 2004].

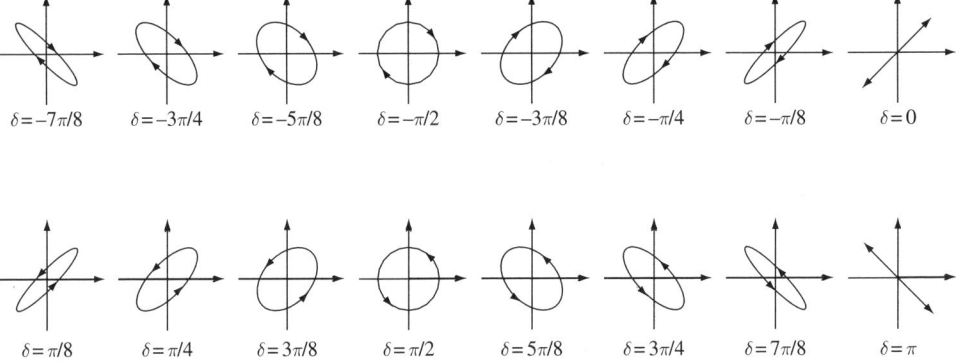

Figure 3.2 Polarization ellipses at various phase angles under assumption of $A_x = A_y$ in equation (3.4)

3.2.2 Complex Number Representation [Yeh P., 1988; Azzam R. M. A., 1989; Yeh P., 1999]

From equation (3.7), a general SOP, which is independent of overall intensity, is completely defined if we know the ratio of A_x to A_y, and relative phase δ. A convenient two-dimensional representation of the SOP uses a single complex number χ, which is defined as:

$$\chi = e^{i\delta} \tan \Psi = \frac{A_y}{A_x} e^{i(\delta_y - \delta_x)} \tag{3.11}$$

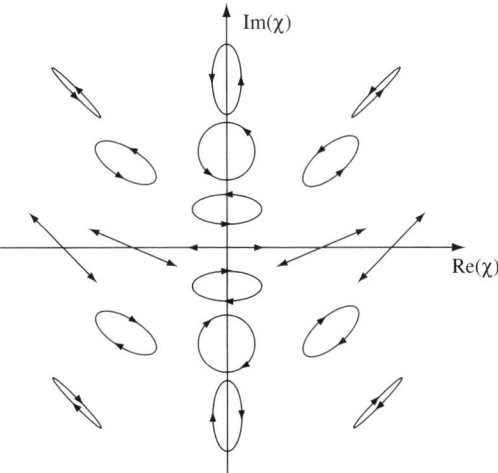

Figure 3.3 Polarization states to different points in complex plane of polarization

where the angle Ψ is confined to the range between 0 and $\pi/2$. Once again, the SOP can be expressed entirely by δ and Ψ, or simply the complex number χ. Figure 3.3 illustrates various polarization states in the complex plane. It can be seen that each point in the complex plane represents a unique polarization state. Left-handed elliptical polarization states are in the upper half of the plane, while right-handed ones are in the lower half. The states along the real axis are linearly polarized with various inclination angles.

The relationship between the two SOP representations can be determined from simple mathematical manipulation. From equation (3.7), the inclination angle ϕ and the ellipticity angle $\varepsilon (\varepsilon = \tan^{-1}(e))$ are simply related to χ by:

$$\tan(2\phi) = \frac{2\mathrm{Re}(\chi)}{1-|\chi|^2} = \tan(2\Psi)\cos\delta \qquad (3.12)$$

$$\sin(2\varepsilon) = -\frac{2\mathrm{Im}(\chi)}{1+|\chi|^2} = -\sin(2\Psi)\sin\delta \qquad (3.13)$$

Orthogonal polarization states satisfy the relationships:

$$\chi^*\chi_{orth} = -1 \qquad \chi_{orth} = \frac{-\chi}{|\chi|^2} = -\frac{e^{i\delta}}{\tan\Psi} \qquad (3.14)$$

3.2.3 Jones' Vector Representation [Yeh P., 1988; Azzam R. M. A., 1989; Yeh P., 1999]

Whilst the representations of the SOP so far presented are very concise, they do not contain information regarding intensity. To investigate the brightness and contrast of an LCD projection system it is clearly necessary to calculate intensities. The Jones' vector, introduced by R. C. Jones [Jones R. C., 1941], is a more convenient means of describing both the SOP and the intensity of a plane wave. Jones' matrices that act to propagate Jones' vectors are

52　POLARIZATION BASICS

a very effective way to determine the polarization state evolution as the light propagates through a birefringent system. A Jones' vector describing a polarization state of a plane wave propagating along z in a conventional Cartesian coordinate system is expressed as:

$$J = \begin{pmatrix} A_x e^{i\delta_x} \\ A_y e^{i\delta_y} \end{pmatrix} = A_x e^{i\delta_x} \mathbf{x} + A_y e^{i\delta_y} \mathbf{y} \quad (3.15)$$

The intensity of light is proportional to $J \cdot J^*$, which equals $A_x^2 + A_y^2$. If we are only interested in the polarization state of a wave, it is convenient to use the normalized Jones' vector which satisfies the condition that $J \cdot J^* = 1$. It is straightforward to show that an arbitrary SOP defined using the complex notation of the previous section can be represented by the following Jones' vector:

$$J(\Psi, \delta) = A \begin{pmatrix} \cos \Psi \\ e^{i\delta} \sin \Psi \end{pmatrix} = A \left(\cos \Psi, \, e^{i\delta} \sin \Psi \right) \quad (3.16)$$

where $A = \sqrt{(A_x^2 + A_y^2)}$.

Jones' vectors representing light with linear polarization along the x- and y-axes are $(1, 0)$ and $(0, 1)$ respectively, while Jones' vectors for a right-hand (R) and left-hand (L) circular polarized light are $(1, -i)$ and $(1, i)$. A Jones' vector can be expressed by any two arbitrary orthogonal eigenstates. In conventional Cartesian coordinates, a Jones' vector is defined typically with its two basis states corresponding to linear polarization along the x- and y-axes. In certain scenarios, it is more convenient to use the R/L circular polarization states, which are called circular Jones' vectors. We can use the following transformation to convert from Cartesian to circular basis states:

$$\begin{pmatrix} x \\ y \end{pmatrix} = \frac{1}{\sqrt{2}} \begin{pmatrix} 1 & 1 \\ i & -i \end{pmatrix} \begin{pmatrix} R \\ L \end{pmatrix} \quad (3.17)$$

For the orthogonal polarization state, the Jones' vector satisfies $J_{orth}^+ \cdot J = 0$.

An equivalent Jones' vector for a polarization state with inclination angle ϕ and ellipticity e can be obtained by a two-step synthesis procedure as shown Figure 3.4. Mathematically the Jones' vector is derived by:

$$J = A \cdot R(\phi) \begin{pmatrix} \cos \varepsilon \\ -i \sin \varepsilon \end{pmatrix} = A \cdot \begin{pmatrix} \cos \phi \cos \varepsilon + i \sin \phi \sin \varepsilon \\ \sin \phi \cos \varepsilon - i \cos \phi \sin \varepsilon \end{pmatrix} \quad (3.18)$$

where $R(\phi)$ is a rotation matrix and A is the amplitude.

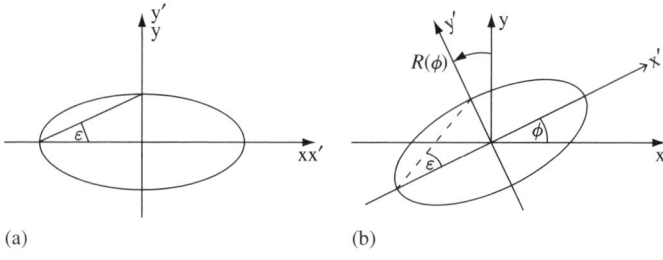

(a)　　　　　　　　　　(b)

Figure 3.4 Two-step synthesis procedure of the Jones' vector of an arbitrary SOP

The equivalent complex number representation is:

$$\chi = \frac{\sin\phi\cos\varepsilon - i\cos\phi\sin\varepsilon}{\cos\phi\cos\varepsilon + i\sin\phi\sin\varepsilon} = \frac{\tan\phi - i\tan\varepsilon}{1 + i\tan\phi\tan\varepsilon} = e^{i\delta}\tan\Psi \quad (3.19)$$

The relationship between (ϕ, ε) and (δ, ψ) is shown in equations (3.12) and (3.13).

3.2.4 Stokes' Parameters [Yeh P., 1988; Azzam R. M. A., 1989; Yeh P., 1999]

A monochromatic plane wave by definition has a defined SOP. In general, however, light is made up of a superposition of plane waves whose exact makeup is often unknown. To represent such a quasi-monochromatic light beam, Stokes' parameters were introduced, whose amplitude $A(t)$ and relative phase $\delta(t)$ can vary with time. If the polarization state changes more rapidly than the speed of observation, the light is partially polarized or unpolarized, as determined by the time average behavior of the polarization state. Assuming a detection time constant of τ_D, which is greater than the reciprocal of the frequency spread of the light, $1/\Delta\omega$, $A(t)$ may change significantly in a time interval τ_D. A practical measurement under these conditions would reflect only the time-averaged behavior. To describe the polarization state of this type of radiation, the following time-averaged Stokes' parameters are used:

$$\begin{aligned} S_0 &= \langle A_x^2 + A_y^2 \rangle \\ S_1 &= \langle A_x^2 - A_y^2 \rangle \\ S_2 &= 2\langle A_x A_y \cos\delta \rangle \\ S_3 &= 2\langle A_x A_y \sin\delta \rangle \end{aligned} \quad (3.20)$$

It can be shown that Stokes' parameters satisfy the relation:

$$S_0^2 \geq S_1^2 + S_2^2 + S_3^2 \quad (3.21)$$

where the equality sign holds only for polarized waves. For unpolarized light, $S_1 = S_2 = S_3 = 0$ and $S_0 = 1$, with the degree of polarization defined as:

$$\gamma = \frac{(S_1^2 + S_2^2 + S_3^2)^{1/2}}{S_0} \quad (3.22)$$

The Stokes' parameters of a plane wave whose complex polarization is $e^{i\delta}\tan\Psi$ are:

$$\begin{aligned} S_0 &= 1 \\ S_1 &= \cos(2\Psi) \\ S_2 &= \sin(2\Psi)\cos\delta \\ S_3 &= \sin(2\Psi)\sin\delta \end{aligned} \quad (3.23)$$

54 POLARIZATION BASICS

In general, S_0 is proportional to the intensity, whereas the parameter S_1 represents the extent to which light is linearly polarized with respect to the x and y Cartesian axes. The proportion of light that is linearly polarized along the x-axis is $(1+S_1)/2$, and along y, $(1-S_1)/2$. S_2 describes the extent to which light is linearly polarized along directions at a 45° angle ($\phi = \pm 45°$) to the x-axis; $(1+S_2)/2$ and $(1-S_2)/2$ are the proportions of light that are linearly polarized along these orthogonal rotated axes. Finally, the parameter S_3 represents the degree of circular polarization. The fraction of light that can be said to have either right-handed or left-handed circular polarization is $(1-S_3)/2$ and $(1+S_3)/2$ respectively.

In summary, there are various representations describing the SOP of a monochromatic plane wave, all of which can be related as described above. When one description is known, all others can be derived. Table 3.1 lists some very typical polarization states in the various representations outlined above.

3.2.5 Poincaré Sphere [Yeh P., 1988; Azzam R. M. A., 1989; Yeh P., 1999]

The polarization state of a monochromatic plane wave is uniquely defined by three of the four Stokes' parameters (S_1, S_2, and S_3). Using $S_0 = 1$, all points with coordinates (S_1, S_2, S_3) are confined to the surface of a unit sphere in three-dimensional space, which is called the Poincaré sphere (Figure 3.5). Each point on the surface of the sphere represents a unique polarization state. Based on equations (3.12), (3,13), and (3.23), the major axis inclination ϕ and ellipticity angle ε are related to Stokes' parameters by:

$$\tan(2\phi) = \frac{S_2}{S_1} \text{ and } \sin(2\varepsilon) = -S_3 \qquad (3.24)$$

Treating the sphere as a globe, equation (3.24) specifies that all points on the equator correspond to a unique orientation of linear polarization. Furthermore, the north pole (0, 0, 1) and south pole (0, 0, −1) represent left-handed circular (LHC) and right-handed circular (RHC) polarization states, respectively. The longitudinal circles, where the ratio S_2/S_1 remains constant, describe polarization states that have the same inclination angle ϕ, but different ellipticities. Similarly, S_3 = constant represents a circle of latitude where polarization states have the same ellipticity, but different inclination angles.

The Poincaré sphere gives a geometrical representation of the SOP transformation produced by standard polarization optics, such as retarders. Complex polarization problems can be solved geometrically, in a way that appeals to intuition, rather than resorting to mathematics. It can be proven that the output SOP from a general birefringent wave plate is related to the input SOP by a simple rotation on the Poincaré sphere [Yeh P., 1999, chapter 2]. Assuming a wave plate of retardance Γ ($=\Delta nd$ in nm) and orientation angle Ω (rad), the output SOP **O** in Figure 3.6 can be obtained by rotating the input SOP **I** about an axis on the equator with orientation 2Ω, by an angle of $2\pi\Gamma/\lambda$ rad, where λ is the wavelength of the light in nm. The Poincaré sphere can also be employed to analyze polarization state evolution

Table 3.1 Various polarization state representations

Polarization ellipse	Jones' vector	(δ, Ψ)	(ϕ, ε)	Stokes' parameter
↔	$\begin{pmatrix} 1 \\ 0 \end{pmatrix}$	$(0, 0)$	$(0, 0)$	$\begin{pmatrix} 1 \\ 1 \\ 0 \\ 0 \end{pmatrix}$
↕	$\begin{pmatrix} 0 \\ 1 \end{pmatrix}$	$(\pi, \pi/2)$	$(\pi/2, 0)$	$\begin{pmatrix} 1 \\ -1 \\ 0 \\ 0 \end{pmatrix}$
↗↙	$\frac{1}{\sqrt{2}} \begin{pmatrix} 1 \\ 1 \end{pmatrix}$	$(0, \pi/4)$	$(\pi/4, 0)$	$\begin{pmatrix} 1 \\ 0 \\ 1 \\ 0 \end{pmatrix}$
↘↖	$\frac{1}{\sqrt{2}} \begin{pmatrix} 1 \\ -1 \end{pmatrix}$	$(\pi, \pi/4)$	$(-\pi/4, 0)$	$\begin{pmatrix} 1 \\ 0 \\ -1 \\ 0 \end{pmatrix}$
◯ (RCP)	$\frac{1}{\sqrt{2}} \begin{pmatrix} 1 \\ -i \end{pmatrix}$	$(-\pi/2, \pi/4)$	$(0, \pi/4)$	$\begin{pmatrix} 1 \\ 0 \\ 0 \\ -1 \end{pmatrix}$
◯ (LCP)	$\frac{1}{\sqrt{2}} \begin{pmatrix} 1 \\ i \end{pmatrix}$	$(\pi/2, \pi/4)$	$(0, -\pi/4)$	$\begin{pmatrix} 1 \\ 0 \\ 0 \\ 1 \end{pmatrix}$
vertical ellipse	$\frac{1}{\sqrt{5}} \begin{pmatrix} 1 \\ 2i \end{pmatrix}$	$(\pi/2, \tan^{-1}(2))$	$(-\pi/2, -\tan^{-1}(1/2))$	$\begin{pmatrix} 1 \\ -3/5 \\ 0 \\ 4/5 \end{pmatrix}$
horizontal ellipse	$\frac{1}{\sqrt{5}} \begin{pmatrix} 2 \\ -i \end{pmatrix}$	$(-\pi/2, \tan^{-1}(1/2))$	$(0, \tan^{-1}(1/2))$	$\begin{pmatrix} 1 \\ 3/5 \\ 0 \\ -4/5 \end{pmatrix}$
tilted ellipse	$\frac{1}{\sqrt{2}} \begin{pmatrix} 1 \\ -i\pi \\ e^8 \end{pmatrix}$	$(-\pi/8, \pi/4)$	$(\pi/4, 0.196)$	$\begin{pmatrix} 1 \\ 0 \\ 0.924 \\ -0.383 \end{pmatrix}$

56 POLARIZATION BASICS

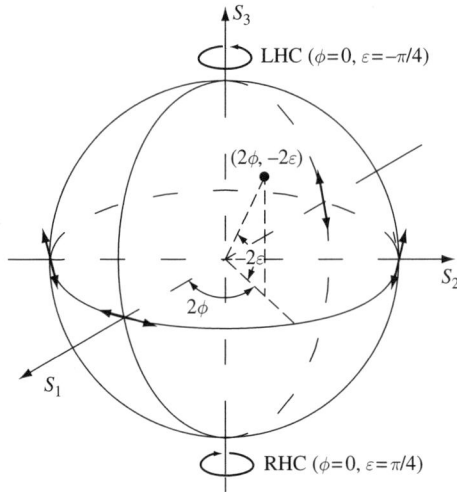

Figure 3.5 Poincaré sphere where each point on its surface represents a unique polarization state

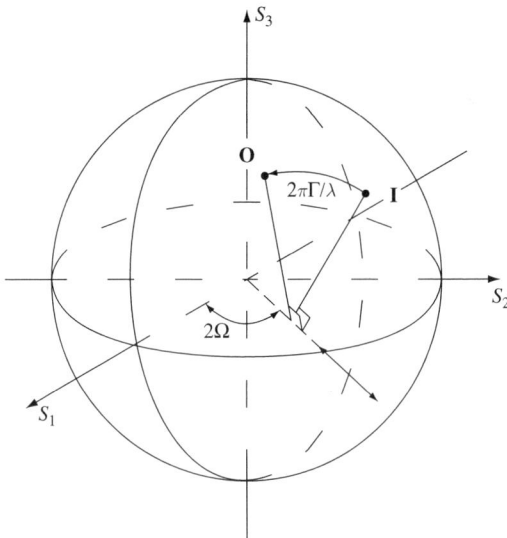

Figure 3.6 Geometrical determination of an output SOP **O** transformed from a given input SOP **I** by a birefringent wave plate of orientation Ω and retardance Γ using the Poincaré sphere

through LCDs, where the LC director is typically inhomogeneous along the device normal. By stratifying the cell into a large number of thin layers, each layer can be considered a homogeneous wave plate, each performing an incremental transformation on the SOP (see Chapter 5).

3.3 Interaction with Media

3.3.1 Reflection and Refraction of Plane Waves [Born M., 1980, chapter 1]

Now consider the reflection and refraction of an optical wave at a plane interface between two media with different dielectric properties. This is a common situation in optical systems and it is inherently polarization sensitive. As shown in Figure 3.7, a plane wave incident on the interface is split into two waves: a transmitted wave propagating into the second medium, and a reflected wave reflecting back into the first medium.

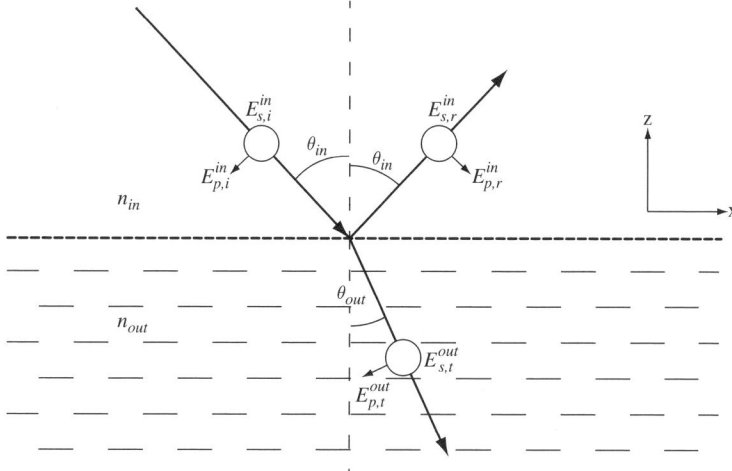

Figure 3.7 Refraction and reflection of a plane wave at the interface between two boundaries

The boundary conditions derived from Maxwell's equations force the tangential componet of the electric field vector to be continuous across the interface. Also, in the absence of any surface currents, the tangential component of the magnetic vector is also continuous across the interface. Hence we can write:

$$E_{x,i}^{in} + E_{x,r}^{in} = E_{x,t}^{out} \qquad E_{y,i}^{in} + E_{y,r}^{in} = E_{y,t}^{out}$$
$$H_{x,i}^{in} + H_{x,r}^{in} = H_{x,t}^{out} \qquad H_{y,i}^{in} + H_{y,r}^{in} = H_{y,t}^{out} \qquad (3.25)$$

which lead to the following equations:

$$\cos\theta_{in}\left(E_{p,i}^{in} - E_{p,r}^{in}\right) = \cos\theta_{out} E_{p,t}^{out}$$
$$E_{s,i}^{in} + E_{s,r}^{in} = E_{s,t}^{out}$$
$$\sqrt{\varepsilon_1}\cos\theta_{in}\left(E_{s,i}^{in} - E_{s,r}^{in}\right) = \sqrt{\varepsilon_2}\cos\theta_{out} E_{s,t}^{out} \qquad (3.26)$$
$$\sqrt{\varepsilon_1}\left(E_{p,i}^{in} + E_{p,r}^{in}\right) = \sqrt{\varepsilon_2} E_{p,t}^{out}$$
$$k_{x,i}^{in} = k_{x,t}^{out}$$

where the conventional notation **s** and **p** has been introduced for the polarization surface eigenmodes, $E_{y,i}^{in}$ being equivalent to $E_{s,i}^{in}$ etc. We can solve equations (3.26) to get the Fresnel formulas and Snell's law:

$$t_p = \frac{E_{p,t}^{out}}{E_{p,i}^{in}} = \frac{2 n_{in} \cos\theta_{in}}{n_{out} \cos\theta_{in} + n_{in} \cos\theta_{out}}$$

$$t_s = \frac{E_{s,t}^{out}}{E_{s,i}^{in}} = \frac{2 n_{in} \cos\theta_{in}}{n_{in} \cos\theta_{in} + n_{out} \cos\theta_{out}}$$

$$r_p = \frac{E_{p,r}^{in}}{E_{p,i}^{in}} = \frac{n_{out} \cos\theta_{in} - n_{in} \cos\theta_{out}}{n_{out} \cos\theta_{in} + n_{in} \cos\theta_{out}} \quad (3.27)$$

$$r_s = \frac{E_{s,r}^{in}}{E_{s,i}^{in}} = \frac{n_{in} \cos\theta_{in} - n_{out} \cos\theta_{out}}{n_{in} \cos\theta_{in} + n_{out} \cos\theta_{out}}$$

$$n_{in} \sin\theta_{in} = n_{out} \sin\theta_{out}$$

The optical power reflectivity R and transmissivity T are obtained by first taking the square of the amplitude reflection and transmssion coefficients (r, t) in equation (3.27), respectively. Second, the optical speed in the two media are factored in. Third, the projected geometrical interface area is incorporated. In this way we get:

$$R = \left(\frac{(E_{p,r}^{in})^2 + (E_{s,r}^{in})^2}{(E_{p,i}^{in})^2 + (E_{s,i}^{in})^2} \right) \quad \text{and} \quad T = \frac{n_{out} \cos\theta_{out}}{n_{in} \cos\theta_{in}} \left(\frac{(E_{p,t}^{out})^2 + (E_{s,t}^{out})^2}{(E_{p,i}^{in})^2 + (E_{s,i}^{in})^2} \right) \quad (3.28)$$

For normal incidence, the reflectance and transmittance for **s**- and **p**-components can be expressed as follows:

$$R_s = R_p = \left(\frac{n_{in} - n_{out}}{n_{in} + n_{out}} \right)^2 \quad (3.29)$$

$$T_s = T_p = \frac{4 n_{in} n_{out}}{(n_{in} + n_{out})^2} \quad (3.30)$$

Two important optical phenomena are derived from equations (3.27): total internal reflection (TIR) and perfect **p**-transmission at Brewster's angle. If the incident medium has a refractive index larger than that of the second medium ($n_{in} > n_{out}$), solutions exist for Snell's law where $\sin\theta_{out} > 1$. Under these conditions no transmitted wave is allowed and lossless TIR occurs. Based on Snell's law ($\sin\theta_{out} = (n_{in}/n_{out}) \sin\theta_{in}$), the critical angle of incidence above which TIR is observed ($\sin\theta_{out} = 1$) is given by:

$$\theta_c = \sin^{-1}\left(\frac{n_{out}}{n_{in}} \right) \quad (3.31)$$

When the incidence angle is larger than the critical angle, the transmitted wave decays exponentially along the surface normal (z-axis) and all of the energy is reflected from the surface. The amplitude of the reflected wave only differs from the incident amplitude by a phase factor. TIR is the phenomenon by which light propagates in optical fibers, and is used in DLP projection systems, where angular discrimination is required between incident and reflected beams.

The Fresnel reflection coefficients r_s and r_p vary differently as a function of the incidence angle. R_s is always greater than the reflectance of R_p, except at normal incidence where they are eqivalent (see Figure 3.8). What is of great interest, though, is the vanishing of the Fresnel reflection coefficient r_p when the incidence angle θ_{in} satisfies the following equation:

$$n_{out} \cos \theta_{in} = n_{in} \cos \theta_{out} \qquad (3.32)$$

Using Snell's law to eliminate θ_{out}, we get the following expression for this particular angle of incidence, known as Brewster's angle:

$$\theta_B = \tan^{-1}(n_{out}/n_{in}) \qquad (3.33)$$

Equation (3.33) is equivalent to the condition $\theta_{in} + \theta_{out} = \pi/2$.

A simple optical component that uses this effect is the so-called 'pile-of-plates' polarizer [Hecht E., 1980, p.245]. Here a stack of glass plates oriented at Brewster's angle can produce linear polarized transmitted light, since the finite s-reflection at each interface gradually

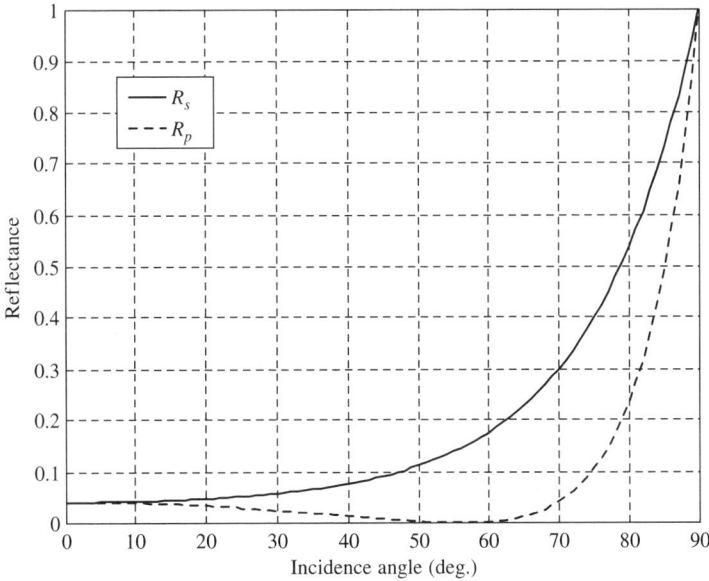

Figure 3.8 The reflectance of R_s and R_p versus the angle of incidence θ for $n_{in} = 1.0$ and $n_{out} = 1.5$. Brewster's angle θ_B is equal to 56°

60 POLARIZATION BASICS

reduces the s-polarized electric field. Windows in glass laser tubes are also typically oriented at Brewster's angle to minimize loss.

3.3.2 Matrix Formulation for Isotropic Layered Media [Yeh P., 1988, chapter 5]

The s- and p-polarization components are independent in layered isotropic media with parallel interfaces and can therefore be treated independently. A two-dimensional description is adequate to describe the coupling between counter-propagating beams of either component. In a simple three-layer dielectric structure as shown in Figure 3.9, E_i^j and E_r^j represent the electric field vector of right (transmitted) and left (reflected) traveling waves in layer j respectively; j can be *in* and *out* for the input and output layers and an additional dash is introduced to distinguish waves at the start, $E_i^{\prime j}$, and end of a layer.

By applying the boundary conditions (3.26), a three-step procedure can lead to relationships between the electric fields of the two counter-propagating waves on either side of each interface. These are:

$$\begin{pmatrix} E_i^{in} \\ E_r^{in} \end{pmatrix} = \mathbf{D}_{in}^{-1} \mathbf{D}_1 \begin{pmatrix} E_i^{\prime 1} \\ E_r^{\prime 1} \end{pmatrix}$$

$$\begin{pmatrix} E_i^{\prime 1} \\ E_r^{\prime 1} \end{pmatrix} = \mathbf{P}_1 \begin{pmatrix} E_i^{\prime 1} \\ E_r^{\prime 1} \end{pmatrix} \quad (3.34)$$

$$\begin{pmatrix} E_i^1 \\ E_r^1 \end{pmatrix} = \mathbf{D}_1^{-1} \mathbf{D}_{out} \begin{pmatrix} E_i^{out} \\ E_r^{out} \end{pmatrix}$$

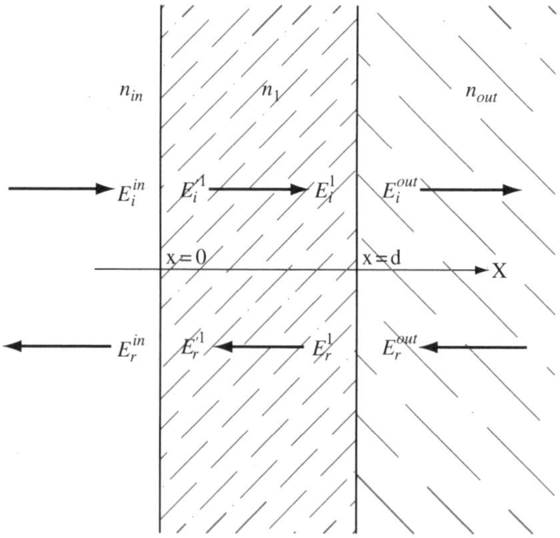

Figure 3.9 A three-layer dielectric structure

where:

$$\mathbf{D}_j = \begin{cases} \begin{pmatrix} 1 & 1 \\ n_j \cos\theta_j & -n_j \cos\theta_j \end{pmatrix} & \text{for s-wave} \\ \begin{pmatrix} \cos\theta_j & \cos\theta_j \\ n_j & -n_j \end{pmatrix} & \text{for p-wave} \end{cases} \quad (3.35)$$

$$\mathbf{P}_j = \begin{pmatrix} e^{-ik_x^j d_j} & 0 \\ 0 & e^{ik_x^j d_j} \end{pmatrix} \quad (3.36)$$

and d_j is the thickness of jth layer, and $k_x^j = n_j \omega \cos\theta_j / c$.

Combining equations (3.34), we get the following relationship between the incident and exiting waves:

$$\begin{pmatrix} E_i^{in} \\ E_r^{in} \end{pmatrix} = \mathbf{D}_{in}^{-1} \mathbf{D}_1 \mathbf{P}_1 \mathbf{D}_1^{-1} \mathbf{D}_{out} \begin{pmatrix} E_i^{out} \\ E_r^{out} \end{pmatrix} \quad (3.37a)$$

If a dielectric structure contains N layers, the extension of the above analysis yields:

$$\begin{pmatrix} E_i^{in} \\ E_r^{in} \end{pmatrix} = \mathbf{D}_{in}^{-1} \left[\prod_{j=1}^{N} (\mathbf{D}_j \mathbf{P}_j \mathbf{D}_j^{-1}) \right] \mathbf{D}_{out} \begin{pmatrix} E_i^{out} \\ E_r^{out} \end{pmatrix} = \begin{pmatrix} M_{11} & M_{12} \\ M_{21} & M_{22} \end{pmatrix} \begin{pmatrix} E_i^{out} \\ E_r^{out} \end{pmatrix} \quad (3.37b)$$

Making E_r^{out} equal to zero allows reflection and transmission coefficients to be determined as follows:

$$r = \left(\frac{E_r^{in}}{E_i^{in}} \right)_{E_r^{out}=0} = \frac{M_{21}}{M_{11}} \quad (3.38)$$

$$t = \left(\frac{E_i^{out}}{E_i^{in}} \right)_{E_r^{out}=0} = \frac{1}{M_{11}} \quad (3.39)$$

This technique is used to calculate the reflection and transmission spectrum of dichroic filters, and anti-reflection coatings.

3.3.3 Matrix Formulation for Anisotropic Layered Media

3.3.3.1 Electromagnetic Propagation in Homogeneous Anisotropic Media [Yeh P., 1988, chapter 3]

LCD video projectors modulate light using the electro-optical effect of optically anisotropic LCs. Despite the inhomogeneity of director profiles in LCDs, we can stratify an LC structure into many sublayers and treat each sublayer as a homogeneous birefringent layer. Understanding electromagnetic propagation in homogeneous anisotropic media is therefore a vital part of developing a mathematical model for propagating the SOP through projection system.

62 POLARIZATION BASICS

To form a mathematical description of light propagation through anisotropic media, it is necessary to include within Maxwell's equations the interaction of the electric field with the material. This interaction is described by the so-called material equations that, in the case of birefringent materials, relate displacement charge to the electric field by a dielectric tensor ε. It is also assumed that neither free charge density nor current density exist. The relevant equations are then:

$$\nabla \times E = -\frac{\partial B}{\partial t} \quad (3.40)$$

$$\nabla \times H = \frac{\partial D}{\partial t} \quad (3.41)$$

$$\nabla \cdot D = 0 \quad (3.42)$$

$$\nabla \cdot B = 0 \quad (3.43)$$

with material equations:

$$D = \varepsilon_0 \varepsilon E \quad (3.44)$$

$$B = \mu_0 H \quad (3.45)$$

Without loss in generality, an optical ray can be defined with propagation vector k within the xz plane of a conventional Cartesian coordinate system (Figure 3.10). A dielectric tensor ε can then be defined that describes the interaction of light with a material. Although biaxial materials are most general [Yeh P., 1999, chapter 3], a uniaxial material is considered here, with optic axis specified by polar and azimuthal angles (θ, ϕ) with respect to the z-axis. The components of the dielectric tensor can be written in the following form:

$$\varepsilon_{\alpha\beta} = \varepsilon_\perp \delta_{\alpha\beta} + (\varepsilon_\parallel - \varepsilon_\perp) n_\alpha n_\beta \quad (3.46a)$$

which yields the following symmetrical dielectric tensor:

$$\begin{pmatrix} n_o^2 + (n_e^2 - n_o^2)\sin^2\theta\cos^2\phi & (n_e^2 - n_o^2)\sin^2\theta\sin\phi\cos\phi & (n_e^2 - n_o^2)\sin\theta\cos\theta\cos\phi \\ (n_e^2 - n_o^2)\sin^2\theta\sin\phi\cos\phi & n_o^2 + (n_e^2 - n_o^2)\sin^2\theta\sin^2\phi & (n_e^2 - n_o^2)\sin\theta\cos\theta\sin\phi \\ (n_e^2 - n_o^2)\sin\theta\cos\theta\cos\phi & (n_e^2 - n_o^2)\sin\theta\cos\theta\sin\phi & n_o^2 + (n_e^2 - n_o^2)\cos^2\theta \end{pmatrix} \quad (3.46b)$$

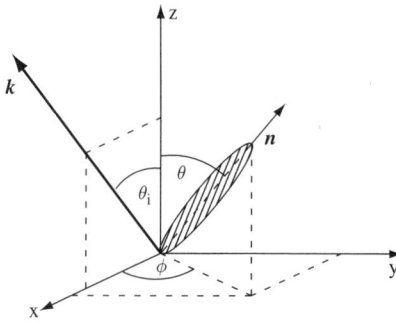

Figure 3.10 Coordinate arrangement for both the optical ray propagation vector and the optic axis of the uniaxial material

To solve Maxwell's equations for propagating wave solutions, with the above dielectric tensor, we first assume plane wave solutions for the electric and magnetic fields of the form:

$$E(r, t) = E_0 e^{i(\omega t - k \cdot r)} \tag{3.47}$$

$$H(r, t) = H_0 e^{i(\omega t - k \cdot r)} \tag{3.48}$$

Using equation (3.40)–(3.45) and (3.47)–(3.48), we then get:

$$k \times E_0 = \omega \mu_0 H_0 \tag{3.49}$$

$$k \times H_0 = -\omega \varepsilon_0 \varepsilon \cdot E_0 \tag{3.50}$$

When H_0 is eliminated from the two equations (3.49)–(3.50), we obtain:

$$k \times (k \times E_0) + \omega^2 \mu_0 \varepsilon_0 \varepsilon \cdot E_0 = 0 \tag{3.51}$$

For non-trivial solutions, the determinant of the matrix in equation (3.51) must vanish, which gives four solutions, $k_{z1}, k_{z2}, \ldots, k_{z4}$, for the normalized z-component of k [Yeh P., 1988, chapter 3; Lien A., 1990]:

$$k_{z1,3}/k_0 = \pm \sqrt{n_o^2 - \left(\frac{k_x}{k_0}\right)^2} \tag{3.52}$$

$$k_{z2,4}/k_0 = -\frac{\varepsilon_{xz}}{\varepsilon_{zz}} \frac{k_x}{k_0} \pm \sqrt{\frac{n_o^2 n_e^2}{\varepsilon_{zz}} - \frac{(\varepsilon_{xx}\varepsilon_{zz} - \varepsilon_{xz}^2)}{\varepsilon_{zz}^2} \left(\frac{k_x}{k_0}\right)^2} \tag{3.53}$$

where $k_x = k_0 \sin(\theta_i)$ and $k_0 = 2\pi/\lambda = \omega/c$.

Of the four solutions, the positive k_z solutions represent the ordinary (**o**) and extraordinary (**e**) transmitted modes, while the negative solutions are the **o** and **e** reflected modes. For each solution the associated electric field vectors are the eigenvectors or normal modes of equation (3.51). As for all linear systems, a general forward propagating solution in this case would consist of a linear sum of the two **o** and **e** modes.

It can be proven that the electric field vector of the **o**-mode is perpendicular to both the wavevector k_o and the material optic axis n, whereas the **e**-mode electric field vector, though perpendicular to the electric field of the **o**-mode, is not orthogonal to k_e. However, the deviation from 90° is very slight and for most practical purposes, where the birefringence is small (often called the small birefringent approximation [Yeh P., 1988, chapter 8]), we can assume orthogonality (see later Section 3.3.3.3 on extended Jones' matrix methods). The displacement vector D of both **o**- and **e**-modes is, however, always exactly perpendicular to the wavevectors k_o and k_e respectively, and can be written explicitly as:

$$D_o = \frac{k_o \times n}{|k_o \times n|} \tag{3.54}$$

$$D_e = \frac{D_o \times k_e}{|D_o \times k_e|} \tag{3.55}$$

and, furthermore k_e, D_e, and E_e are coplanar.

3.3.3.2 Jones' Matrix Formulation

The previous section showed that light propagating within a birefringent medium can be considered a linear superposition of two normal modes. These modes have well-defined phase velocity and polarization directions. The modes are mutually orthogonal and their electric field vectors are called the "slow" and "fast" axes of the medium for that direction of propagation. The calculation necessary to determine the change in the polarization state as light travels through a succession of birefringent layers can become very involved. One very powerful and instructive technique for this calculation uses Jones' matrices.

Consider a system where normally incident light propagates through a series of birefringent elements, where the optic axes are orthogonal to the ray direction, with negligible interface reflection. This system is shown schematically in Figure 3.11. The approach involves projecting the incident field onto the normal modes of propagation for each layer. These normal modes then propagate through the layer with different phase velocities yielding the complex electric field vector incident on the next layer.

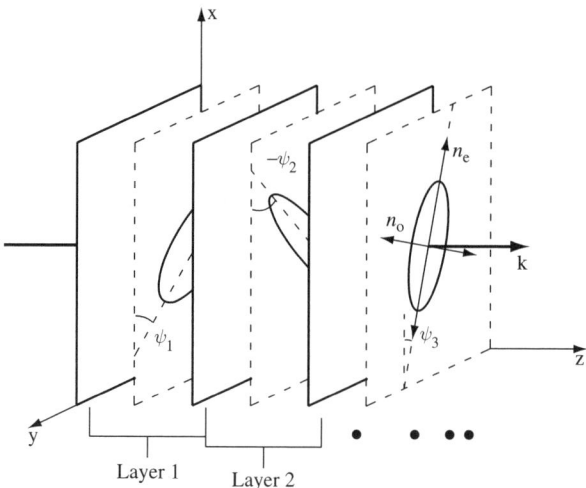

Figure 3.11 Coordinate system of Jones' matrix algebra

Consider the polarization of a light beam as it passes through the system shown in Figure 3.11. Initially, the electric field can be described by the following Jones' vector:

$$E^{in} = \begin{pmatrix} A_x \\ A_y \end{pmatrix} \quad (3.56)$$

where A_x and A_y are the complex electric field amplitudes in the (x, y) coordinate system. The electric field of the output beam is then calculated by expressing the electric field vector in the orthogonal coordinate system defined by the o- and e-polarization vectors (e, o). These vectors are parallel and orthogonal to the material optic axis since here it is assumed normal to the propagation. A matrix representing a simple geometrical rotation re-expresses the electric field in these rotated coordinates. The components of the rotated electric field then represent the amplitudes of the **e**- and **o**-modes launched into the material. In traversing the material these amplitudes remain the same, but their phase differs dependent on their propagation speeds. Finally, to re-express the emerging electric field in the original coordinate system a further rotation of the opposite sense is required. Mathematically, the output electric field having passed through layer 1 of the stack is related to the input by:

$$\begin{pmatrix} A'_x \\ A'_y \end{pmatrix} = R(-\psi) P R(\psi) \begin{pmatrix} A_x \\ A_y \end{pmatrix} \tag{3.57}$$

where $R(\psi)$ is the rotational matrix:

$$R(\psi) = \begin{pmatrix} \cos\psi & \sin\psi \\ -\sin\psi & \cos\psi \end{pmatrix} \tag{3.58}$$

and P is the propagation matrix for the ordinary and extraordinary modes:

$$P = \begin{pmatrix} e^{-ik_e d} & 0 \\ 0 & e^{-ik_o d} \end{pmatrix} = \begin{pmatrix} e^{-\frac{2i\pi n_e d}{\lambda}} & 0 \\ 0 & e^{-\frac{2i\pi n_o d}{\lambda}} \end{pmatrix}$$

$$= e^{-i\gamma} \begin{pmatrix} e^{-i\delta/2} & 0 \\ 0 & e^{i\delta/2} \end{pmatrix} \tag{3.59}$$

and δ is the phase retardation equal to $2\pi d(n_e - n_o)/\lambda$; $\gamma (= (n_e + n_o)\pi d/\lambda)$ is an arbitrary phase factor and can be ignored.

If the system contains a series of birefringent elements, the exiting electric field can be expressed as:

$$\begin{pmatrix} A'_x \\ A'_y \end{pmatrix} = \prod_{j=1}^{j=N} R(-\psi_j) P_j R(\psi_j) \begin{pmatrix} A_x \\ A_y \end{pmatrix} \tag{3.60}$$

In real systems, polarizers are used at various angles. Jones' matrix representation of an ideal sheet polarizer oriented at ψ is simply:

$$P_{pol}(\psi) = R(-\psi) \begin{pmatrix} 1 & 0 \\ 0 & 0 \end{pmatrix} R(\psi) \tag{3.61}$$

Jones' matrix formulation is almost universally used to determine the optical properties of transmissive and reflective LC systems [Wu S. T., 2001], and is also the preferred method by which retarder stack filters are modeled.

3.3.3.3 Extended Jones' Matrix Method [Yeh P., 1982; Gu C. 1993; Yeh P., 1999, chapter 8]

In the previous section, basic Jones' matrix formulation was addressed, which is limited to normal incidence rays where no refraction occurs at material boundaries. In LCD projection systems, light propagates through optical components at various incidence angles.

66   POLARIZATION BASICS

Off-axis optical properties therefore play a very important role in determining overall system performance. Projection system design is frequently based on an FOV (Field Of View) compromise, preserving performance such as contrast, while making efficient use of the lamp emission by reducing $f_{/\#}$. An extended Jones' matrix method is able to address off-axis propagation in a stratified medium which consists of a stack of birefringent layers as shown in Figure 3.12.

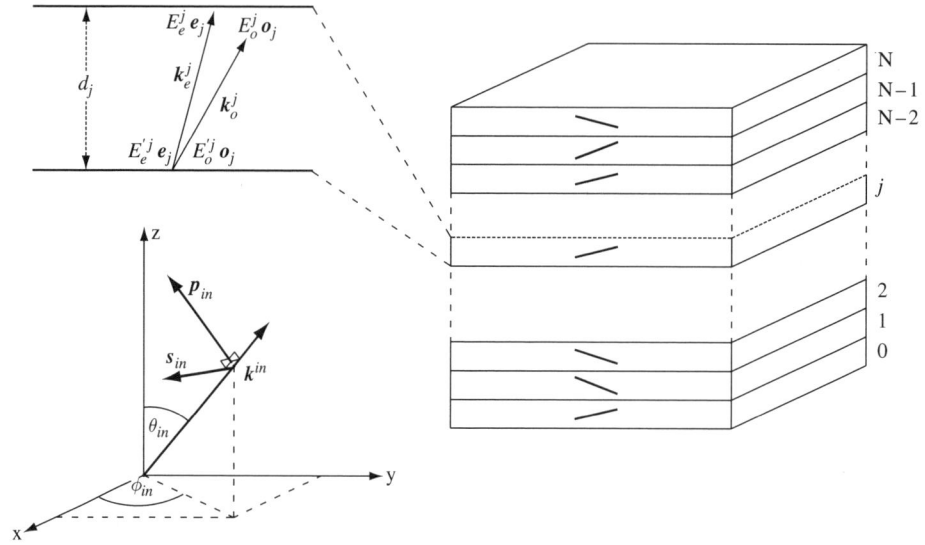

Figure 3.12   A stratified medium, consisting of a stack of birefringent layers

Again, reflections at internal layer interfaces are assumed negligible. Input/output Fresnel reflections can be considered power losses without giving rise to interference. Under the small birefrigence approximation ($|n_e - n_o| \ll n_e, n_o$), we can assume the electric field of both e- and o-modes to be parallel to their displacements. According to equations (3.54) and (3.55), we can decribe the polarization directions of each mode as:

$$o = \frac{k_o \times n}{|k_o \times n|} \quad (3.62)$$

$$e = \frac{o \times k_o}{|o \times k_o|} \quad (3.63)$$

The propagation matrix **M** through the whole stack can be obtained as the product of four types of matrices:

$$M = D_{out} P_N D_{N,N-1} P_{N-1} D_{N-1,N-2} \cdots D_{3,2} P_2 D_{2,1} P_1 D_{in} \quad (3.64)$$

and the electric field of the exiting light beam is equal to:

$$\begin{pmatrix} E_s^{out} \\ E_p^{out} \end{pmatrix} = M \begin{pmatrix} E_s^{in} \\ E_p^{in} \end{pmatrix} \quad (3.65)$$

The individual types of matrices in equation (3.64) include the following.

(a) **Input Dynamic Matrix D_{in}**
The input dynamic matrix D_{in} is the transition matrix from a semi-infinite input space to the first layer, which simultaneously transforms the electric field between the input (s,p) coordinate system to that of the first layer e and o polarization vectors: e_1 and o_1:

$$\begin{pmatrix} E_e^1 \\ E_o^1 \end{pmatrix} = \begin{pmatrix} t_s(s_1 \cdot e_1) & t_p(p_1 \cdot e_1) \\ t_s(s_1 \cdot o_1) & t_p(p_1 \cdot o_1) \end{pmatrix} \begin{pmatrix} E_s^{in} \\ E_p^{in} \end{pmatrix} \quad (3.66)$$

Here, t_s and t_p are the Fresnel transmission coefficients of equation (3.27) relating to the interface between the first layer and the bounding semi-infinite material.

(b) **Output Dynamic Matrix D_{out}**
The output dynamic matrix D_{out} is the transition matrix from the last layer of the stack to semi-infinite space. The transformation is effectively the inverse of the input dynamic matrix relating to the interface between the last layer and the bounding material:

$$\begin{pmatrix} E_s^{out} \\ E_p^{out} \end{pmatrix} = \begin{pmatrix} t_s'(e_N \cdot s_N) & t_s'(o_N \cdot s_N) \\ t_p'(e_N \cdot p_N) & t_p'(o_N \cdot p_N) \end{pmatrix} \begin{pmatrix} E_e^N \\ E_o^N \end{pmatrix} \quad (3.67)$$

where t_s' and t_p' are again the Fresnel transmission coefficients for the final boundary as given by equation (3.27).

(c) **Propagation Matrix P_j**
P_j is the propagation matrix inside the jth birefringent layer:

$$P_j = \begin{pmatrix} e^{-ik_{e,z}^j d_j} & 0 \\ 0 & e^{-ik_{o,z}^j d_j} \end{pmatrix} \quad (3.68)$$

and:

$$\begin{pmatrix} E_e^j \\ E_o^j \end{pmatrix} = P_j \begin{pmatrix} E_e'^j \\ E_o'^j \end{pmatrix} \quad (3.69)$$

where $k_{e,z}^j$ and $k_{o,z}^j$ are the z-components of the e- and o-mode wavevectors in the jth layer. They can be derived from the positive solutions of equations (3.52)–(3.53).

(d) **Dynamic Matrix $D_{j+1,j}$**
Dynamic matrix $D_{j+1,j}$ performs the transition from layer j to j+1, since in general the eigenpolarizations differ between layers. Mathematically it is expressed as:

$$D_{j+1,j} = \begin{pmatrix} (e_j \cdot e_{j+1}) & (o_j \cdot e_{j+1}) \\ (e_j \cdot o_{j+1}) & (o_j \cdot o_{j+1}) \end{pmatrix} \quad (3.70)$$

and:

$$\begin{pmatrix} E_e'^{(j+1)} \\ E_o'^{(j+1)} \end{pmatrix} = D_{j+1,j} \begin{pmatrix} E_e^j \\ E_o^j \end{pmatrix} \quad (3.71)$$

Though the extended Jones' matrix method described here assumes uniaxial materials with small birefringence, the method works for high birefringence and biaxial systems

[Yeh P., 1999, chapter 8]. The method by which the eigenvectors are determined for any given $\boldsymbol{k}$-vector is, however, more complex, and often detracts from the supposed simplicity of this approach. A slightly different extended Jones' matrix approach was developed in 1990, in which the laboratory coordinate system is used instead of principal coordinates [Lien A., 1990; 1996].

3.3.3.4 Berreman 4 × 4 Approach [Berreman D. W., 1970; Teitler S., 1970; Berreman D. W., 1972]

The Jones' matrix approach does not include the reflection of light between material layers, and is therefore only an approximation. Consequently, it cannot, for example, simulate interference phenomena. To include reflections, a four-dimensional approach is necessary (two for transmission and two for reflection). A convenient and established rigorous method of modeling wave propagation in anisotropic materials is the "so-called" Berreman calculus.

To derive this approach we will continue to use the coordinate system shown in Figure 3.10. Assuming the optical anisotropy only varies along the z-direction (1D problem), the electric and magnetic fields can be written as:

$$\boldsymbol{E}(x, y, z) = \boldsymbol{E}(z) e^{i(k_x x - \omega t)} \tag{3.72}$$

$$\boldsymbol{H}(x, y, z) = \boldsymbol{H}(z) e^{i(k_x x - \omega t)} \tag{3.73}$$

From Maxwell's equations and the material equations (3.40)–(3.45), and for convenience of expression choosing Gaussian units ($\mu_0 = 1$ and $\varepsilon_0 = 1$), the following equations can be derived:

$$\begin{pmatrix} 0 & -\frac{d}{dz} & 0 \\ \frac{d}{dz} & 0 & -ik_x \\ 0 & ik_x & 0 \end{pmatrix} \begin{pmatrix} E_x \\ E_y \\ E_z \end{pmatrix} = i \frac{\omega}{c} \begin{pmatrix} H_x \\ H_y \\ H_z \end{pmatrix} \tag{3.74}$$

and:

$$\begin{pmatrix} 0 & -\frac{d}{dz} & 0 \\ \frac{d}{dz} & 0 & -ik_x \\ 0 & ik_x & 0 \end{pmatrix} \begin{pmatrix} H_x \\ H_y \\ H_z \end{pmatrix} = -i \frac{\omega}{c} \vec{\varepsilon} \begin{pmatrix} E_x \\ E_y \\ E_z \end{pmatrix} \tag{3.75}$$

By eliminating E_z and H_z between the above two expressions the following relationship is obtained:

$$\frac{d}{dz} \begin{pmatrix} E_x \\ H_y \\ E_y \\ -H_x \end{pmatrix} = ik_0 \begin{pmatrix} -m \frac{\varepsilon_{zx}}{\varepsilon_{zz}} & 1 - \frac{m^2}{\varepsilon_{zz}} & -m \frac{\varepsilon_{zy}}{\varepsilon_{zz}} & 0 \\ \varepsilon_{xx} - \frac{\varepsilon_{zx}\varepsilon_{xz}}{\varepsilon_{zz}} & -m \frac{\varepsilon_{xz}}{\varepsilon_{zz}} & \varepsilon_{xy} - \frac{\varepsilon_{xz}\varepsilon_{zy}}{\varepsilon_{zz}} & 0 \\ 0 & 0 & 0 & 1 \\ \varepsilon_{yx} - \frac{\varepsilon_{zx}\varepsilon_{yz}}{\varepsilon_{zz}} & -m \frac{\varepsilon_{zy}}{\varepsilon_{zz}} & \varepsilon_{yy} - \frac{\varepsilon_{zy}\varepsilon_{yz}}{\varepsilon_{zz}} - m^2 & 0 \end{pmatrix} \begin{pmatrix} E_x \\ H_y \\ E_y \\ -H_x \end{pmatrix} \tag{3.76}$$

where $m = n_{in} \sin \theta_{in}$ and $k_0 = \omega/c$ (wavevector in air).

INTERACTION WITH MEDIA

By defining:

$$\Psi(z) = \begin{pmatrix} E_x \\ H_y \\ E_y \\ -H_x \end{pmatrix} \quad (3.77)$$

equation (3.76) can be written simply as:

$$\frac{d\Psi(z)}{dz} = ik_0 \Delta(z) \Psi(z) \quad (3.78)$$

where $\Psi(z)$ is the Berreman vector.

It is convenient to select these four tangential components of electric and magnetic fields (E_x, E_y, H_x, and H_y) as a description of the electromagnetic field since it is naturally continuous at boundaries between materials. By integrating equation (3.78), a propagation matrix $P(d)$ can be derived where:

$$\Psi(d) = P(d)\Psi(0) \quad (3.79)$$

and:

$$P(d) = e^{ik_0 \Delta((N)\Delta z)\Delta z} \cdot e^{ik_0 \Delta((N-1)\Delta z)\Delta z} \cdot \ldots \cdot e^{ik_0 \Delta(0)\Delta z} = e^{ik_0 \int_0^d \Delta(z)dz} \quad (3.80)$$

Considering the electric field vectors entering and exiting a layered system, in which light impinges on the input side only, we can write:

$$\Psi_t = P(d)(\Psi_i + \Psi_r) \quad (3.81)$$

where Ψ_i, Ψ_r, and Ψ_t are the incident, reflective, and transmissive Berreman vectors as shown in Figure 3.13.

In equation (3.81) there appears to be only four relationships, and yet eight variables: four for reflected and four for transmitted electromagnetic waves. However, since the external medium is isotropic, by definition, the magnetic field within the bounding medium can be

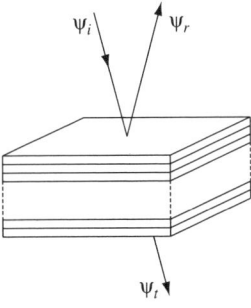

Figure 3.13 Incident, reflective, and transmissive Berreman vectors

70 POLARIZATION BASICS

expressed in terms of its electric field. The remaining four unknown variables can then be obtained from equation (3.81), allowing a solution.

Computation of the propagation matrix $P(d)$ can be tedious if a simple Taylor series expansion technique is used. A faster approach uses Cayley and Hamilton's theorem that equates $e^{ik_0\Delta(z)\Delta z}$ to a finite power series with n terms, where $n \times n$ are the dimensions of the $\Delta(z)$ matrix ($n = 4$ in our case) [Berreman D. W., 1972]. That is:

$$e^{ik_0\Delta(z)\Delta z} = \beta_0 I + \beta_1 \Delta(z) + \beta_2 \Delta^2(z) + \beta_3 \Delta^3(z) \tag{3.82}$$

If the eigenvalues of $\Delta(z)$ are equal to λ_i ($i = 1$ to 4) then:

$$e^{ik_0\lambda_i\Delta z} = \beta_0 + \beta_1 \lambda_i + \beta_2 \lambda_i^2 + \beta_3 \lambda_i^3 \tag{3.83}$$

where:

$$\beta_0 = -\sum_{i=1}^{4} \lambda_j \lambda_k \lambda_l \frac{f_i}{\lambda_{ij}\lambda_{ik}\lambda_{il}}$$

$$\beta_1 = \sum_{i=1}^{4} (\lambda_j \lambda_k + \lambda_j \lambda_l + \lambda_k \lambda_l) \frac{f_i}{\lambda_{ij}\lambda_{ik}\lambda_{il}}$$

$$\beta_2 = \sum_{i=1}^{4} (\lambda_j + \lambda_k + \lambda_l) \frac{f_i}{\lambda_{ij}\lambda_{ik}\lambda_{il}}$$

$$\beta_3 = \sum_{i=1}^{4} \frac{f_i}{\lambda_{ij}\lambda_{ik}\lambda_{il}}$$

$$\lambda_{ij} = \lambda_i - \lambda_j; \quad f_i = e^{ik_0\lambda_i\Delta z}$$

For a uniaxial material, the four eigenvalues represent the ordinary and extraordinary transmissive and reflective modes:

$$\lambda_{0\pm} = \pm \sqrt{(\varepsilon_\perp - m^2)}$$

$$\lambda_{e\pm} = -m\frac{\varepsilon_{xz}}{\varepsilon_{zz}} \pm \sqrt{\frac{\varepsilon_\parallel \varepsilon_\perp}{\varepsilon_{zz}} - m^2 \left(\frac{\varepsilon_{xx}\varepsilon_{zz} - \varepsilon_{xz}^2}{\varepsilon_{zz}^2}\right)} \tag{3.84}$$

Though seemingly complex, this method can be computed very easily, allowing accurate numerical simulations to be carried out. One drawback with this method is the interference between highly coherent waves reflecting from interfaces separated by macroscopic distances. In practice, unless laser sources are used, such interference is not apparent. To eliminate these effects from simulation results, anti-reflection layers or index-matched boundaries must be included (see Section 3.5 on modeling).

3.4 Index Ellipsoid Visualization [Yeh P., 1999, chapter 3]

From Section 3.3, electromagnetic propagation in homogeneous anisotropic media can be complicated, especially when biaxial materials are involved. To visualize eigenmode polarization directions and the magnitude of any birefringence for wave propagation within general anisotropic materials, the index ellipsoid geometrical construct is often very useful.

In the principal coordinates of the dielectric tensor (i.e., when it is diagonal), an index ellipsoid can be defined as:

$$\frac{x^2}{n_x^2} + \frac{y^2}{n_y^2} + \frac{z^2}{n_z^2} = 1 \tag{3.85}$$

To find the normal propagation modes along a general direction s, the vector s is first drawn from the origin of the ellipsoid. Then an ellipse can be formed between the intersection of the ellipsoid surface and the plane normal to s containing the origin. The two principal axes of this intersection ellipse are then equal in length to n_1 and n_2, which are the two-eigenmode indices of refraction. This is illustrated in Figure 3.14. Furthermore, the two principal axes are parallel to the directions of the displacement vectors D_1 and D_2 of the normal modes.

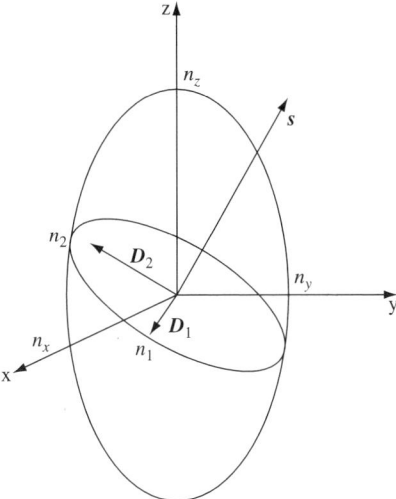

Figure 3.14 Index ellipsoid geometrical construct for visualizing eigenmode polarizations and their propagation speeds; n_1 and n_2 are the refractive indexes for waves with propagation vector s

As an example, consider the case of a plane wave in a uniaxial material with propagation vector given by:

$$s = (\sin\theta\cos\phi, \sin\theta\sin\phi, \cos\theta) \tag{3.86}$$

From Figure 3.14, the following relationships can be derived geometrically:

$$D_1 = (\sin\phi, -\cos\phi, 0) \quad \text{and} \quad D_2 = -(\cos\theta\cos\phi, \cos\theta\sin\phi, -\sin\theta) \tag{3.87}$$

$$n_1 = n_o \quad \text{and} \quad n_2 = \frac{n_e n_o}{\sqrt{n_e^2 \cos^2\theta + n_o^2 \sin^2\theta}} \tag{3.88}$$

The index ellipsoid method provides an alternative and instructive approach for determining the direction of polarization of the normal modes. This approach is particularly beneficial when dealing with biaxial media [Yeh P., 1999, chapter 3].

3.5 Modeling Techniques

Optical simulation has become vital to understanding and optimizing system performance of LCD projectors. In this section we identify certain problems that are encountered when implementing the mathematics into a software model, and their solution.

(a) Eliminating Fresnel Interference Patterns in the Berreman 4×4 Approach

Berreman's 4×4 algorithm is a rigorous solution for stratified media. Since light is usually incident from (and exits into) air, there is typically a big refractive index discontinuity at the incident and exit boundaries. A side effect when using the 4×4 method directly is to produce fine interference spectral patterns in any calculated spectrum, which is rarely seen in a real measurement due to component thickness irregularity and the incoherence of the distributed light used in projection systems. One solution is to smooth the calculated spectrum. Typically, 10-point averaging is needed, which can slow down the calculation speed significantly [Lien A., 1996]. A second method is to use the extended 2×2 Jones' matrix approach, which avoids interference by ignoring reflections.

Another effective solution is to immerse the system in index-matched media. The Fresnel loss can be calculated separately as a simple bulk property. For the oblique incidence case, the incidence angle must be scaled according to Snell's law ($\theta'_{in} = \sin^{-1}(\sin\theta_{in}/n)$) as shown in Figure 3.15.

Yet another three-stage solution treats the first and last elements, such as a sheet polarizer or analyzer, as semi-infinite media as shown in Figure 3.16. Instead of calculating the propagation matrix for a whole system, we break it down into three subsystems. The advantage of this approach is that no extra steps are needed, and any Fresnel loss has been taken into account. In this case, we ignore the interference among the three reflected beams Ψ_{r1}, Ψ_{r2}, and Ψ_{r3}.

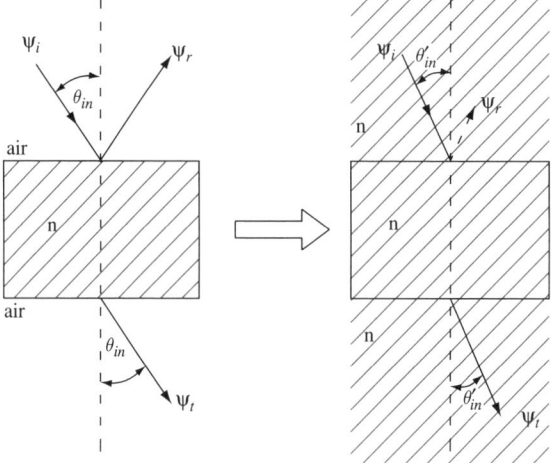

Figure 3.15 Index-matching approach

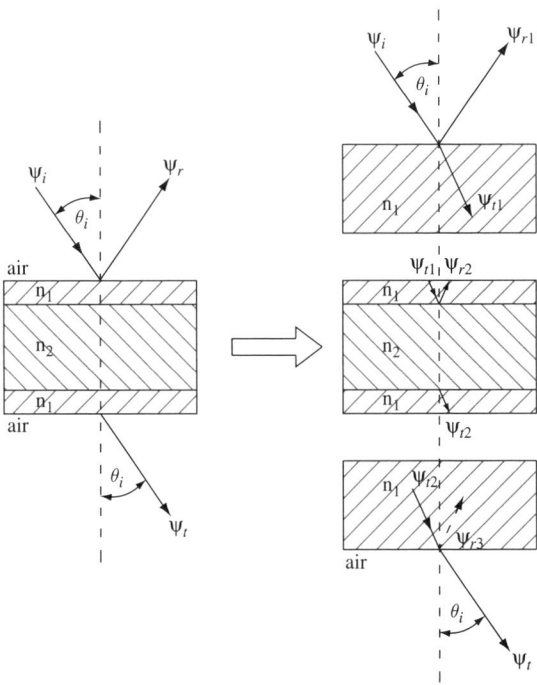

Figure 3.16 Three-stage calculation using the 4 × 4 matrix method

(b) Geometrical Coordinate Transformations

In some circumstances, the optical system contains components with interfaces that are not parallel to one another. For example, a dichroic beam splitter or PBS coating is usually oriented at 45° to the light beam. Since the optical calculation performs the simulation of one component at a time, a coordinate transformation can be carried out before and after any oblique interface (see Figure 3.17) before proceeding with the calculation.

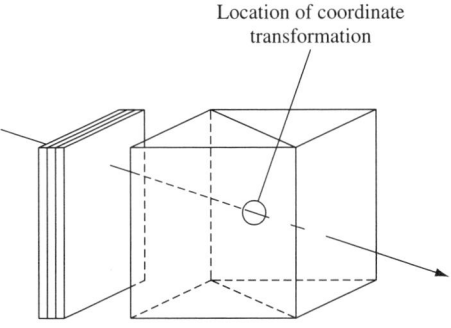

Figure 3.17 Situation for stratified planes not parallel to each

74 POLARIZATION BASICS

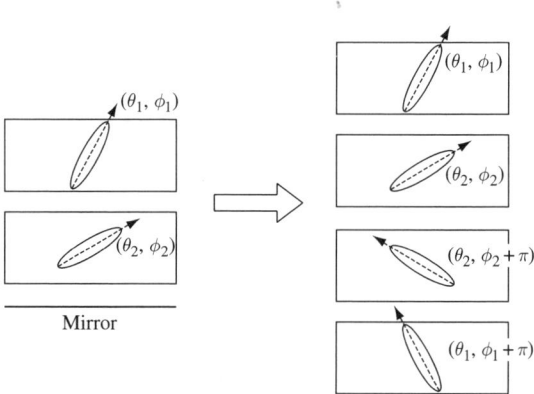

Figure 3.18 A reflective system can be simulated as an unfolded transmissive system

(c) Reflective Systems

A reflective system can be simulated as an unfolded transmissive system as shown in Figure 3.18. While the Berreman 4×4 approach can simulate the reflective system directly, the true characteristics of the mirror (n, k, and thickness d) can be difficult to model and can often create a misleading result. With a Jones' matrix calculation, the inability to deal with reflected light is overcome. By unfolding a reflective system, modeling results are based on an ideal mirror. This technique is particularly useful in investigating the contrast of LCOS panels based on LC director distribution only

References

[Azzam R. M. A., 1989] R. M. A. Azzam and N. M. Bashara, *Ellipsometry and polarized light*, North-Holland, Elsevier Science Publishers, Amsterdam, 1989.

[Berreman D. W., 1970] D. W. Berreman and T. J. Sheffer, Bragg reflection of light from single-domain cholesteric liquid crystal films, *Phys. Rev. Lett.*, 25, p.577, 1970.

[Berreman D. W., 1972] D. W. Berreman, Optics in stratified and anisotropic media: 4×4-matrix formulation, *J. Opt. Soc. Am.*, 62, pp.502–510, 1972.

[Born M., 1980] M. Born and E. Wolf, *Principles of Optics* (6th edition), pergamon Press, Oxford, 1980.

[Chen J., 2004] J. Chen, M. G. Robinson, and G. D. Sharp, General methodology for LCoS panel compensation, SID'04, Digest, pp.990–993, 2004.

[Gu C., 1993] C. Gu and P. Yeh, Extended Jones matrix method II, *J. Opt. Soc. Am.*, A10, pp.966–973, 1993.

[Hecht E., 1980] E. Hecht and A. Zajac, *Optics*, Addison-Wesley World Student Series Edition, 5th printing, Addison-Wesley, Reading, MA, 1980.

[Jones R. C., 1941] R. C. Jones, New calculus for the treatment of optical systems, *J. Opt. Soc. Am.*, 31, p.488, 1941.

[Lien A., 1990] A. Lien, Extended Jones matrix representation for the twisted nematic liquid-crystal display at oblique incidence, *Appl. Phys. Lett.*, 57, pp.2767–2769, 1990.

[Lien A., 1996] A. Lien and C.-J. Chen, A new 2×2 matrix representation for twisted nematic liquid crystal displays at oblique incidence, *Jpn. J. Appl. Phys*, 35B, pp.L1200–1203, 1996.

[Teitler S., 1970] S. Teitler and B. W. Henvis, Refraction in stratified, anisotropic media, *J. Opt. Soc. Am.*, 60, p.830, 1970.

[Wu S. T., 2001] S. T. Wu and D. K. Yang, *Reflective Liquid Crystal Displays*, John Wiley & Sons, Ltd., Chichester, chapter 4, pp.89–111, 2001.

[Yeh P., 1988] P. Yeh, *Optical Waves in Layered Media*, a Wiley-Interscience, New York, 1988.

[Yeh P., 1999] P. Yeh and C. Gu, *Optics of liquid crystal displays*, a Wiley-Interscience, New York, 1999.

[Yeh P., 1982] P. Yeh, Extended Jones matrix method, *J. Opt. Soc. Am.*, 72, pp.507–513, 1982.

4

System Components

4.1 Introduction

A projection system consists of a series of components whose individual performances complement each other to achieve good overall system performance. This is particularly true for LC projection systems that demand good polarization control. Subtle polarization effects from optical components must often be corrected or compensated by subsequent components. Methods for integrating a system from individual components without losing polarization control will be discussed in later chapters. Introduced here are the properties of isolated components in order that their individual effect on polarization is clearly understood, particularly for off-axis optical rays. The components will be considered separately and in no particular order, although a large section will be devoted to interference filters, which include dichroic mirrors, AR coatings, MacNeille cube polarizing beam splitters (PBSs), etc. Retardation stack filters (ColorSelect®) are left for more detailed discussion in Chapter 6.

4.2 Retarders

Retarders are commonly used in LCD-based projectors, particularly in LCOS systems, primarily to improve contrast. In general, a retarder (or wave plate) is a component that modifies, or transforms, the state of polarization (SOP). It is in principle lossless, and therefore does not increase the degree of polarization. A retarder therefore relies on a polarized input to be effective and relies on its birefringence to transform the SOP. A linear birefringent material (as distinct from, say, optically active materials) has an intrinsic material coordinate system, where a different refractive index can exist along each axis in three-dimensional space.

78 SYSTEM COMPONENTS

Birefringent (or anisotropic) material can be characterized by three refractive indices (n_x, n_y, n_z), where the most general case is termed biaxial. With many common materials, termed uniaxial, two of the three refractive indices are identical or degenerate. When light passes through birefringent material, the field is in general subdivided into two discrete waves or modes, resulting from the projection of the optical polarization onto each material axis (see Chapter 3). Each wave that passes through a retarder experiences a different optical path length as the waves propagate with different speeds according to the difference, Δn, of their projected extraordinary, n_e, and ordinary, n_o, refractive indices. For uniaxial materials, the ordinary index is always equal to the value of its two degenerate refractive indices. A phase delay therefore exists between each wave exiting a retarder, which depends upon the difference in refractive index and the physical thickness, and can be used to control the SOP.

A retarder is characterized by its retardation, given by:

$$\Gamma(\lambda) = (\Gamma_0 + 2\pi m)\frac{\lambda_0}{\lambda}\frac{\Delta n(\lambda)}{\Delta n(\lambda_0)}$$

where Γ_0 is the retardation of order m, at wavelength λ_0. The retardation is inversely proportional to wavelength, but also depends upon the dispersion of the birefringence. In display, the wavelength dependence of retardation is generally minimized, calling for zero-order ($m = 0$) retarders. Zero-order retarders are conventionally formed by crossing two multi-order retarders, to yield a difference retardance that is less than the wavelength. Such retarders formed from like materials, however, exhibit strong angle sensitivity not shown to the same extent in true zero-order retarders made from single films. When zero-order retardation is not adequate, achromatic retarders are used.

A retarder is further characterized by material properties. Some common terms to characterize retarders are as follows:

1. A uniaxial retarder can have either positive $(n_e - n_o) > 0$ or negative $(n_e - n_o) < 0$ birefringence. Negative retarders are better suited to compensating LC devices, which are generally positive birefringent.

2. A uniaxial retarder with the optic axis contained in the device aperture is termed an a-plate. These are very common, used to compensate direct-view panels, and to eliminate the in-plane retardation of an LC cell.

3. A uniaxial retarder with the optic axis normal to the device aperture is termed a c-plate. Negative c-plates are commonly used to compensate the FOV of a homeotropic LC device.

4. A uniaxial retarder with the optic axis inclined at any other angle is termed an o-plate (i.e., oblique). These can be used to compensate an LC device with an inhomogeneous director profile.

5. The most general homogeneous retarder is a biaxial o-plate, which has a three-dimensional refractive index, and arbitrary orientation with respect to the device.

6. LC polymer retarders are generally uniaxial, but are capable of forming o-plates, with the additional feature that the optic axis can be inhomogeneous in the thickness direction.

Retarders can be made using either inorganic (crystalline) or organic materials. Naturally occurring birefringent materials include crystals such as quartz, mica, and calcite [Yeh P., 1988, chapter 10] which exhibit birefringence on an atomic scale. Crystalline retarders are frequently made by grinding and polishing, resulting in components of high optical quality and durability. The manufacturing process is, however, limited to small batches, and is time consuming, frequently resulting in expensive components. Additionally, the small phase delays required for projection demand layers that are often impractically thin, both for the polishing process and for the mechanical stability of the finished optics. Unfortunately, there are no crystalline materials with the practical benefits of quartz, and a low enough birefringence to yield a free-standing compensator.

Recent improvements to thin-film coating processes have allowed retarders based on form birefringence to be realized [Eblen J. P., 1994; 1997]. Form birefringence [Yeh P., 1988, chapter 6] is a property that emanates from small anisotropic structures formed from isotropic materials, the simplest structure being one made from alternate layers of two materials as shown in Figure 4.1.

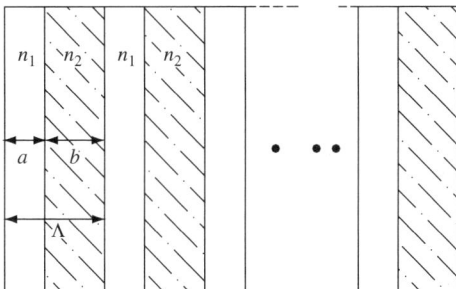

Figure 4.1 Periodic structure of form birefringent material

Mathematically, the superposition of propagating electromagnetic waves at any given wavelength cannot conform to the structure's features when the wavelength is large relative to the structure, and only solutions based on suitably averaged material properties exist. Formally we can consider solutions to Maxwell's equations within this two-material periodic layer system. When the periodicity Λ is close to that of the wavelength of radiation, banded Bloch wave solutions typical of many periodical physical systems [Kronig R. de L., 1931] are obtained. In the long-wavelength limit (equivalent to small layer periodicity), the wave solutions become effectively continuous and the material behaves to first order as a uniaxial material with refractive indices given by:

$$n_o^2 = \frac{a}{\Lambda}n_1^2 + \frac{b}{\Lambda}n_2^2 \quad \text{and} \quad \frac{1}{n_e^2} = \frac{a}{\Lambda}\frac{1}{n_1^2} + \frac{b}{\Lambda}\frac{1}{n_2^2} \qquad (4.1)$$

where n_1 and n_2 are the refractive indices of the two layers, a and b the layer thicknesses, and Λ the periodicity of the structure ($= a + b$). The direction of the optic axis is normal to the plane of the layers, which yields a c-plate retarder.

Assuming a stack of alternating materials with a 50% duty cycle (i.e., equal thickness layers) then the ratio of the extraordinary index to the ordinary index is:

$$\frac{n_e}{n_o} = \frac{2n_1 n_2}{(n_1^2 + n_2^2)} \qquad (4.2)$$

Since the material refractive indices are >1 then $n_e < n_o$ making the stack negative uniaxial. Though somewhat restrictive, negative c-plates are very effective at compensating LC panels (see Chapter 7) offering superior durability to organic polymers. Typical retardance values for panel compensation lie between a quarter and half wave (130–280 nm). A multilayer component formed from common coating materials, TiO_2 ($n \sim 2.4$) and SiO_2 ($n \sim 1.45$), would require a total thickness of $\sim 0.8\,\mu m$ or about 10–20 $\lambda/10$ layers.

Most LCOS compensators require an additional 5–20 nm in-plane retardance, which can be introduced into a cast polymer c-plate by stretching. In the case of a multilayer component, oblique evaporation can produce some in-plane retardance [Eblen J. P., 1994; 1997].

More recently, organic retarders, exhibiting birefringence on a molecular scale, were developed. These retarders can be manufactured using continuous processes, and are therefore inexpensive. These include coated polymers [Wu S. T., 1994], polymer casting [Cheng S. Z. D., 1997], LC polymers (LCPs) [Mori H., 1997a; Seiberle H., 2003], and stretched polymer films [Fujimura Y., 1991]. The mass manufacturing of retarder film was brought about by the direct-view display industry. Retarder films are used to correct for inadequacies of the display, including color, head-on contrast, and field of view. While LC layers are generally inhomogeneous, it was initially found that first-order correction by a homogeneous retarder is often sufficient. This is significant because it is not practical through stretching to produce inhomogeneous films that more closely match the LC characteristics. Additional improvements in display compensation have come about via compensation with multiple films, and through more exotic stretching. However, for the large angles involved in direct view, an inhomogeneous film matched to the LC characteristics is often required, such as LCPs. An LCP film can be aligned on a single substrate and cross-linked with UV light to form a compensating retarder that, in the case where discotic negative birefringent LC material is used, perfectly compensates the LC layer in the dark state.

Compensation requirements in projection are generally quite different. Light levels are much higher, as are contrast requirements, but incidence angles are much smaller. While this often leads to a unique design solution, the volume demands of projection are relatively insignificant, forcing the industry to 'make the most' of the materials available.

A stretched polymer retardation film is formed by coarsely aligning the long-chain molecules of isotropic polymeric materials through controlled stretching. By virtue of this process, the material local electronic transport properties at optical frequencies become directional, and birefringence is induced. Initially, the polyvinyl alcohol (PVA) required for manufacturing polarizer was also used for retarder film. But PVA is not stable environmentally and requires encapsulation. Today, the workhorse of the display industry is polycarbonate, which forms a positive uniaxial retarder when stretched. Retardation values as little as 20 nm, and as large as 2000 nm have been demonstrated. Low retardation values are difficult to control because the degree of stretching is insufficient to produce a film with uniform spatial statistics.

With simple one-dimensional stretching, essentially uniaxial birefringence is formed with associated optical properties. More recently, manufacturers have developed three-dimensional stretching of polycarbonate (PC) to address LCD contrast and FOV enhancement requirements. The more complex 2D stretching, which includes shearing, can form layers that exhibit biaxiality. By controlling the extent of biaxiality, improvements in critical off-axis performance can be achieved. The extent to which off-axis performance is improved can be readily calculated for varying degrees of biaxiality in a viewing plane containing two of the film's three orthogonal optic axes (n_x, n_z). Figure 4.2 shows the plane in question and represents the birefringence of the film as an index ellipsoid (see Chapter 3).

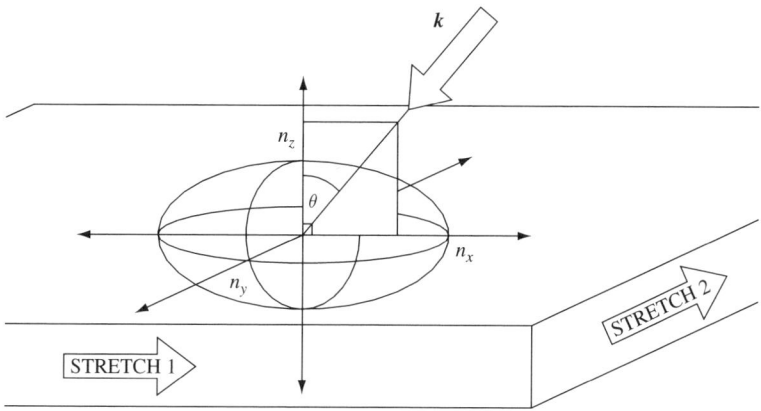

Figure 4.2 Biaxial film geometry

The optical properties of a biaxial film can be characterized by the so-called Nz factor [Fujimura Y., 1991], where $Nz = (n_x - n_z)/(n_x - n_y)$. Figure 4.3 shows the various biaxial retardances currently achievable with stretching.

By rigorous calculation described in Chapter 3, it can be proven that the retardation in this particular incidence plane is independent to first order in angle when $Nz = 0.5$ as shown in Figure 4.4.

Polymer retarders are typically manufactured by coating the resin on a metal casting belt. As such, the optical properties (e.g., transmitted wavefront distortion) of a polymer retarder can adversely affect the image quality when incorporated into a projection system. One solution is to mount the polymer film between optically flat substrates (such as glass) using an optical adhesive. Film casting is followed by stretching. The stretching is accomplished in a continuous fashion via an accurately controlled oven with a pair of rollers turning at different rotation speeds. The precision of this process determines the spatial statistics of the retarder film, including retardation and optic axis. Stretched polymer retarders are low-cost birefringent elements that can be manufactured with large apertures, good cosmetic quality, high durability, and with varying, prescribed retardances.

82 SYSTEM COMPONENTS

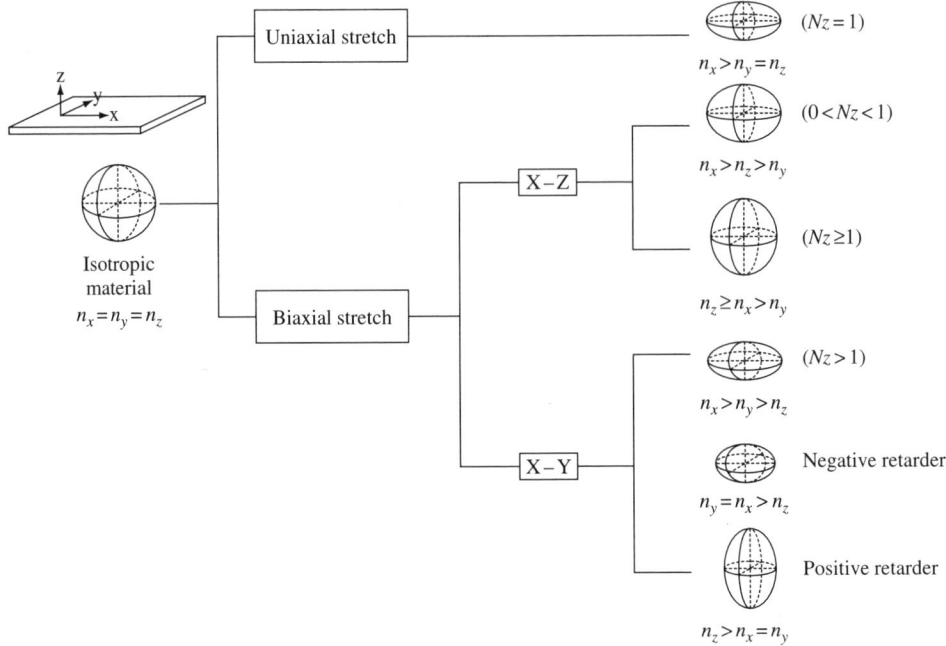

Figure 4.3 Film birefringence attainable by polymer stretching

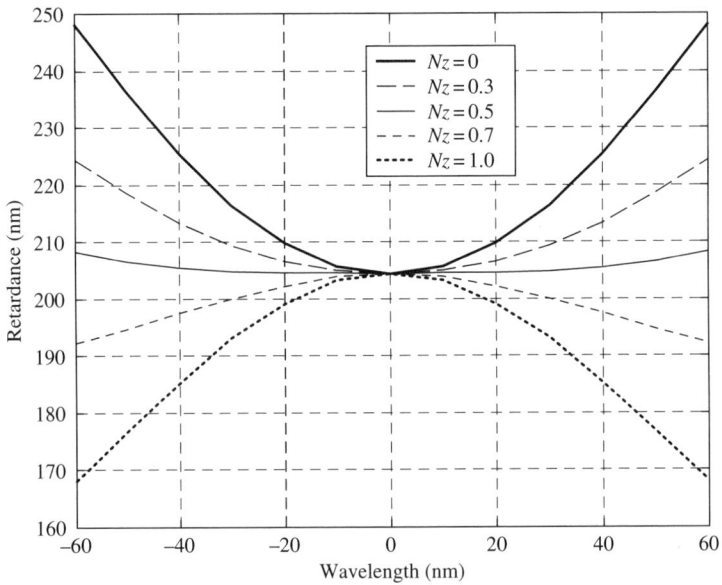

Figure 4.4 Retardation versus polar angle for a 205 nm film with varying Nz factors

4.3 Polarizers

A polarizer is a component that increases the degree of polarization and must therefore extinguish a portion of transmitted light, using mechanisms that include absorption, reflection, or scatter. A polarizer can in principle have arbitrary eigenpolarizations, meaning that the orthogonal transmitted/extinguished polarizations can be any elliptical SOP. Cholesteric LCs, for instance, reflect one handedness of circular polarization, while transmitting the orthogonal SOP. By far the most common polarizer, however, is the linear polarizer, illustrated in Figure 4.5. The linear polarizer used in projection takes two forms: the absorptive polarizer and the reflective polarizer. These components will be referred to as sheet polarizers from now on.

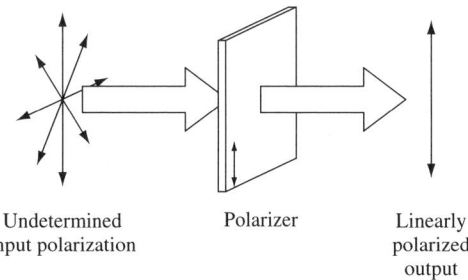

Figure 4.5 Linear polarizer

4.3.1 Absorptive Polarizer

Absorptive polarizers consist of absorbing molecules oriented by a stretched polymer, such that light of a single linear polarization is allowed to pass through the material without significant loss, whereas the orthogonal polarization is strongly absorbed. There are two common types: iodine and dye-type polarizer.

The first absorptive polarizer commercially manufactured was iodine based, developed by Edwin H. Land in 1941 [Land E. H., 1941; 1943]. It is made by first heating and stretching a sheet of PVA which is then imbibed in an iodine solution of chromophore formation, before being stabilized with a boron complex. By stretching PVA, its polymeric molecular structure becomes aligned, which acts to order the chemically bonded iodine molecules. The PVA sheet is structurally weak (approximately 35μm thick) and will decompose in a moist environment, so the polarizer must be laminated between substrates providing a moisture barrier. Today, the stretched polarizer film is chemically bonded between two films of triacetyl cellulose (TAC) for support. Its structure is shown in Figure 4.6.

Dye-type polarizers simply replace the iodine molecules with more durable dye molecules, which make it more suitable for the high luminance levels experienced in LCD projection systems. Companies such as Polatechno Corp. (SHC Series) and Sumitomo currently offer dye-type polarizers for LC projection systems.

For a given concentration and order parameter of the absorptive molecules, absorption polarizers can be made to have higher contrast by increasing by the thickness of the absorption

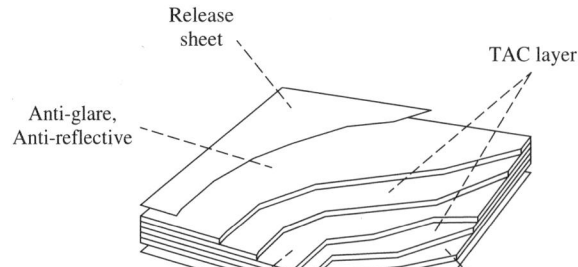

Figure 4.6 Absorptive polarizer structure

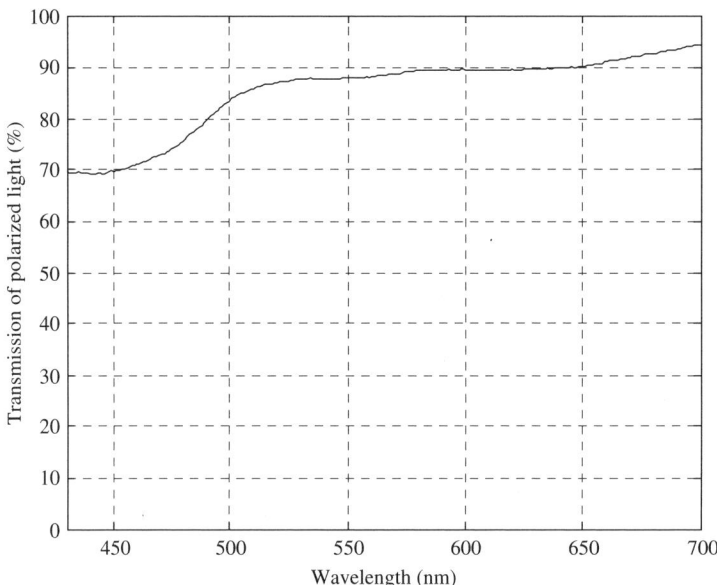

Figure 4.7 Transmission of linearly polarized light through Polatechno's dye-type polarizer SHC125U

layer. Unfortunately increased thickness results in lower transmission. For high-durability dye-type polarizers with typical contrast $\sim 1000:1$, internal transmission of blue (450 nm) light is only $\sim 76\%$ (see Figure 4.7).

It is important to realize that the substrates surrounding the absorbing layer are usually birefringent, which can influence the transmitted SOP. For example, TAC film has a negative c-type birefringence ($\Delta nd \sim -80$ nm) [Mori H., 1997b]. Often, polarizer film is mounted on sapphire, which possesses high thermal conductivity, thus improving reliability. But sapphire is negative uniaxial, and in typical (a-cut) substrate thicknesses has a large retardation

($\Delta nd \sim -10\,000$nm). To minimize the impact on the SOP, the optic axis is aligned parallel or orthogonal to the polarization axes of the polarizer. It is furthermore positioned upstream (i.e., closer to the source than the PVA) when the component acts as a polarizer, or downstream when analyzing.

Optically, an absorptive polarizer can be described as a birefringent film which absorbs one of the normal modes. If the ordinary mode is transmitted then it is called **o**-type. Typical iodine and dye-type polarizer are **o**-type, while more recently **e**-type polarizers have been commercialized based on lyotropic LC films [Paukshto M., 2001; Lazarev P., 2001; Paukshto M., 2002]. The FOV properties of **e**- and **o**-type polarizers have subtle differences, such that when used in polarizer/analyzer combinations the off-axis leakage can be significantly affected, as quantified in Chapter 7.

4.3.2 Reflective Polarizers

4.3.2.1 Multilayer Birefringent Polarizers

Multilayer birefringent polarizers consist of alternate layers of different uniaxial birefringent materials with a common optic axis. In one direction, the refractive indices of the materials are matched. Because the structure appears homogeneous, light polarized along this direction is transmitted without loss. In the orthogonal direction, the structure is inhomogeneous, taking the form of a thin-film multilayer mirror. The indices are mismatched such that light polarized along this direction is affected by the layer boundaries, resulting in reflection if:

$$d_1 n_1 + d_2 n_2 = \frac{\lambda_0}{2} \qquad (4.3)$$

where d_1, d_2, n_1, and n_2 are the alternate layer thickness and indices respectively, and λ_0 is the center wavelength of the reflected band [Welford W. T., 1991]. This relationship is derived from isotropic interference calculations introduced in Chapter 3.

The structure of the generic stacked birefringent polarizer is shown in Figure 4.8.

Increasing the number of layers in the stack increases the reflection of the center wavelength according to:

$$R_{\lambda_0} = \left(\frac{1 - \left(\frac{n_1}{n_2}\right)^{4N}}{1 + \left(\frac{n_1}{n_2}\right)^{4N}} \right)^2 \qquad (4.4)$$

where N is the number of layers, ignoring external reflections from the first and last interfaces, and $n_1 \sim n_2$ [Born M., 1980, chapter 1]. To achieve $>99\%$ reflectivity with a typical material system, where $\Delta n = n_1 - n_2 \ll n_2 (= n)$, the number of layers $N > 6n/\Delta n$. For typical polymer systems where $\Delta n > 0.2$ are difficult to produce through stretching, over 50 layers are necessary. Increasing the number of layers beyond that required for $\sim 100\%$ at λ_0 results in increasing reflection bandwidth given by [Welford W. T., 1991]:

$$\delta\lambda = \frac{4\lambda_0}{\pi} \sin^{-1}\left(\frac{\Delta n}{2n}\right) \qquad (4.5)$$

86 SYSTEM COMPONENTS

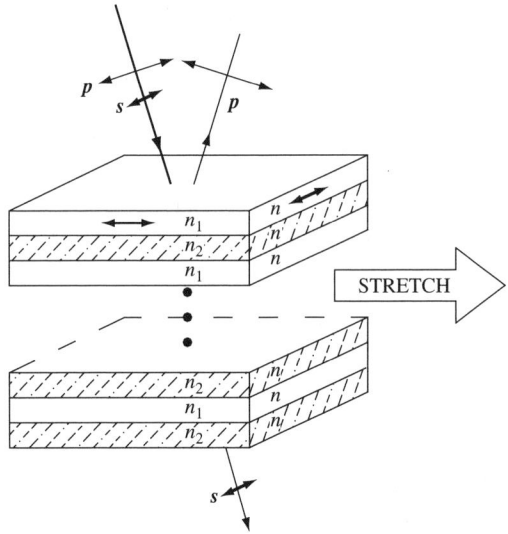

Figure 4.8 Schematic of the stacked birefringent polarizer

Assuming $\Delta n \sim 0.002$, a stack centered at 550 nm would have a maximum width of ~ 0.5 nm. To produce a polymer-based stacked birefringent polarizer with reflectivity over the visible spectrum would then require about 800 layers in either discrete or graded steps in layer thickness.

Such a polarizing film was developed as a brightness enhancement film for direct-view LCDs. Marketed commercially as DBEF, this material has more than 800 layers of alternate polymers [Wortman D. L., 1997]. Made from thermally processed extruded multilayer polymer, the entire layered structure is stretched to form the necessary birefringence within the layers. In this way the extraordinary index is controlled to achieve high reflectivity, whereas the ordinary axis is matched. The optic axes of all layers are essentially parallel to the stretch direction and can therefore be considered the optic axis of the composite film. Since the unaffected transmitted wave is orthogonal to this optic axis, the polarizer is effectively an **o**-type polarizer in transmission, but **e**-type in reflection. This material is most relevant to projection through its use as a polarizing beam splitter, as will be discussed later in this chapter.

4.3.2.2 Wire Grid Polarizers

Wire grid polarizers (WGPs) are effectively 1D metal gratings, ideally with a pitch $<1/10$ of the shortest illuminating wavelength. Wire grid elements have been used extensively as polarizers for long-wavelength infrared and microwave radiation [Hecht E., 1980]. More recently, high-resolution lithography has allowed gratings with progressively smaller feature sizes. Today, it is practical to manufacture polarizers with acceptable polarizing efficiency and transmission throughout the visible. The structure of a metal grid polarizer is shown in Figure 4.9.

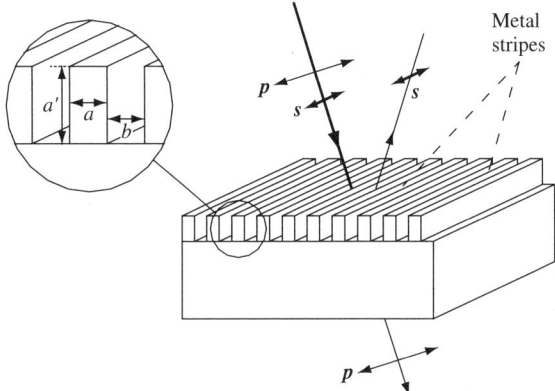

Figure 4.9 Schematic structure of a WGP

The performance of a high-aspect-ratio ($a'/a \gg 1$) WGP can modeled as a form–birefringence element consisting of alternate metal/air layers normal to the substrate (see Section 4.2 above). Assuming an aluminum wire grid structure of the type shown in Figure 4.9, with a 50% fill factor ($a = b$) and a pitch much smaller than visible light wavelengths ($\ll 0.5\,\mu$m), we have the following uniaxial refractive indexes:

$$n_o^0 = \sqrt{\frac{n_{air}^2 + n_{Al}^2}{2}} = 0.696 + 4.71i \quad \text{and} \quad n_e^0 = \frac{\sqrt{2}n_{air}n_{Al}}{\sqrt{n_{air}^2 + n_{Al}^2}} = 1.43 + 0.0045i \quad (4.6)$$

taking the refractive index of the aluminum, n_{Al}, to be $0.974 + 6.73i$ at $\lambda = 550$ nm [www.gsolve.com].

The large imaginary index of the ordinary ray (**o**-mode) (polarized in the plane of the substrate and parallel to the metal stripes) produces severe attenuation in transmission and high reflectivity at boundaries. On the other hand, the **e**-mode (with electric field perpendicular to the metal stripes) experiences minimal attenuation and reflectivity. Hence the structure acts as an **e**-type polarizer in transmission and an **o**-type in reflection. In the more practical case, however, $a'/a \sim 1$ making the structure resemble more closely a cylindrical wire grid, which from symmetry would be an **o**-type polarizer in transmission with its optic axis along the wires. Currently available WGP components are best described as predominantly **o**-type polarizers with slight biaxiality in the plane normal to the stripes.

Visible WGPs are commercially available today [Pentico C., 2001], and are commonly used as pre-polarizers, or as PBSs in projection systems. Normal-incidence transmission of polarized light through a typical WGP can vary between 85% and 92% depending on the thickness of the metal [www.Moxtek.com, Normal Incidence Polarizers, May 2004]. For most practical purposes the WGP can be assumed angularly tolerant and essentially achromatic, unlike absorptive polarizers, which tend to perform significantly better with increasing wavelength. For blue-starved projection systems, the high blue transmission and durability of a WGP make it an ideal choice for a pre-polarizer.

Polarizer contrast, being the inverse of polarizing efficiency, is defined as the ratio of transmission of light polarized perpendicular to the wires, to that parallel to the wires. It depends upon wavelength and metal thickness, but not to any large degree on incidence angle.

88 SYSTEM COMPONENTS

Published data [www.Moxtek.com, Normal Incidence Polarizers, May 2004] indicates a linear increase in contrast with wavelength and includes a thickness-dependent offset, *Off*. Empirically, this published data can be described by:

$$Contrast \approx 2.4\,(\lambda - 450) + Off \tag{4.7}$$

where the values of the *Off* for commercially available products range between 200 and 800.

WGPs are routinely used as pre-polarizers in LC projection systems, but more recently, a PBS version has been used in high-contrast LCOS systems, which will be considered later in the PBS section of this chapter.

4.4 Interference Filters

An interference filter typically consists of a stack of optically thin (sub-wavelength) dielectric layers deposited onto a glass substrate. The refractive indices, thicknesses, and number of layers are all selected in order to control the net reflectivity as a function of wavelength. The mathematics that describes this phenomenon are discussed in Chapter 3. The simplest designs involve few layers, which include anti-reflection coatings. Increasing the number of layers allows chromatic reflection filters to be designed that can operate as beam splitters. The design and fabrication of interference filters has been described in detail elsewhere [Macleod H. A., 1969]. Here we will consider specific examples of interference designs in order to highlight their general properties, particularly the polarization effects in chromatic beam splitters.

4.4.1 Anti-reflection Coatings

Equation (3.34) of Chapter 3 explicitly describes the reflection and transmission from a single layer coated onto a substrate material. A minimum reflectivity, anti-reflection (AR) solution to this equation is obtained when the layer thickness is a quarter wave at the design wavelength, with refractive index n_1, where:

$$n_1 = \sqrt{n_{in} n_{out}} \tag{4.8}$$

This solution, though simple in construction, is often not sufficiently achromatic, and materials with the correct refractive index are often not available to produce zero reflection even at the design wavelength. By increasing the number of layers, a lower reflectivity can be achieved, with broader spectral coverage, as shown in Figure 4.10.

Broadband AR coatings are used to minimize unwanted reflections, while increasing the transmission of free-standing components. Precise designs depend on available coating materials, specific substrate indexes, and tolerance to manufacturing limitations. The typical four-layer AR coatings reduce surface reflections to <0.5% over the visible (430–680 nm) spectrum. In certain narrowband channels, as in the RGB paths of three-panel projection systems, the coatings can be modified to achieve closer to 0.25% reflectivity, which can be important when reflected light affects system contrast (see Chapter 7). Their broadband

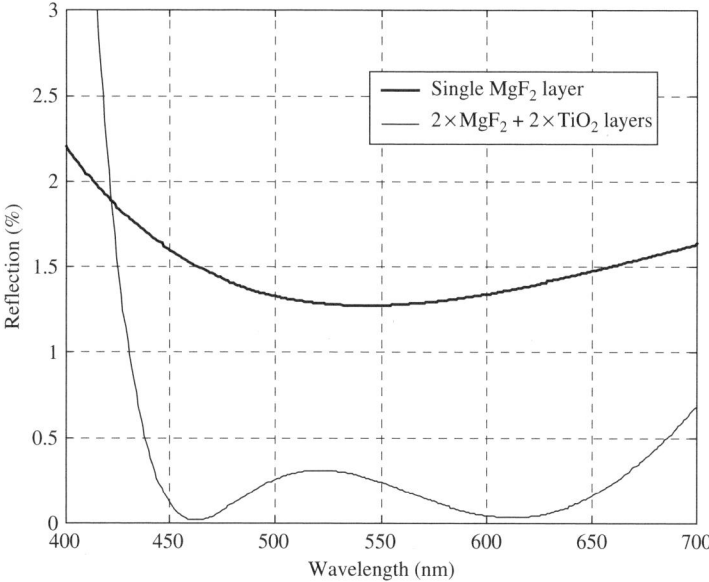

Figure 4.10 Normal incidence reflection spectra for a single and four-layer AR coating on glass

nature together with their near-normal incidence operation does not introduce any significant polarization or chromatic effects for off-normal rays.

The design of multilayer coatings is carried out in general with multivariable optimization algorithms. Many commercially available software packages exist [www.sspectra.com]. As with all optimization methods of this type, however, final designs can depend heavily on initial conditions and stopping criteria. For this reason there are few generic coatings, one exception being the quarter-wave stack. It is useful to look at this coating in some detail since it shows quite clearly how increasing the number of layers produces high reflectivity over finite color bands; a first example of a color filter.

4.4.2 Quarter-wave Stack

A quarter-wave stack (QWS) consists of alternate quarter-wave-thick layers of two different materials, which creates constructive interference between successive interface reflections, forming a highly reflecting mirror coating. Increasing the number of layers acts both to increase the reflectivity (at the design wavelength) and to widen the spectral coverage of high reflection. Figure 4.11 clearly shows this effect, where reflection spectra of SiO_2/TiO_2 QWSs are plotted versus number of layers.

This approach is generally sufficient when designing broadband mirror coatings. However, a spectral filter requires high reflectivity in one portion of the spectrum, with high transmission in the complementary portion of the spectrum. With this additional constraint, it is necessary to optimize the thickness of each layer, as for the AR coating above. These so-called dichroic filters are a key component of projection systems.

90 SYSTEM COMPONENTS

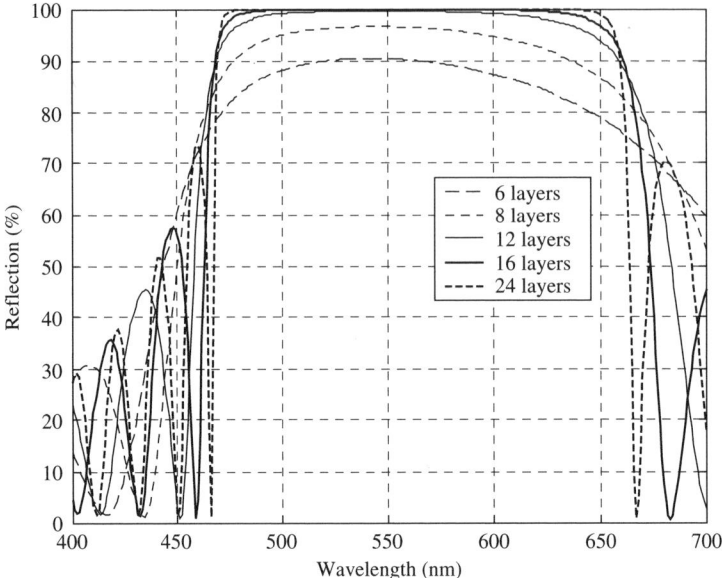

Figure 4.11 Reflectivity of normal incidence light on a stack of alternate, quarter-wave-thick, TiO_2/SiO_2 layers on BK7 glass for different number of layers

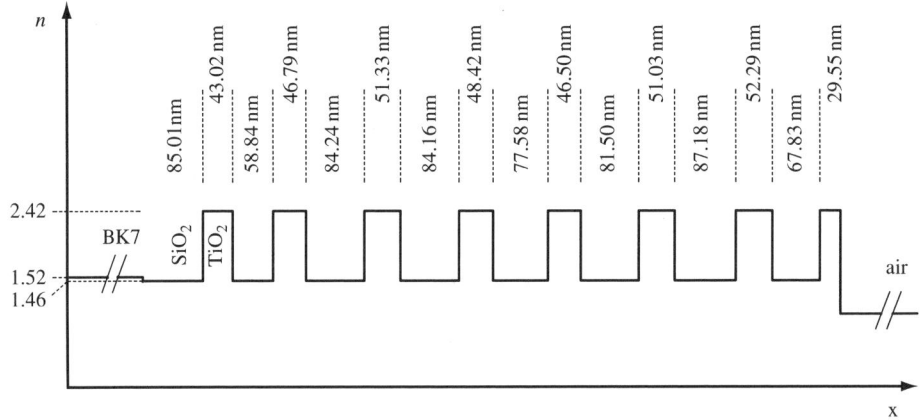

Figure 4.12 Blue reflecting dichroic mirror design

4.4.3 Normal Incidence Dichroic Filters

Using a typical optimization algorithm from the initial 16-layer QWS design above, it is possible to arrive at the blue reflecting/red transmitting dichroic design shown in Figure 4.12.

The normal incidence transmission of this filter is shown in Figure 4.13, together with the transmission for light incident at off-normal angles.

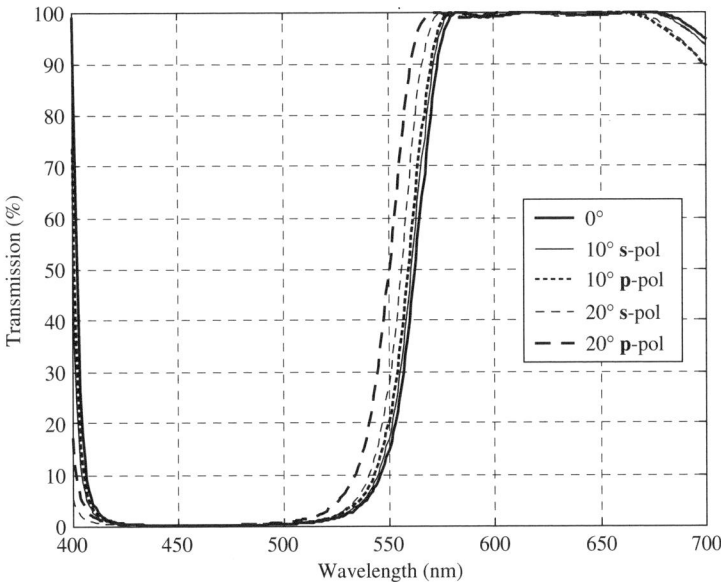

Figure 4.13 Transmission spectra for off-normal incidence of a blue reflecting dichroic interference filter

Off-normal light incident on a dichroic filter experiences two phenomena. First, the reduction in optical path length results in a shift of the filter spectrum toward a shorter wavelength. Second, and as predicted by the polarization-dependent Fresnel reflection coefficients of equation (3.27), the change in spectral profile differs for **s**- and **p**-rays.

In general, the phase delay decreases as ray angles deviate further from normal incidence. The result is a shift of the filter transition band to shorter wavelengths, as clearly seen in Figure 4.13. The shift can be estimated from the geometrical decrease in phase delay between reflections from successive layer boundaries, assuming the interface reflections remain constant. This approach yields the following well-known expression describing the incident angular dependent shift in the 50% transmission wavelength [Melles Griot, 1997]:

$$\lambda_{50\%}(\theta) \approx \lambda_{50\%}(0) \sqrt{1 - \frac{\sin^2 \theta}{n_{eff}^2}} \tag{4.9}$$

where $\lambda_{50\%}(\theta)$ is the 50% transmission wavelength for rays propagating at $\theta°$ with respect to normal incidence in air, and n_{eff} is an average of the layer refractive indices [Stupp E. H., 1999, p. 96].

For this particular filter at 10° incidence, the 50% point is predicted to shift from 562.6 nm to 560.3 nm, which is almost exactly the simulated average shift of the **s**- and **p**-polarizations.

Both **s**- and **p**-polarizations experience a change in interface reflectivity, with **p**-polarized light generally seeing a reduction in reflectivity as a function of incidence angle. Any difference in transmission between **s**- and **p**-polarizations implies an increase in the degree of polarization. Consequently, a dichroic filter can be considered a partial polarizer, with characteristics dependent upon ray direction and wavelength. Not surprisingly, dichroic

92 SYSTEM COMPONENTS

filters therefore present challenges in situations requiring polarization control in convergent/divergent light. Fortunately, at normal incidence and with illumination cone angles typical of projection systems these effects are small, allowing, for example, dichroics to be used as trim filters within the transmissive system described in Chapter 9.

In a typical telecentric LC projection system, light that is incident on a dichroic filter has a set of rays defined by the $f_{/\#}$ of the system. The net transmission of the filter in this case is an average over the ray cone. Averaging over the blue shift in filter response with angle produces an effective transmission spectrum with a shallower transition slope, centered at a lower wavelength, as shown in Figure 4.14.

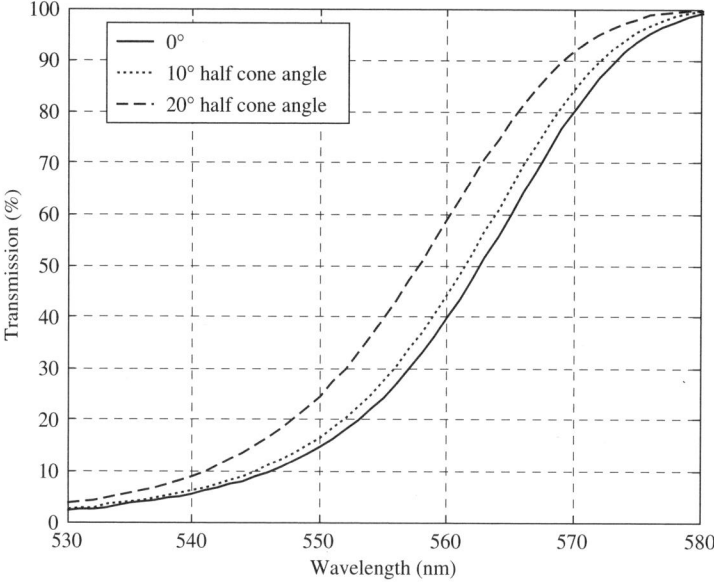

Figure 4.14 Cone-averaged normal incidence transmission of a 16-layer blue reflecting interference filter

Again, at normal incidence the effect of a finite illumination cone is very small. Both polarization and chromatic effects are therefore much more significant in dichroic beam splitters.

4.4.4 Dichroic Beam Splitters

A simple tilted dichroic plate does not in general yield a good dichroic beam splitter, since the 45° incidence spectrum is dramatically different from the normal incidence spectrum. Optimization is necessary, but can only be carried out effectively by considering **s**- and **p**-polarizations separately. Due to the higher reflectivity of **s**-polarized light at coating boundaries, fewer layers are required for **s**-designs. Hence they are favored from a manufacturing and cost standpoint.

Chromatic effects characteristic of dichroic splitters include a significant difference in the transition wavelengths between the **s**- and **p**-components, and the magnified 50% point shift

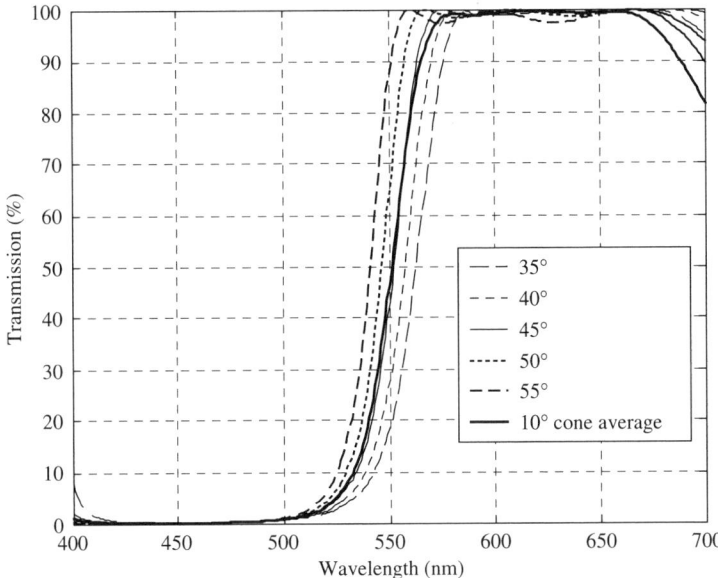

Figure 4.15 The s-polarization transmission spectra for incidence angles about 45°. Also included is the 10° ($\approx f_{/2.8}$) cone average

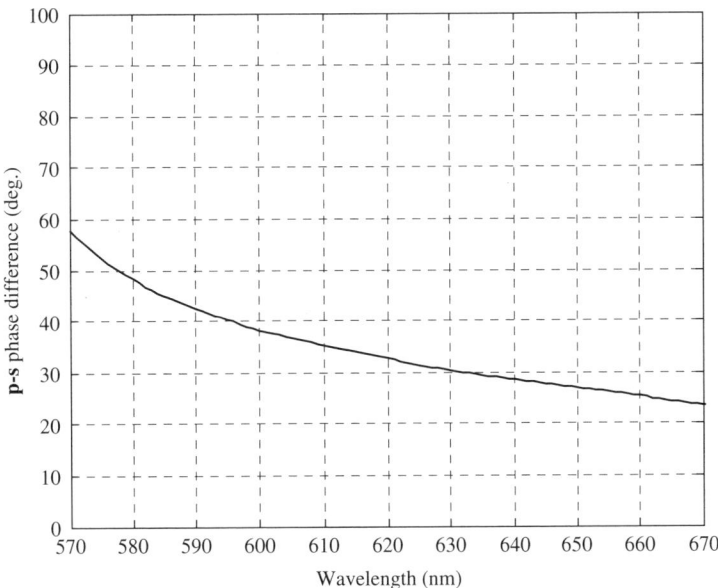

Figure 4.16 Phase difference in degrees between transmitted s- and p-polarized beams incident at 45° to a blue/red dichroic beam splitter

94 SYSTEM COMPONENTS

as a function of incidence angle. The former can be ignored to a great extent in systems that use essentially one polarization. The latter, however, results in a reduction in the transition slope for increasing illumination cone angles (see Figure 4.15).

Polarization effects are significant in dichroic beam splitters due to the large reflection amplitude difference between **s** and **p** in the filter transition region, and the large **s** and **p** phase difference in the transmitted and reflected wavelength bands (see Figure 4.16). At large incidence angles, there is sufficient separation of the **s**- and **p**-spectra that PBS behavior occurs at intermediate wavelengths. Near the transition band (where some portion of **s** and **p** are transmitted/reflected) is a large phase difference, which causes significant transformation in the SOP. Again, this transformation depends upon incidence angle and wavelength, which tends to reduce the cone-angle-averaged degree of polarization. The extent of polarization mixing for skew rays will be discussed next in the PBS section.

4.5 Polarizing Beam Splitters (PBSs)

4.5.1 Dichroic Cube PBS

Embedding a dichroic coating within a high-index medium at 45° enhances the PBS behavior. Invented by S. M. MacNeille in 1946 [MacNeille S. M., 1946], the cube PBS consists of alternate layers of high- and low-index materials coated onto the hypotenuse of a triangular glass prism, which is then bonded to a second prism to form a cube. It works by choosing coating material such that Brewster's angle condition is met at all interfaces, satisfying the Banning relation between refractive indices [Mouchart J., 1989]:

$$n_{sub}^2 = \frac{2n_L^2 n_H^2}{n_L^2 + n_H^2} \tag{4.10}$$

If this relation is met, then **p**-polarized light is always transmitted perfectly (i.e., $T_p = 100\%$, $R_p = 0\%$) regardless of the layer thickness. To efficiently reflect **s**-polarization (i.e., $T_s \sim 0$), the thickness of the layers must then be optimized. In practice, the Banning relation only holds at the 45° incidence angle and for a single wavelength, assuming the materials have non-matched dispersion. To minimize the angular variation, high-index substrate materials are preferred. To achieve good PBS performance, both **p**- and **s**-polarizations must be optimized simultaneously. Figure 4.17 shows R_p as a function of wavelength for a good high-index (PBH56, $n_{sub} = 1.85$) MacNeille PBS. The same PBS has $T_s < 0.05\%$ over the entire visible spectrum for similar incidence angles. In a system, the PBS performance can be represented by an average over the illumination cone angle, which for the case of Figure 4.17 yields $\bar{R}_p < 5\%$ (i.e., $\bar{T}_p > 95\%$) for $> f_{/2.8}$ between 450 and 650 nm.

A key property of cube PBSs relevant to projection systems is geometrical skew ray polarization mixing. It is a consequence of the **s**- and **p**-polarization axes rotating as a function of ray direction and is purely geometrical. Mathematically:

$$s = \frac{k \times n}{|k \times n|} \quad \text{and} \quad p = \frac{k \times s}{|k \times s|} \tag{4.11}$$

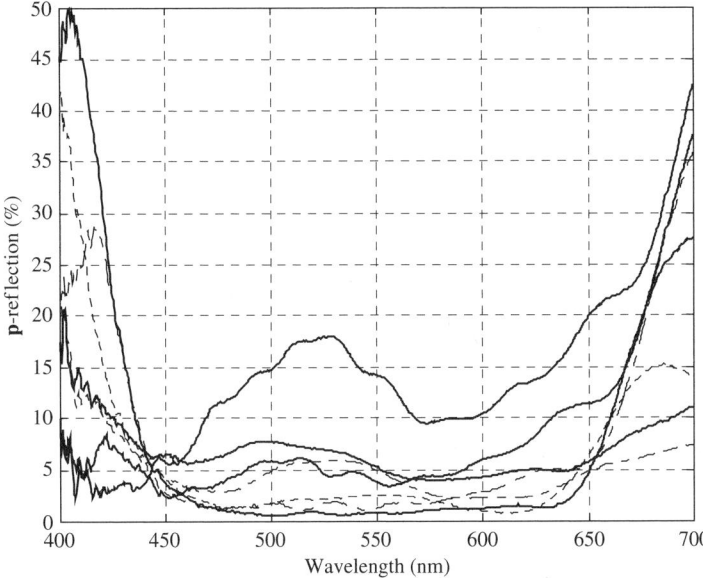

Figure 4.17 Measured reflection of **p**-polarized light, $R_p(\lambda, \theta)$, for a range of incidence angles in air on the face of a PBH56 high-angle-tolerant cube PBS

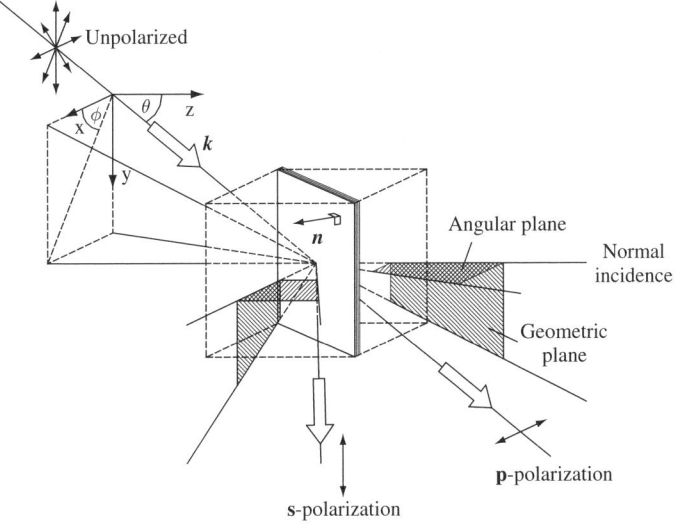

Figure 4.18 General beam splitting geometry

where k is the wavevector of a ray, and n is the reflecting plane normal as shown in Figure 4.18.

For rays contained within the angular plane, there is no rotation of the **s**- and **p**-polarization axes, which remain parallel to coordinate axes defined by the cube. For all other rays the **s**- and **p**-polarization axes are rotated.

96 SYSTEM COMPONENTS

Taking a conventional right-handed Cartesian coordinate system, where the light propagates at small angles θ to the z-axis with respect to a PBS that acts to reflect s-polarized rays toward the x-axis, we can express the vectors $\mathbf{k}$ and $\mathbf{n}$ as follows:

$$\mathbf{k} = (\sin\theta\cos\phi, \sin\theta\sin\phi, \cos\theta) \quad \text{and} \quad \mathbf{n} = \left(\tfrac{1}{\sqrt{2}}, 0, -\tfrac{1}{\sqrt{2}}\right) \quad (4.12)$$

which makes the projection of $\mathbf{s}$ onto the xy plane equal to:

$$(\sin\theta\sin\phi, -\cos\theta - \sin\theta\cos\phi) \quad (4.13)$$

This in turn yields a rotation angle relative to the normal incidence polarization axes of:

$$\beta = \tan^{-1}\left(\frac{-\sin\theta\sin\phi}{\cos\theta + \sin\theta\cos\phi}\right) \quad (4.14)$$

For the particular set of rays in the geometric ($\phi = 90°/270°$) plane, the rotation of the polarization β (shown in Figure 4.19) is equivalent to the polar angle θ of the ray for typical small projection angles ($<15°$).

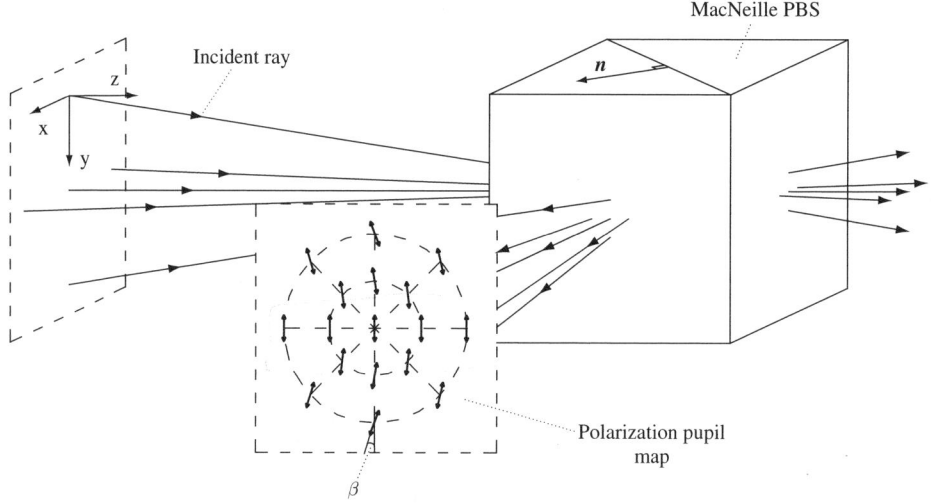

Figure 4.19 Illustration of polarization axes rotation of skew rays propagating through a cube PBS. The degree of polarization rotation of s-polarization is a function of incidence angle

Consequences of the axis rotation are seen when a cube PBS is used either as a polarizer or analyzer in conjunction with other polarizing components such as sheet polarizers, retarders, or other PBSs. A well-known problem resulting from this is the very low contrast ($<100:1$) in uncompensated LCOS systems that use a MacNeille PBS as both polarizer and analyzer [Rosenbluth A. E., 1998]. Solutions to this and related issues are covered in detail in Chapter 7.

Asymmetries in coating performance, and the described geometrical effects, make biasing the angle of the reflecting surface to $<45°$ attractive. From a system standpoint, angles shallower than $45°$ can cause increases in optical path lengths. Nevertheless, a $3 \times$ PBS/X-cube architecture has been successfully demonstrated (see Chapter 10).

4.5.2 Multilayer Birefringent Cube PBS (MBC PBS)

Incorporating the wideband multilayer birefringent reflecting polarizing material of Section 4.3.2.1 between glass prisms produces a cube PBS. For narrowband polarizing films there is a shift to shorter wavelength as for any multilayer interference filter. Unlike the MacNeille case, the orientation of the optic axis defines the transmitting and reflecting polarizations, as for a conventional sheet polarizer. This in principle allows transmission of either **s**- or **p**-polarizations. Figure 4.20 shows the two configurations and their respective optic axes.

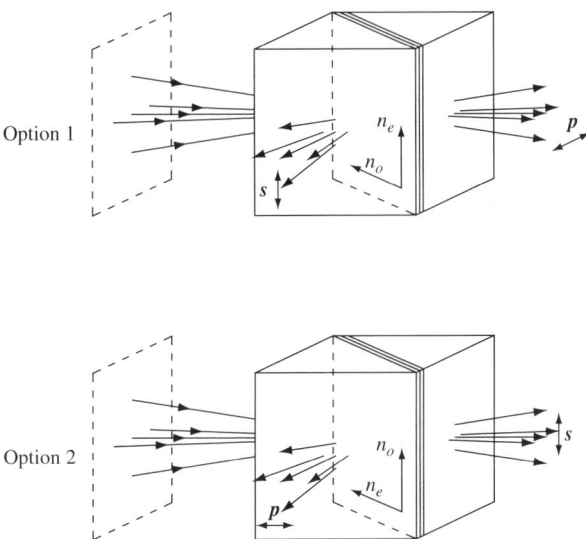

Figure 4.20 Two options for MBC PBSs

Although the two options in Figure 4.20 work well for on-axis rays (i.e., those incident at 45° to the reflecting material), off-axis effects are significantly different due to geometry. In both cases the polarization axis of the transmitted **o**-mode can be written:

$$o = \frac{k \times e}{|k \times e|} \quad (4.15)$$

If **e** is at 45° to the optic axis (option 2), then rotation of the polarization axis similar to the MacNeille cube will result. This leads to low contrast in the absence of compensation. Option 1 has the same geometrical effect as an **o**-type sheet polarizer, with minimal rotation of axes (see Section 4.5.1). In a telecentric LCOS projection system, where the MBC PBS acts as both polarizer and analyzer, there is no geometrical leakage (i.e., self-compensation occurs).

MBC PBSs of the option 1 type have been commercially developed [Bruzzone C. L., 2003] with typical contrast shown in Figure 4.21. Its transmission is >95% over the entire visible with the same $f_{/2.3}$ illumination.

Compared to the conventional MacNeille PBS, this type offers lower $f_{/\#}$ operation with higher transmission and minimal geometrical affects. However, one concern with any organic

98 SYSTEM COMPONENTS

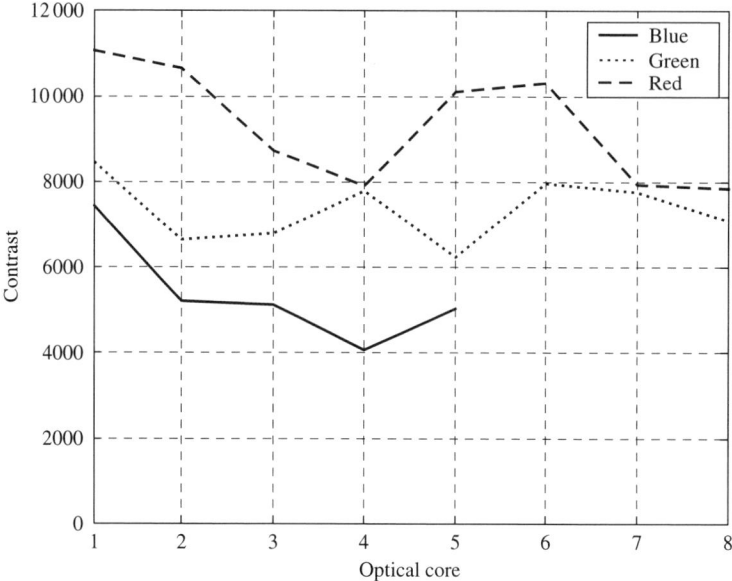

Figure 4.21 Measured contrast of several MBC PBSs for RGB wavelength bands with perfect input polarization at $f_{/2.3}$ using quarter-wave plates (QWPs) and mirrors (Courtesy C. Bruzzone)

material is stability under high-intensity, short-wavelength radiation. This issue is currently being addressed [Bruzzone C. L., 2004]. Another concern is stress-induced birefringence in both the polymer and the surrounding glass, as will be discussed later in this chapter.

4.5.3 Wire Grid Plate PBS

Simply tilting the wire grid plate (WGP) of Section 4.3.2.2 produces a plate PBS with improved contrast performance in the blue. Since astigmatism must be avoided in the image path of a projection system, light from the panel is always analyzed in reflection. Of the two input polarization options, geometrical considerations dictate that the wire orientation should be normal to the system optic axis (equivalent to option I in Figure 4.20) since the component tends to behave predominantly as an o-type polarizer with its axis along the wires. Figure 4.22 shows the only practical configuration for telecentric LCOS projection.

Commercial ProFlux™ WGP PBS components [Arnold S., 2001] have their transmission and reflection properties summarized in Figure 4.23.

Similar to the normal-incidence polarizer, the WGP PBS's intrinsic contrast varies between 100 and 500, depending primarily upon metal thickness.

Compared to a MacNeille PBS, the WGP PBS is very angularly insensitive and can operate well in low-$f_{/\#}$ systems. It also has significantly reduced geometrical effects, which yield very high contrast in systems using clean-up polarizers [Pentico C., 2001]. Stress-induced birefringence is also reduced, since the critical projection optical path does not pass through the substrate. On the downside, transmission is low relative to the McNeill PBS, which includes a reflection loss of ~10%. Also, to achieve very high contrast, polarizers

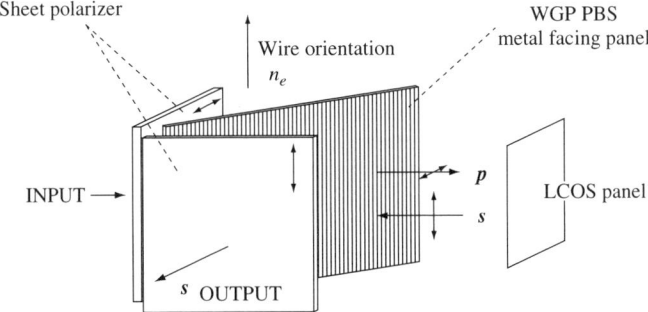

Figure 4.22 The only practical wire grid LCOS configuration

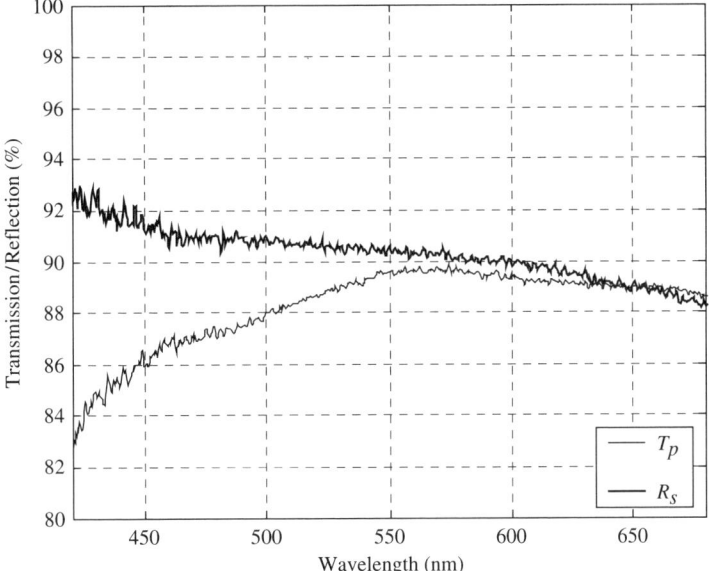

Figure 4.23 Measured T_p and R_s for WGP PBS

are necessary in both entrance and exit ports; the exit polarizers best being the absorptive type. Absorptive polarizers are used to avoid low ANSI contrast in typical small-bfl RPTV systems. The low refractive index of air compared to MacNeille PBS glass prisms results in longer path lengths and larger angular divergence. As a result, systems can be relatively large. Mechanically, the plates are more difficult to align and to maintain flatness during operation. Protection of the open metal layer against moisture and dust is also a concern.

More detailed measurement of the wire grid PBS shows that its description in terms of a simple **o**-type polarizer is not entirely adequate. Direct measurements indicate significant induced ellipticity of skew ray polarization in transmission consistent with small biaxiality. Figure 4.24 shows a typical pupil plot of transmitted (**p**-polarized) light exiting the exposed metal. If the wire grid is considered an **o**-type polarizer with a negative c-plate component

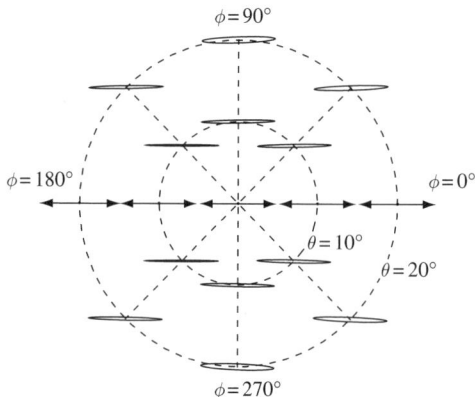

Figure 4.24 WGP PBS polarization pupil plot in transmission. At $(\theta, \phi) = (20°, 90°)$ the transmitted polarization state $(\phi, \varepsilon) \sim (1°, 3°)$

typical of thin-film structures, the measured ellipticity is consistent with a -50 nm, c-plate retardance.

To understand the polarization performance of the wire grid component in detail requires more sophisticated analysis based on rigorous diffraction theory [www.Gsolver.com].

4.6 Other Components

4.6.1 Mirrors

Front surface mirrors are used routinely to fold the optical path in systems to reduce their physical size (see for example Figure 1.15). It should be noted that reflection off a metal surface imparts not only angle-dependent amplitude variation between the **s**- and **p**-modes, but also a significant phase difference. The reflectivity and phase of a metal mirror can be calculated directly from the Fresnel equations using the appropriate complex refractive index associated with the metal (see Chapter 3). For example, a bare aluminum-coated mirror has the amplitude and phase-dependent reflectivity shown in Figure 4.25.

In most cases, only mirror reflectivity is of concern for throughput reasons. However, depending upon position within a polarization-based system, subtle differences in **s**- and **p**-reflectance can cause problems. For example, most projection systems use folding mirrors between the projection lens and the screen. When RGB colors are projected with different polarizations, as in conventional LCD projectors, color non-uniformities occur in the corners.

Commercially available mirrors often have additional deposited layers, which act both to protect and enhance reflectivity. Polarization effects are still an issue, however.

4.6.2 Light-pipe

Light-pipes are used for efficiently homogenizing the illumination (see Chapter 2). They are rectangular in cross-section, and can be made either from individual coated mirror surfaces,

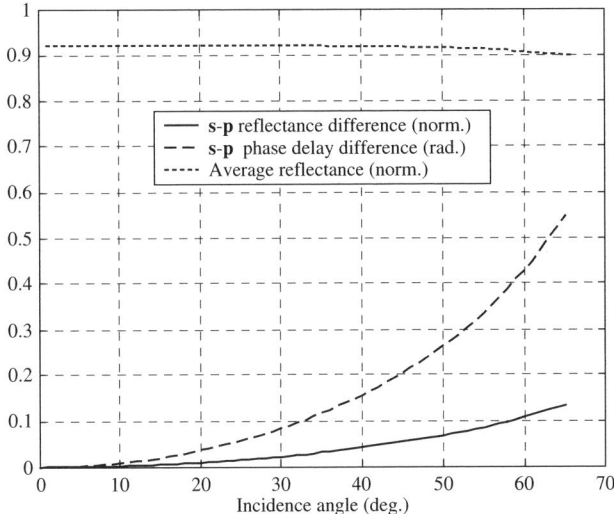

Figure 4.25 Average **s**- and **p**-reflectance off a bare aluminum front surface mirror and their difference as a function of incidence angle. Their phase difference is plotted in radians on the same axis

or from solid glass relying on total internal reflection (TIR). In both cases, polarization is not conserved due to **s** and **p** reflection phase and amplitude differences. Mirrored surface polarization effects are similar to that of a mirror (see preceding section), whereas TIR **s** and **p** phase changes on reflection are given by [Yeh P., 1988, chapter 3]:

$$\delta_s = 2\tan^{-1}\left(\frac{\sqrt{n^2\sin^2\theta - 1}}{n\cos\theta}\right) \quad (4.16)$$

$$\delta_p = -\pi + 2\tan^{-1}\left(\frac{n\sqrt{n^2\sin^2\theta - 1}}{\cos\theta}\right) \quad (4.17)$$

The multiple reflections within a light-pipe cause polarization mixing which prevents efficient polarization recycling.

4.6.3 Substrates

The most common substrate is glass, which in general suffers from stress-induced birefringence. Any intrinsic or induced birefringence alters the polarization state of light, causing non-uniform system performance [Ou C. R., 2001]. The extent to which glass exhibits this phenomenon is related to its photoelastic constant, β, which is defined as the change in birefringence (Δn) as a function of applied pressure. The lower the value, the more tolerant the glass is to applied stress. Typical substrate glasses have the β values listed in Table 4.1.

Induced birefringence can result from several conditions. The first is internal stress due to the forming of glass. This can be seen in glass prisms that have been formed without sufficient annealing. BK7 is annealed by slowly reducing the temperature of the raw mold

Table 4.1 Typical substrate glass properties

Glass	Photoelastic constant β (nm/cm/10^5 Pa)	Young's modulus E (10^8 N/m^2)	Expansion coefficient α (10^{-7}/°C)	Absorption (%/10 mm) at 460 nm
PBH56	0.09	592	84	0.5
SF57HHT	0.02	540	83	0.9
SF6	0.65	550	81	1.1
SF1	1.80	560	81	0.6
SF2	2.62	550	84	0.5
N-BK7	2.77	820	71	0.3

over a period of several days. Second, bonding and mounting glass components must be done carefully to avoid stress. Both UV and thermoset adhesives shrink when cured. Slow curing is beneficial to minimize induced birefringence. Finally, thermally induced birefringence can be controlled through careful system thermal management. Induced birefringence derives from non-uniform expansion of glass by thermal gradients and mismatched material thermal coefficients. The extent to which these thermal effects are seen is related not only to the glass photoelastic constant, but also to absorption, thermal expansion coefficient, and Young's modulus (listed in Table 4.1). High-lead-content glasses such as PBH56 and SF57 have low β values [Shimomura H., 1998] and exhibit minimal non-uniform birefringence, and are thus favored for components like MacNeille PBSs in projection systems. However, the future demand for lower lead-content glasses could favor a compromise solution with glasses such as a wider transmission band SF6 glass (SF6W), for example.

References

[Arnold S., 2001] S. Arnold, E. Gardner, D. Hansen, and R. Perkins, An improved polarizing beam-splitter LCOS projection display based on wire-grid polarizers, SID'01, Digest, p.1282, 2001.

[Born M., 1980] M. Born and E. Wolf, *Principles of Optics*, Pergamon Press, Oxford, 1980.

[Bruzzone C. L., 2003] C. L. Bruzzone, J. Ma, D. J. W. Aastuen, and S. K. Eckhardt, SID'03, Digest, p.126, 2003.

[Bruzzone C. L., 2004] C. L. Bruzzone, J. J. Schneider, and S. K. Eckhardt, SID'04, Digest, p.60, 2004.

[Cheng S. Z. D., 1997] S. Z. D. Cheng, F. Li, E. P. Savitsky, and F. W. Harris, Uniaxial negative birefringent aromatic polyimide thin films as optical compensators in liquid crystalline displays, *Trends Polym. Sci.*, 5, p.51, 1997.

[Eblen J. P., 1994] J. P. Eblen, Jr., W. J. Gunning, J. Beedy, D. Taber, L. Hale, P. Yeh, and M. Khoshnevisan, Birefringent compensators for normally white TN-LCDs, SID'94 Digest, p.245, 1994.

[Eblen J. P., 1997] J. P. Eblen, Jr., L. G. Hale, B. K. Winker, D. B. Taber, P. Kobrin, and W. J. Gunning, Advanced gray-scale compensator for TN-LCDs for avionics applications, SID'97, Digest, p.683, 1997.

[Fujimura Y., 1991] Y. Fujimura, T. Nakatsuka, H. Yoshimi, and T. Shimomura, Optical properties of retardation films for STN-LCDs, *SID Tech. Dig.*, 22, p.739, 1991.

[Hecht E., 1980] E. Hecht and A. Zajac, *Optics*, Addison-Wesley World Student Series Edition, 5th printing, Addison-Wesley, Reading, MA, 1980.

REFERENCES

[Kronig R. de L., 1931] R. de L. Kronig and W. G. Penney, Quantum mechanics of electrons in crystal lattices, *Proc. R. Soc.*, 130, pp.499–513, 1931.

[Land E. H., 1941] E. H. Land, Light polarizer and process of manufacture, US Patent 2,237,567, 1941.

[Land E. H., 1943] E. H. Land, Light polarizer and process of manufacture, US Patent 2,328,219, 1943.

[Lazarev P., 2001] P. Lazarev and M. Paukshto. Low leakage off-angle in Epolarizers, *J. SID*, 9, pp.101–105, 2001.

[Macleod H. A., 1969] H. A. Macleod, *Thin film optical filters*, Adam Hilger, Bristol, 1969.

[MacNeille S. M., 1946] S. M. MacNeille, Beam splitter, US Patent 2,403,731, 1946.

[Melles Griot, 1997] Melles Griot Inc., 1997–1998 Catalogue, p.199, Irvine, CA, 1997.

[Mori H., 1997a] H. Mori, Y. Itoh, Y. Nishlura, T. Nakamura, and Y. Shinagawa, Novel optical compensation film for AMLCDs, SID'97, Digest, p.941, 1997.

[Mori H., 1997b] H. Mori and P. J. Bos, Application of a negative birefringence film to various LCD modes, Proceedings of International Display Research Conference (IDIC), Toronto, Canada, M88, 1997.

[Mouchart J., 1989] J. Mouchart, J. Begel, and E. Duda, Modified MacNeille cube polarizer for wide angular field, *Appl. Phys.*, 28, pp.2847–2853, 1989.

[Ou C. R., 2001] C. R. Ou, W. T. Lee, C. F. Yaung, H. S. Lin, Y. L. Wu, and J. C. Yoo, Experiments on the thermal induced effect of PBS for LCOS system, SID'01, Digest, p.926, 2001; C. R. Ou, W. T. Lee, S. C. Chung, C. C. Lin, and L. C. Lin, Thermal-optical simulations of ColorQuad for image quality, SID'01, Digest, p.922, 2001.

[Paukshto M., 2001] M. Paukshto and L. D. Silverstein, Viewing angle enhancement of TN LCD using E-type polarizers, SID'01, Digest, p.902, 2001.

[Paukshto M., 2002] M. Paukshto and L. D. Silverstein, Two novel applications of thin-film E-type polarizers, SID'02, Digest, p.722, 2002.

[Pentico C., 2001] C. Pentico, E. Gardner, D. Hansen, and R. Perkins, New, high performance, durable polarizers for projection displays, SID'01, Digest, p.1287, 2001.

[Rosenbluth A. E., 1998] A. E. Rosenbluth, D. B. Dove, F. E. Doany, R. N. Singh, Y. H. Yang, and M. Lu, Contrast properties of reflective liquid crystal light valves in projection displays, *IBM J. Res. Dev.*, 42, p.359, 1998.

[Seiberle H., 2003] H. Seiberle, C. Benecke, and T. Bachels, Photo-aligned anisotropic optical thin films, SID, Digest, 1162, 2003.

[Shimomura H., 1998] H. Shimomura, K. Numazaki, Y. Oikawa, N. Shimamura, M. Ueda, and T. Hasegawa, Projection type display apparatus, US Patent 5,808,795, 1998.

[Stupp E. H., 1999] E. H. Stupp and M. S. Brennesholtz, *Projection Displays*, John Wiley & Sons, Ltd, Chichester, 1999.

[Welford W. T., 1991] W. T. Welford, *Useful Optics*, p.78, The University of Chicago Press, Chicago, 1991.

[Wortman D. L., 1997] D. L. Wortman, A recent advance in reflective polarizer technology, Proceedings of IDRC'97, p.M-98–106, 1997.

[Wu S. T., 1994] S. T. Wu, Film compensated homeotropic liquid crystal cell for direct view display, *J. Appl. Phys.*, 76, p.5975, 1994.

[Yeh P., 1988] P. Yeh, *Optical Waves in Layered Media*, John Wiley & Sons, Inc., New York, 1988.

5

Liquid Crystal Displays (LCDs)

5.1 Description and Brief History

LCs represent an intermediate state of matter, existing between a highly ordered crystalline solid state and an isotropic liquid. As a material possessing intermediate order, LCs have long been the subject of scientific study, but more recently have proven useful in display applications [Demus D., 1990; Chandrasekhar S., 1992; de Gennes P. G., 1993]. Many organic materials pass through phases of intermediate order, called mesophases, as they are heated from the crystalline to the isotropic liquid phase. In the crystalline phase, the component molecules that comprise the material have both long-range positional and orientational order in three dimensions. Conversely, isotropic liquids possess no orientational or positional order. The partial ordering of molecules in LCs, intermediate between these extremes, can be either translational, orientational, or both.

Liquid crystalline materials can be divided into two categories [Chandrasekhar S., 1992; de Gennes P. G., 1993], consisting of lyotropic and thermotropic materials. They are distinguishable by primary physical parameters that can be modified to produce a change in material phase. For lyotropic LCs, the variable parameter is the concentration of a surfactant in a solvent, while for thermotropic LCs it is temperature. Depending upon the detailed molecular structure, the system may pass through one or more mesophases before becoming isotropic. An essential requirement for LC phases to exist is that the molecule must be geometrically anisotropic in shape, e.g., rod-like or disk-like. While there are many LCs which comprise disk-like molecules, by far the most prevalent in commercial applications are thermotropic LCs, composed of rod-like molecules. Following the nomenclature proposed originally by Friedel [Friedel G., 1922], they are classified broadly into three categories: nematic, cholesteric, and smectic.

Most LC applications are based on the unique optical response of thermotropic LCs to an electric field, particularly nematic LCs. The nematic LC has a high degree of long-range orientational order, but no long-range translational order (Figure 5.1). This differs from the isotropic liquid in that the molecules are spontaneously oriented with their long axes approximately parallel. The preferred direction usually varies from point to point in the medium, but a uniformly aligned specimen is strongly optically birefringent with positive uniaxiality. The physical properties of nematics clearly show that the order is axial, not polar, and hence the orientational order in a uniaxial nematic can be described by a second-rank tensor. For practical purposes, it is often convenient to define a scalar order parameter $S = \langle \frac{3}{2} \cos^2 \theta - \frac{1}{2} \rangle$, where θ is the polar angle made by the long axis of the molecule with respect to a preferred direction, and the angular brackets indicate a statistical average [Tsvetkov V., 1948]. Note that S becomes zero for an isotropic distribution of molecules, and is unity when all of the molecules are strictly parallel to one another.

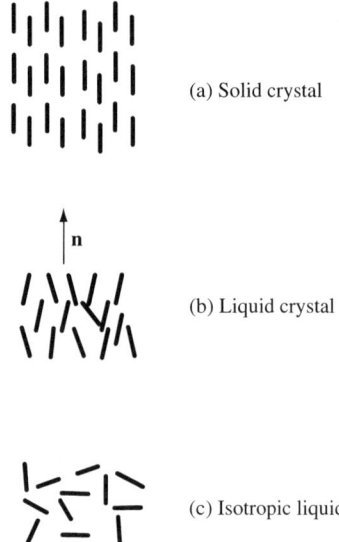

Figure 5.1 Schematic diagram of molecular order in the crystalline, nematic, and isotropic phases

The cholesteric mesophase is related to the nematic, except that it is composed of optically active molecules. As a consequence, the structure acquires a spontaneous twist about an axis normal to any alignment layer. The twist may be right-handed or left-handed depending upon the molecular conformation. The pitch, P_0, of the cholesteric is the distance along the axis required to trace out a complete period of the twist. The spiral arrangement of the molecules in the cholesteric is responsible for its unique optical properties, e.g., selective reflection of circularly polarized light and rotatory power roughly a thousand times greater than that of ordinary optically active substances. Selective reflection is strongest when the pitch is equivalent to the wavelength in the medium. But other effects (e.g., optical rotation) can be produced when the pitch is substantially larger than the incident wavelength. LCs used in display applications frequently have a chiral dopant added to create a modulation mode capable of fast switching that exhibits high resistance to electric field fringing of the addressing structure and mechanical stability within LC display configurations.

Smectic LCs have stratified structures, but a variety of molecular arrangements are possible within each stratification [Birgeneau R. J., 1978; Pindak R., 1981; Gray G. W., 1984; Leadbetter A. J., 1987; Persham P. S., 1988]. In the smectic A phase, the molecules are upright in each layer with their centers irregularly spaced in a liquid-like fashion (Figure 5.2(a)). The interlayer attractions are weak when compared to the lateral forces exerted between molecules. As a consequence, the layers are able to slide over one another relatively easily. The smectic C phase is a tilted form of the smectic A, where the molecules are inclined with respect to the layer normal. Many other smectic phases have been identified and they possess higher order. To describe order in the smectic phase, in addition to the orientational order parameter S of the nematic phase, we need a translational order parameter $S_t = \langle \cos(2\pi z/d) \rangle$ to describe the interlayer spacing order.

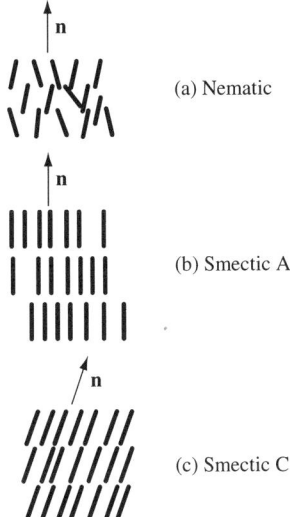

(a) Nematic

(b) Smectic A

(c) Smectic C

Figure 5.2 Schematic diagram of the molecular arrangement in nematic, smectic A, and smectic C LCs

A typical rod-like LC molecule has the general molecular structure depicted in Figure 5.3 [Coates D., 1990; Chandrasekhar S., 1992; de Gennes P. G., 1993; Yeh P., 1999],

X and Y are terminal chemical groups and A is a linking group between two or more ring systems. Z and Z' are lateral substituent chemical groups. Each group plays a different

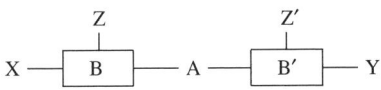

Figure 5.3 General molecular structure for a rod-like LC molecule

role in determining the physical properties of the LC. The rings (B, B') can be either benzene rings or cyclohexanes, or a combination of both. The presence of rings provides the short-range intermolecular forces needed to form the nematic phase. They affect the absorption, dielectric anisotropy, birefringence, electric constants, and viscosity. They also have profound influence on the UV or near-UV stability of LCs, which is essential for projection applications.

The terminal group X, often called the side chain, has three common groups: (a) alkyl chain C_nH_{2n+1}, (b) alkoxy chain $C_nH_{2n+1}O$, and (c) alkenyl chain that contains a double bond. The length of the chain has a strong influence on elastic constants, as well as the phase transition temperature. Due to the geometry factor [Persham P. S., 1988], the clearing temperature (or temperature at which the LC becomes isotropic) shows a zigzag pattern versus the number n, which is known as the "odd–even" effect. LC compounds with a longer chain length possess a lower melting point, but usually have a higher viscosity. Any terminal group that extends the molecular long axis, without increasing the molecular width, will increase the LC stability.

The terminal group Y plays an important role in determining the dielectric constant and the anisotropy, $\Delta\varepsilon$, and can be either a polar or non-polar group. A non-polar group has little impact on the dielectric anisotropy, while LC compounds containing a CN terminal group can have a strong contribution to $\Delta\varepsilon$. The common polar terminal groups are CN, F, and Cl. CN has the highest polarity, and hence a high dielectric anisotropy, but this is often coupled with high viscosity, insufficient resistivity (resulting from purification difficulties), and instability under UV illumination. On the other hand, LC molecules with an F terminal group exhibit low viscosity, high resistivity, and relatively good UV stability. Due to the relatively low polarity, these LC materials exhibit relatively smaller dielectric anisotropy and optical birefringence.

In most cases, the lateral substituents Z and Z' will widen the molecule, tending to reduce lateral attractions and lowering the nematic and smectic stability. If the lateral substituents are polar groups, LC materials with negative dielectric anisotropy can result, which is required for switching the vertically aligned (VA) LC modes. The VA mode is poised to become the industry standard for LCOS projection applications, due to high contrast ratio and large cell gap tolerance. However, such LC materials are much more difficult to synthesize, because a lateral dipole moment is only about half as effective at inducing negative dielectric anisotropy as a terminal dipole is in producing a positive one. Nevertheless, significant progress has been made in this area.

The common linking groups, A, are a single bond, ring, C_2H_4, C_2H_2, and COO. The principal objective of the linking group between ring systems is to lengthen the molecule to increase its aspect ratio and also influence its polarizability and molecular flexibility. These modifications affect the bulk optical and electrical properties of the LC material.

In practice, there is no single LC compound that meets all LC specifications for most display applications. To achieve optimum performance for any one application, different LC materials are typically mixed together. An example property that is readily tailored by LC mixing is phase transition temperatures. It is well known that the melting point of a binary mixture of compounds is less than either of its constituent compounds, with the melting point dependent on the mixture ratio. At the eutectic point, the melting point reaches its minimum. The clearing point of the LC mixture is usually the linear average of the composition. As a result, a mixture of two LC compounds can offer a much larger temperature range in the nematic phase. In addition to the temperature range of the nematic phase, many other physical properties such as dielectric constants, elastic constants, birefringence, and viscosity also depend upon mixing ratios. In practice, the synthesis of LC materials necessary to satisfy LCD application requirements typically involves 20 to 30 LC compounds. For more information on LC material composition, the reader is advised to consult more detailed reference books on the topic [Coates D., 1990].

5.2 Anisotropic Properties of Liquid Crystals [Chandrasekhar S., 1992; de Gennes P. G., 1993]

Mechanical, electrical, magnetic, and optical anisotropy are a direct consequence of the macroscopic ordering of anisotropic molecules. In the uniaxial nematic phase, the dielectric constants differ for electric fields applied parallel or perpendicular to the molecular orientation axis n ($\varepsilon_{\parallel}$ and $\varepsilon_{\perp}$ respectively). The dielectric anisotropy is defined as:

$$\Delta\varepsilon = \varepsilon_{\parallel} - \varepsilon_{\perp} \tag{5.1}$$

Since a nematic LC is uniaxial, the mean dielectric permittivity $\bar{\varepsilon}$ for a nematic LC can be described as:

$$\bar{\varepsilon} = (\varepsilon_{\parallel} + 2\varepsilon_{\perp})/3 \tag{5.2}$$

The dielectric anisotropy of an LC can be positive or negative, depending upon the permanent dipole orientations and on the electrical drive frequency. For certain LCs, the dielectric anisotropy can change sign with frequency [Blinov L. M., 1994]. The sign and magnitude of $\Delta\varepsilon$ are of the utmost importance in determining the electro-optical response.

As in the previous electric field case, a uniaxial LC also shows anisotropy in the refractive index. The ordinary wave with electric field vector perpendicular to the LC orientation direction n (i.e., the material optic axis) experiences a refractive index n_o, while the extraordinary wave, with the electric field vector parallel to the director direction, experiences a refractive index n_e. The optical anisotropy is defined as:

$$\Delta n = n_e - n_o \tag{5.3}$$

The mean value of the refractive indices is then given by the relationship [de Gennes P. G., 1993]:

$$\overline{n^2} = (n_e^2 + 2n_o^2)/3 \tag{5.4}$$

Unlike magnetic susceptibility, where anisotropy is directly proportional to the order parameter, S, there is no simple relationship between dielectric (optical) anisotropy and order parameter. This is because the effective electric field seen by one molecule is given as the superposition of the field due to the external source with the field generated by the neighboring dipoles. The latter contribution is large [de Gennes P. G., 1993].

5.3 Frank Free Energy and Electromagnetic Field Contribution to Free Energy

In an ideal nematic phase, the molecules are (on average) aligned along a common direction n. In an actual LCD, both surface alignment forces and the external field determine a spatial distribution for n. This so-called director distribution will vary from point to point. If the variation of the LC orientation direction is significantly larger than the molecular dimensions, the deformation can be described by a continuum theory, which was first stated by Oseen [Oseen C., 1933] and Zocher [Zocher H., 1938], and completed by Frank [Frank F. C.,

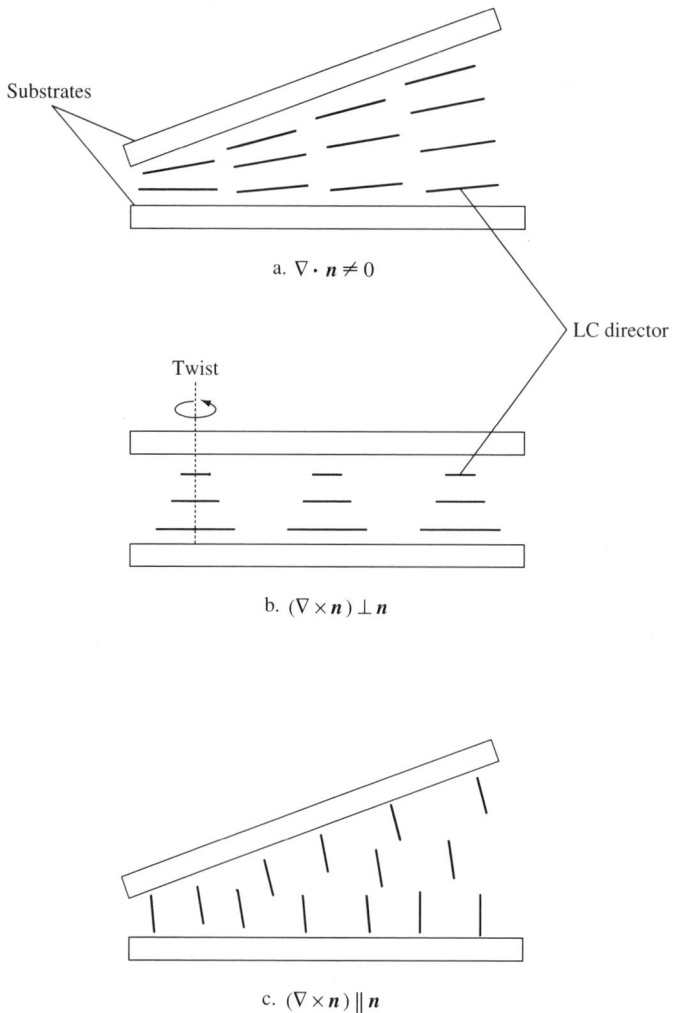

Figure 5.4 Three typical distortion configurations in a nematic LC: (a) splay, (b) twist, and (c) bend. The mathematical expression relating to the free energy is given in each case below the relevant diagram

1958]. After considering certain symmetries and eliminating surface terms, the Frank free energy density of the nematic medium with a curvature deformation in its director field is [Chandrasekhar S., 1992; de Gennes P. G., 1993]:

$$F_B = \frac{1}{2}\left[K_1(\nabla \cdot \boldsymbol{n})^2 + K_2(\boldsymbol{n} \cdot (\nabla \times \boldsymbol{n}))^2 + K_3(\boldsymbol{n} \times (\nabla \times \boldsymbol{n}))^2\right] \quad (5.5)$$

The three terms correspond to splay, twist, and bend deformation respectively, whose geometries are shown in Figure 5.4. K_1, K_2, and K_3 are the splay, twist, and bend elastic constants. These constants are weak compared to those in solids. If the LC has a spontaneous helical twist, the Frank free energy density will become:

$$F_B = \frac{1}{2}\left[K_1(\nabla \cdot \boldsymbol{n})^2 + K_2(\boldsymbol{n} \cdot (\nabla \times \boldsymbol{n}) - q_0)^2 + K_3(\boldsymbol{n} \times (\nabla \times \boldsymbol{n}))^2\right] \quad (5.6)$$

where q_0 is the wavevector equal to $2\pi/P_0$, and P_0 is the intrinsic pitch of the LC. The positive and negative values of q_0 correspond to left- and right-handed helices, respectively.

Under an external electromagnetic field, the contribution of field to the free energy density can be represented by:

$$U_{EM} = \tfrac{1}{2}\boldsymbol{D} \cdot \boldsymbol{E} + \tfrac{1}{2}\boldsymbol{B} \cdot \boldsymbol{H} \quad (5.7)$$

where $\boldsymbol{D}$ and $\boldsymbol{B}$ are constant. If the electromagnetic fields ($\boldsymbol{E}$ and $\boldsymbol{H}$) are constant, then the free energy density due to contributions from the electromagnetic field can be obtained by the Lagrange transformation, expressed as:

$$F_{EM} = U_{EM} - \boldsymbol{D} \cdot \boldsymbol{E} - \boldsymbol{B} \cdot \boldsymbol{H} = -\tfrac{1}{2}\boldsymbol{D} \cdot \boldsymbol{E} - \tfrac{1}{2}\boldsymbol{B} \cdot \boldsymbol{H} \quad (5.8)$$

The equilibrium LC director profile can be obtained by calculus of variations. It leads to a balance between the elastic restoring force and that due to the applied electric field (see Appendix A).

5.4 Alignment Layer and LC Pretilt Angle [Cognard J., 1982; Uchida T., 1990]

A uniform surface alignment is a prerequisite for producing a high-quality LC display. A special substrate treatment induces an orientation to the LC alignment at the interface of the alignment substrate. This influence propagates through the bulk by elastic restoring forces described by the Frank theory. The molecular alignment of nematic LCs is believed to depend on: (a) physicochemical interactions such as van der Waals, steric, or dipole–dipole interactions between LC and alignment layer molecules, such as rubbed polyimide (PI); and (b) mechanical interactions related to the surface topography and anisotropic elasticity of the LC medium, such as an SiO_x evaporated surface. Which mechanism plays the dominant role in LC alignment depends upon circumstances [Taylor G. N., 1973]. In some circumstances, both mechanisms simultaneously contribute to the LC alignment.

Figures 5.5(a) and (b) show two typical geometries of uniform LC alignment (e.g., homogeneous (tangential) and homeotropic (normal) alignment). In practice, a small "pretilt" angle at the LC boundary is necessary to break the symmetry in order to yield domain-free

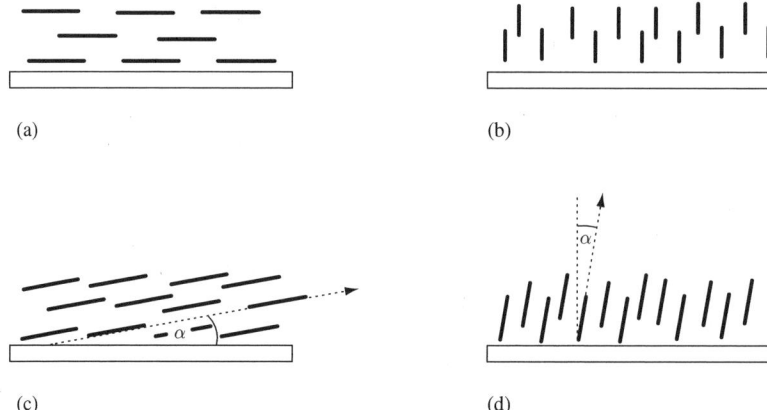

Figure 5.5 Fundamental molecular alignments on substrate surfaces. α is the LC pretilt angle. (a) Parallel (homogeneous), (b) perpendicular (homeotropic), (c) small tilt from homogeneous, (d) small tilt from homeotropic

molecular orientation when an electric field is applied (see Figures 5.5(c) and (d)). The specification for pretilt angle depends upon the display configuration. For example, the pretilt angle can be 2° to 5° in a conventional twisted nematic (TN) display, while a higher pretilt angle (>5°) is preferred in a super twisted nematic (STN) display to avoid stripe instability. A higher pretilt angle is also important to contain the disinclination lines induced by the fringing field in a TFT structure, which can degrade the contrast in HTPS and LCOS projection systems.

The method for producing a homeotropically aligned LC device is to treat the substrates with lecithin [Cognard J., 1982], silane [Taylor G. N., 1973], and PIs with a long alkyl chain. A small pretilt angle (with respect to the boundary normal) can be introduced by then rubbing the surfaces. Unfortunately, these PIs are soft, making it difficult to avoid rubbing marks. The pretilt angle is also not sufficiently stable for projection applications. Perfect homeotropic alignment can also be achieved on a single oblique [Lu M., 2000] or omni-azimuthal (rotating the plate during evaporation) oblique SiO_x evaporated plates [Hiroshima K., 1982; 1984]. A unidirectional evaporation can be added to break the azimuthal symmetry to obtain a finite pretilt angle on an omni-azimuthal SiO_x evaporated surface [Vithana H., 1995]. A small pretilt angle can also be generated by applying a negative uniaxial LC fluid material on a single evaporation surface with a defined evaporation angle [Lu M., 2000]. The most popular method to achieve homogeneous or tilted LC alignment is to unidirectionally rub PI films. Due to the low cost and high yield of this method, the alignment layers for almost all LCDs are obtained in this way [Cognard J., 1982]. The pretilt angle can be controlled by the choice of PI material, rubbing pressure, rotation speed, and substrate translation rate [Geary J. M., 1987].

Since alignment layers have a profound influence on the performance of an LCD, including voltage holding ratio, image sticking, and pretilt angle stability, stringent reliability tests must be performed to qualify a material. Due to the high temperature and light flux found in microdisplay projectors, evaporated inorganic SiO_x alignment layers are gaining popularity relative to traditional PIs.

5.5 Rotational Viscosity

The viscosity of a fluid is an internal resistance to flow, defined as the ratio of shearing stress to the rate of shear. It results from the intermolecular forces in the fluid. The viscous behavior of a particular LC material has a profound effect on the dynamical behavior of the resulting display system. This behavior is further influenced by the (exponential) temperature dependence of viscosity. The rotational viscosity coefficient γ_1, which provides a resistance to rotational motion of LC molecules under external torque, is an important parameter in determining the response time of an LCD. The response time associated with energizing the display, and that given when the field is removed, are given, respectively, by [Jakeman E., 1972]:

$$\tau_{on} = \frac{\gamma_1}{\left(\varepsilon_0 \Delta \varepsilon E^2 - \frac{\pi^2 K}{d^2}\right)} \qquad (5.9)$$

$$\tau_{off} = \frac{d^2 \gamma_1}{K \pi^2} \qquad (5.10)$$

where K is the effective elastic constant, d is the thickness of the LC layer, E is the electric field, and $\Delta \varepsilon$ is the dielectric anisotropic constant. The turn-on time is controlled by the electric field as well as the rotation viscosity γ_1. However, the turn-off time is dominated by $\gamma_1 d^2$. Small cell gaps are desired, but there are thickness limitations imposed by manufacturing methods. This is why low-viscosity LC fluids are so vital for high-speed LCDs, such as those used in sequential-color displays.

For most nematic materials used in display applications, the range of rotational viscosity is approximately 0.02 to 0.5 Pa s (N s/m^2). This is relatively high compared to water, which, at 20° C, has a viscosity of ~0.001 Pa s. LC materials with an increased number of rings, or with longer alkyl chains, tend to have higher viscosity. Due to the strong polar interaction, a large $\Delta \varepsilon$ may also contribute to high viscosity.

5.6 Electro-optical Effect of LCs

Virtually all LCD applications are based on electro-optical effects of LCs. As such, we can rewrite the Gibbs energy (5.7) in the following form assuming that only an electric field is applied:

$$\begin{aligned} F_{EM} &= -\tfrac{1}{2} \mathbf{D} \cdot \mathbf{E} = -\tfrac{1}{2} \varepsilon_0 (\mathbf{E} + \mathbf{P}) \cdot \mathbf{E} \\ &= -\frac{\varepsilon_0}{2}(E^2 + \varepsilon_{ij} E_i E_j) = -\frac{\varepsilon_0}{2}\left[(1+\varepsilon_\perp)E^2 + \Delta\varepsilon(\mathbf{E} \cdot \mathbf{n})^2\right] \end{aligned} \qquad (5.11)$$

The result indicates that in order to reduce the Gibbs energy, the LC director tends to orient parallel to the electric field if $\Delta \varepsilon$ is positive (shown schematically in Figure 5.6), while the orientation is perpendicular to the electric field if $\Delta \varepsilon$ is negative. The electric field influences

114 LIQUID CRYSTAL DISPLAYS

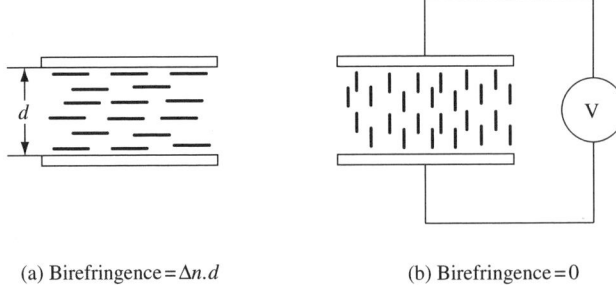

(a) Birefringence = $\Delta n.d$ (b) Birefringence = 0

Figure 5.6 Electro-optical effect of a positive uniaxial LC. The normal incidence birefringence of a homogeneous LC cell without voltage applied is Δnd and it approaches zero as a high voltage is applied

the director profile, which dictates the bulk anisotropic properties of the LC film. Polarized light transmitted through the LC film is thus affected by the applied electric field, producing the electro-optical effect used in LCDs. By spatially varying the electric field applied to an LC layer, the polarization of transmitted light can be spatially encoded, which when viewed through a suitable polarization analyzer yields an image.

5.7 LC Modes for Projection

Several physical mechanisms have been developed for modulating light using LCs. These include the electrically controlled birefringence (ECB) effect, the TN effect, mixed mode or hybrid effects, absorption (dye dopant), light scattering, and Bragg reflection. Displays utilizing absorption, light scattering, or Bragg reflection do not require a polarizer and usually yield low contrast. They are therefore less suitable for projection applications. The majority of LC modes used in projection systems fall into the first three categories. If a display appears bright without voltage activation, it is termed a normally white mode. Conversely, if the display is in the dark state with no applied field, it is a normally black mode. Conventionally, the black mode is achieved when the LC layer has no influence on the SOP, and the LCD is placed between crossed polarizers.

In a reflection mode display, such as an LCOS projector, both on-axis and off-axis arrangements are possible. The latter permits different optics in the illumination and image paths, as in transmissive (HTPS) systems, where a true retro-reflecting system is more restrictive (see Figure 5.7). A polarizing beam splitter (PBS) of some form is required for an on-axis system.

5.7.1 Electrically Controlled Birefringence (ECB) Mode [Kobayashi S., 1990]

For more general treatment of LC modes, the reader is referred elsewhere [Scheffer T. 1990; Wu S. T., 2001]. In this section, we will restrict ourselves to preferred LC modes used in projection systems. One case is the ECB mode, where the LC director is contained in a plane normal to the bounding substrates containing the parallel alignment direction vectors. Here

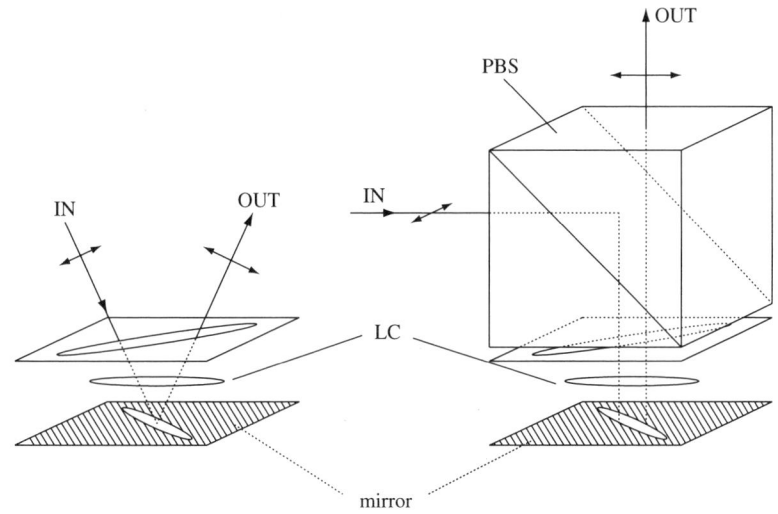

Figure 5.7 Off-axis (left) and on-axis (right) LCOS system

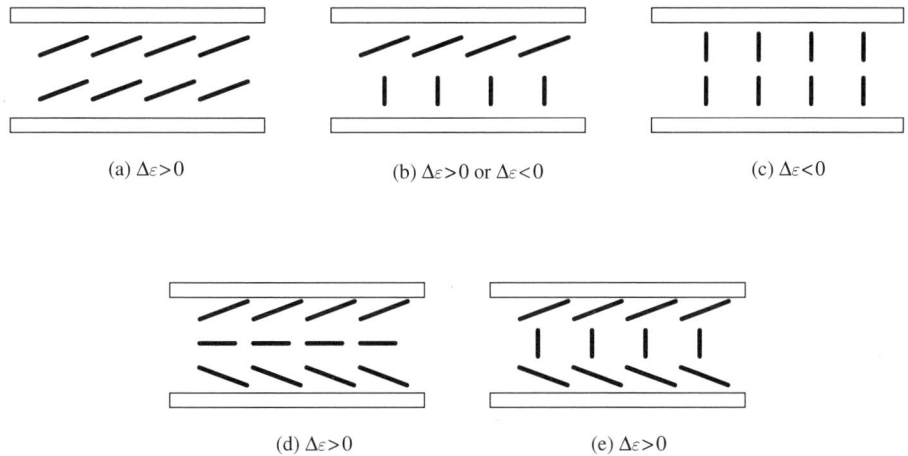

Figure 5.8 ECB LC mode configurations: (a) homogeneous, (b) hybrid, (c) VA, (d) splay, and (e) bend modes

no twist is involved and therefore there is no optical rotation. The homogeneous aligned (parallel rubbed), hybrid, VA, splayed, and bend mode devices all belong to the ECB family (Figure 5.8). Note that the splay and bend modes are often treated as a distinct group, due to different topological symmetry.

When a plane wave is incident normally on an ECB cell sandwiched between two polarizers, the outgoing beam experiences a phase retardation δ due to the different propagation velocities of extraordinary and ordinary rays inside the LC:

$$\delta = \frac{2\pi}{\lambda} \int_0^d \left(\frac{n_e n_o}{\sqrt{n_e^2 \cos^2 \theta(z) + n_o^2 \sin^2 \theta(z)}} - n_0 \right) dz \qquad (5.12)$$

where $\theta(z)$ is the polar angle of the LC relative to the substrate normal, and d is the LC thickness. In order to achieve maximum polarization conversion efficiency (PCE), the LC director plane must be at 45° relative to the polarizer. The polarizer and analyzer can be either parallel or crossed. The normalized light transmittance for the two cases is:

$$T_\parallel = \cos^2(\delta/2) \quad T_\perp = \sin^2(\delta/2) \qquad (5.13)$$

Both homogeneous and hybrid devices are poor candidates for projection display due to their narrow viewing angle and lack of a neutral dark state. The bend mode is also rarely used for projection display due to instability of the bend configuration and large voltage requirements. Due to the ultrafast response time of the bend mode, so-called π-cells are more commonly used to make light shutters for sequential projection applications (see Chapter 11).

The most promising ECB mode for projection is the VA mode. It usually is operated in the normally black mode with a crossed polarizer configuration, ensuring a very high normal incidence contrast. A further advantage is that residual compensation is not required, provided that the pretilt angle is not too large ($<2°$). Figure 5.9 shows the T–V curve of a VA cell, based on Merck MLC-6608 LC fluid, with the cell thickness d optimized for 550 nm. A VA cell can also be operated in reflection mode, with a cell gap half that required for transmission mode.

The VA ECB mode has the merits of high head-on contrast, 100% PCE, easy compensation for FOV, and fast response speed due to thin cell gap. It also has some shortcomings, such

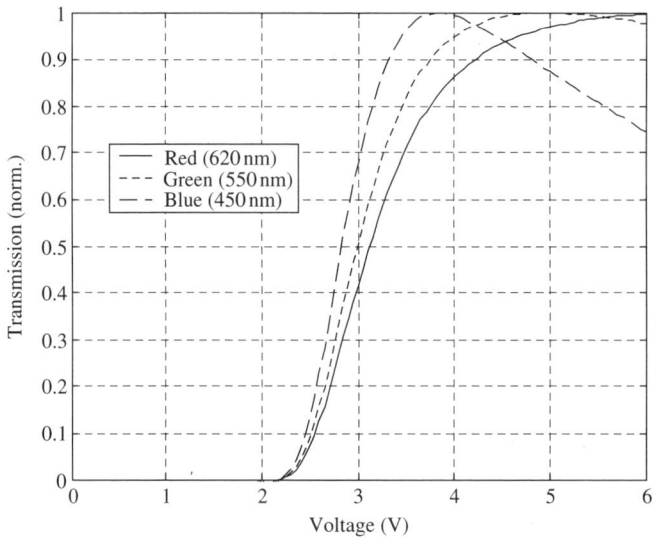

Figure 5.9 T–V curve of a VA LC cell ($\Delta nd = 370$ nm)

as sensitivity to pretilt angle with regard to head-on contrast. Also, gray-scale inversion is observed when viewing in transmission (not significant for the small angles involved for projection).

5.7.2 90° TN and VA 90° TN Mode

The 90° TN mode invented by M. Schadt and W. Helfrich in 1971 [Schadt M., 1971] represents an enormous breakthrough in LC-based displays. Today, nearly all AMLCDs used in laptop and LC light valves for HTPS use the 90° TN mode. The operating principle is illustrated in Figure 5.10, where the LC anchoring direction at the top substrate is perpendicular to that of the bottom one. Thus, in the inactivated state (left), the LC director undergoes a continuous 90° twist. Linearly polarized input light propagates through and, under ideal circumstances, follows the director as it propagates through the LC structure. In this case, light emerges from the LC layer polarized parallel to the transmission axis of the analyzing polarizer. The conventional TN display is therefore normally white. When an electric field is applied across the LC layer, the director distribution in the center of the cell (where the molecules have the greatest freedom to rotate) is predominantly parallel to the electric field. The twisted structure is thus distorted, and, with sufficient voltage, begins to vanish. The polarization direction of the light is no longer rotated as light passes through the cell, so the light is blocked by the analyzing polarizer, and the display appears dark.

In reality, the propagation of light through a TN cell is more complicated than the picture presented above, which is only accurate if the Mauguin condition [Mauguin C., 1948] is satisfied:

$$\Delta nd \gg \frac{\lambda}{2} \tag{5.14}$$

In an actual display this condition is only partially fulfilled, resulting in reduced brightness and contrast, and coloration caused by interference of the extraordinary and ordinary modes

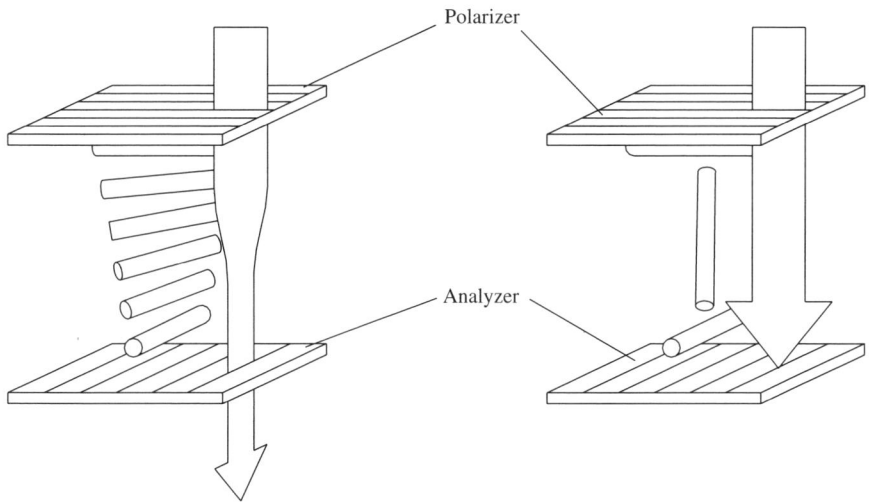

Figure 5.10 Operation principle of 90° TN mode: the inactivated state (left) and activated state (right)

118 LIQUID CRYSTAL DISPLAYS

(analogous to the ECB case). The most favorable operating condition for the TN mode uses $\Delta nd = 0.866\lambda$, the so-called first minimum condition. This LC thickness represents a compromise that provides good FOV, low coloration (achromatic appearance), and high brightness. Figure 5.11 shows the evolution of the polarization state on the Poincaré sphere as light propagates through two thicknesses of a TN display. In the first minimum case, the polarization state evolves from a linear (entrance) to a circular polarization state at the midpoint, and is ultimately converted to an orthogonal linear polarized output. The other case shown is that of a very thick cell ($\Delta nd = 18\lambda$), where the polarization moves along the equator as it is converted to the orthogonal state. It has been shown that the general transformation of a TN device can be described as a 'rolling-cone' path on the Poincaré sphere.

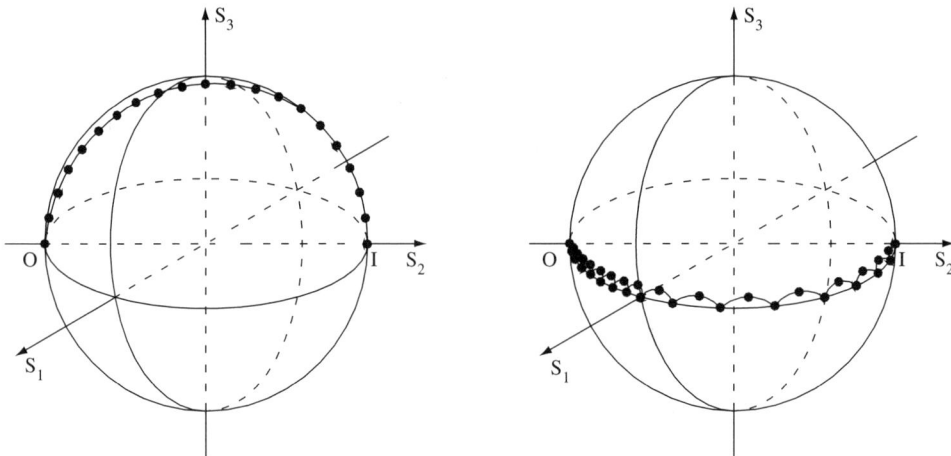

Figure 5.11 Evolution of the polarization state as light propagates through TN display with $\Delta nd = 0.866\lambda$ (left) and 18λ (right)

TN devices are classified according to **e**-mode and **o**-mode, depending on the orientation of the entrance polarizer relative to the rubbing direction of the input substrate. They are further catalogued into normally white or black mode, depending on whether polarizers are crossed or parallel. The optical performance is very similar between **e**-mode and **o**-mode, except that the **e**-mode has a slightly better FOV performance. Figure 5.12 shows T–V curves of the first minimum, normally white TN display at 550 nm, assuming that MLC4972 fluid is used. The normally black TN mode is rarely used in projection applications due to low cell gap tolerance.

The 90° VA TN mode, shown in Figure 5.13, uses negative dielectric ($\Delta \varepsilon < 0$) LC materials [Kashnow R., 1975; Wu S. T., 1997b]. It is usually operated in a normally black **e**-mode (with crossed polarizers). In the inactivated state, the LC is homeotropically aligned and it appears black under a pair of crossed polarizers. In the activated state, the molecules in the center of the cell reorient. At a sufficient voltage, a TN structure is present, resulting in a fully bright state. There are several advantages of the 90° VA TN over the conventional 90° TN, including symmetric and wide FOV, high normal incidence contrast, and large cell gap tolerance. In practice, the Δnd value is slightly larger than that of a conventional TN

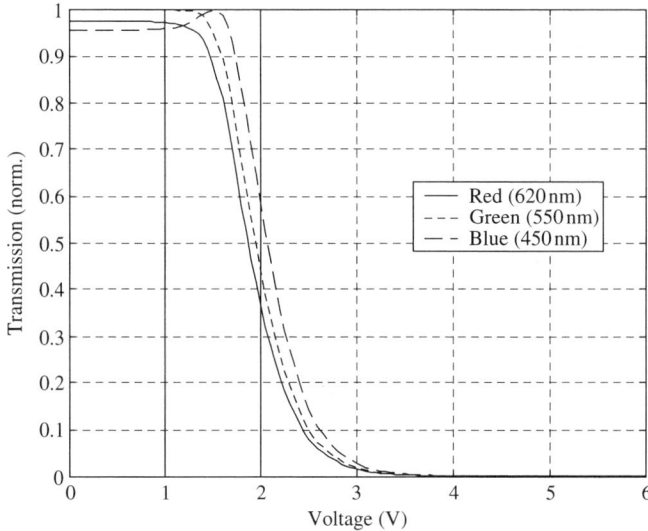

Figure 5.12 *T–V* curves of first minimum, normally white TN mode at 550 nm. The LC material MLC4972 is assumed

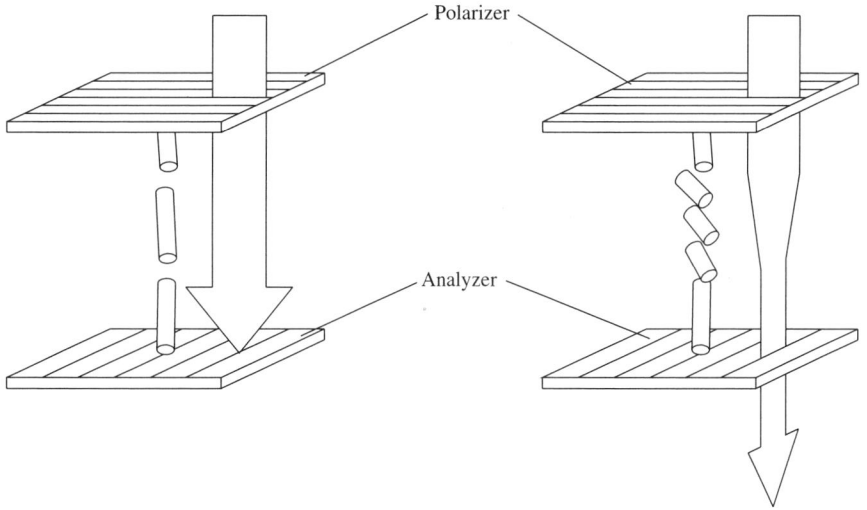

Figure 5.13 Operation principle of 90° VA TN mode: the inactivated state (left) and activated state (right)

display to achieve a fully bright state, in order to compensate for the lack of contribution near the boundaries. Another distinguishing feature of this mode is that the contrast is less sensitive to the magnitude of LC pretilt angle (unlike the VA ECB mode) since the LC is aligned orthogonally at the boundaries.

In practice, an SiO_x evaporation layer is usually used for the VA mode alignment layer of projection panels. This is a challenging process, in that the uniformity of the pretilt angle is very sensitive to evaporation conditions, and thus difficult to control. Such non-uniformities result in dark-state non-uniformity if the VA ECB mode is used, with the LC aligned at ±45° to the polarizer. To obtain a 90° VA TN mode a chiral dopant is used ($d/p = 0.25$), which reduces the LC response time relative to the conventional VA mode, making it a potential candidate for sequential color. It can also be operated in either a transmissive or reflective mode, making it a strong candidate for LCOS projection systems.

5.7.3 45° Reflective TN Mode

The 45° reflective TN [Grinberg J., 1975], shown in Figure 5.14, has the distinction of being the first mode used for LCOS projection displays. More recently, it was employed in an off-axis LCOS projection system [Bone M. F., 1998]. Computer modeling shows that the maximum PCE achievable with the normally black 45° TN mode is only ~93% if the angle, β, between the input LC alignment direction and the input polarizer is zero. However, by setting $\beta = 12°$ in a normally white mode, the PCE can reach 100% for almost all visible wavelengths. Figure 5.15 shows the PCE versus retardance (Δnd) for the two cases. The figure shows that the $\beta = 12°$ case can only really be used in the normally white mode, since the normally black mode has very poor cell gap tolerance. This can be contrasted with the $\beta = 0°$ case, where the normally black mode is favored due to relatively large cell gap tolerance and superior PCE (normally white mode is only about 80%). Assuming the properties of MLC6080 fluid, the cell gap is below 1 micron (0.81 μm) for a normally white mode ($\beta = 12°$). Since the LC relaxation time is proportional to the square of the cell gap, a fast reflective LC display can be realized. The electro-optical performances of a normally white mode ($\beta = 12°$) and normally black mode ($\beta = 0°$) are shown in Figure 5.16. For a three-panel RGB projection display system, the red and green can use the same cell gap, while the blue cell gap should be optimized for the blue band. Without compensation, the normal incidence contrast is about 50:1 with $5V_{rms}$ applied. Using the optimum panel compensation for this mode, much more acceptable (>500:1) system contrasts are possible (refer to Chapter 7).

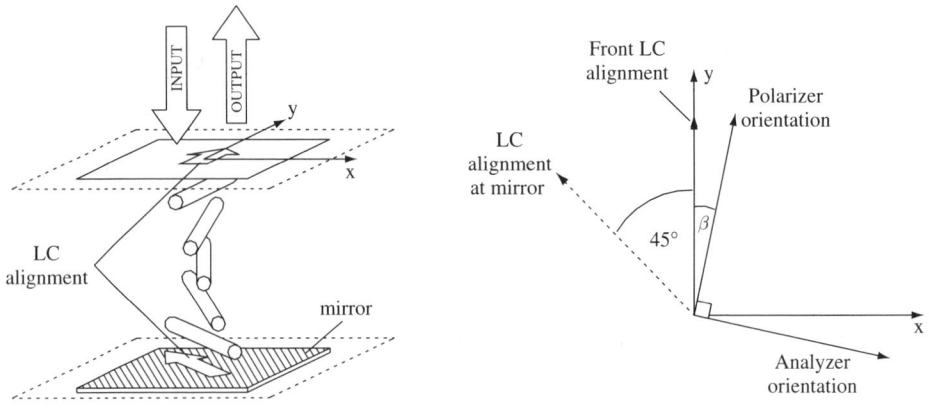

Figure 5.14 Schematic showing the 45° reflective TN mode

LC MODES FOR PROJECTION

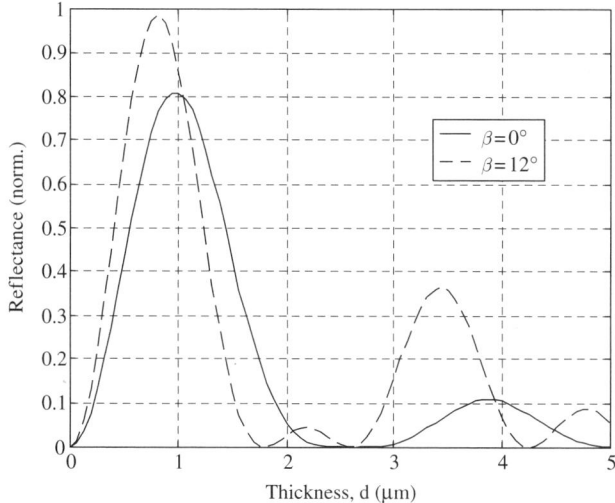

Figure 5.15 Reflectance of 45° TN mode at 550 nm versus cell gap for $\beta = 12°$ and $\beta = 0°$

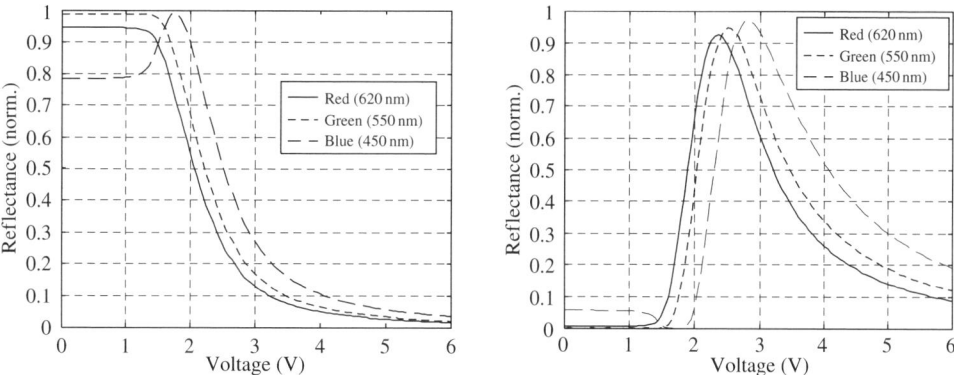

Figure 5.16 R–V response of 45° TN normally white ($\beta = 12°$) and black ($\beta = 0°$) mode. The LC MLC6080 is assumed with cell gaps of 0.81 μm and 2.6 μm respectively

Due to polarization-converted reflections from the interface between the LC and PI (see Chapter 7), which severely degrade the system contrast in LCOS projection systems, the $\beta = 0°$ mode is preferred at the expense of brightness [Chen J., 2004].

5.7.4 63.6° Mixed TN (MTN) Mode

A TN LC mode that achieves 100% PCE with $\beta = 0°$ employs a $\phi = 63.6°$ twist [Sonehara T., 1990a]. This was first demonstrated by Sonehara and Okumura [Sonehara T., 1990b; Wu S. T., 1997b] and is called the TN-ECB or MTN mode. It has the advantage of high PCE (100%) together with low color dispersion and a fast response time. The configuration,

122 LIQUID CRYSTAL DISPLAYS

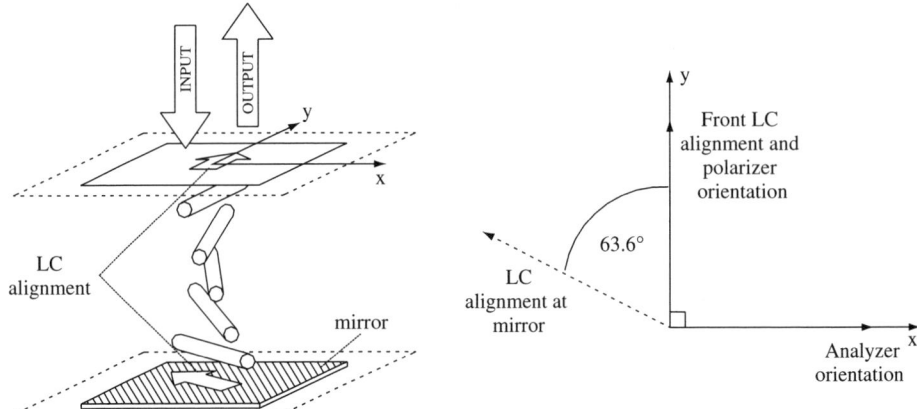

Figure 5.17 63.6° reflective MTN e-mode

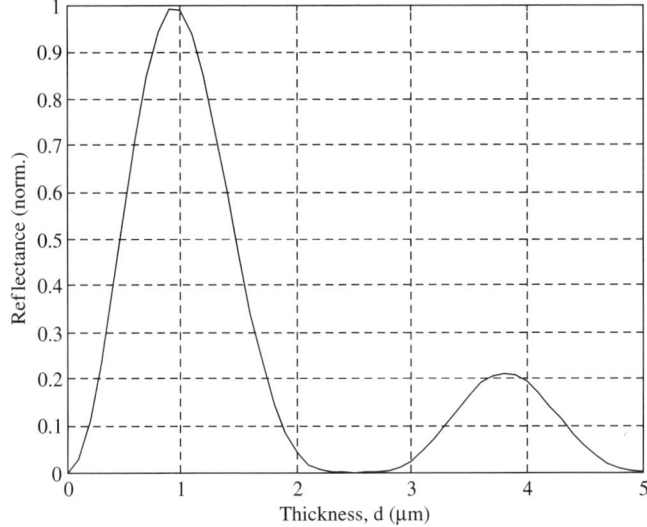

Figure 5.18 Reflectance of 63.7° MTN mode versus Δnd at 550 nm

shown in Figure 5.17, can be operated in either **e**-mode or **o**-mode. The PCE versus Δnd is shown in Figure 5.18 for a cell of the material MLC6080 with $\Delta nd = 192$ nm and a cell gap of 0.95 μm. The R–V curves for R, G, and B are depicted in Figure 5.19. Since the twist angle is not 90° in this mode (i.e., orthogonal orientations at the boundaries), residual retardation exists, which results in dark-state leakage. Without the use of compensation, the normal incidence contrast is only 50:1 at $5V_{rms}$, which is similar to the normally white 45° reflective TN mode. This mode has the advantage of fast response time (<500 μs response time has been reported (Intel Incorporation press release)) and a high PCE. Moreover, the rubbing direction on the input substrate is parallel to the input polarization, eliminating polarization-converted reflections from the PI/LC layer. This is an excellent mode for sequential-color applications.

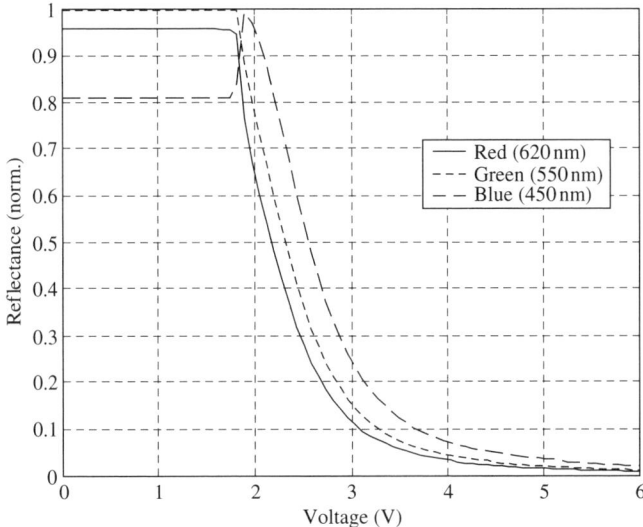

Figure 5.19 *R–V* response of 63.6° MTN normally white mode. MLC6080 fluid is modeled with a cell gap of 0.94 μm

5.7.5 90° MTN Mode

The 90° MTN mode [Wu S. T., 1997a] is also employed in certain LCOS systems [Katayama T., 2001], and has the advantages of low operating voltage, high contrast, and achromatic behavior. However, it suffers from low (~86%) PCE. The configuration is shown in Figure 5.20, and its PCE versus retardance (Δnd) shown in Figure 5.21. This assumes $\Delta nd = 230$ nm and a cell gap of 1.12 μm (again MLC6080 material properties are used to model the mode's response). The *R–V* curves for R, G, and B are graphed in Figure 5.22.

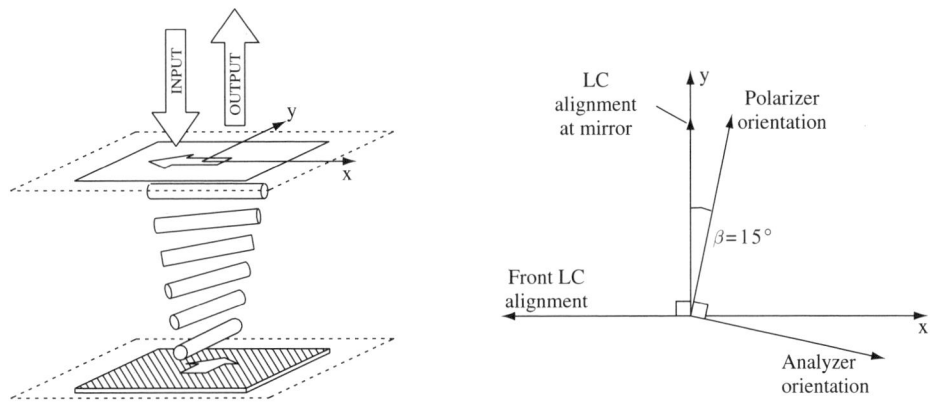

Figure 5.20 90° reflective MTN e-mode

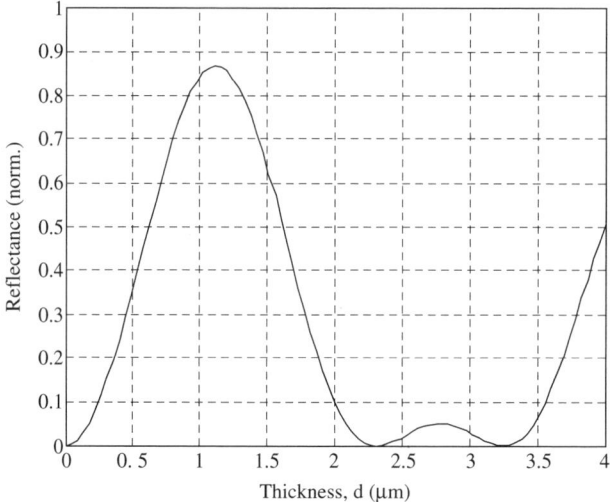

Figure 5.21 Reflectance of the 90° MTN mode versus Δnd at 550 nm

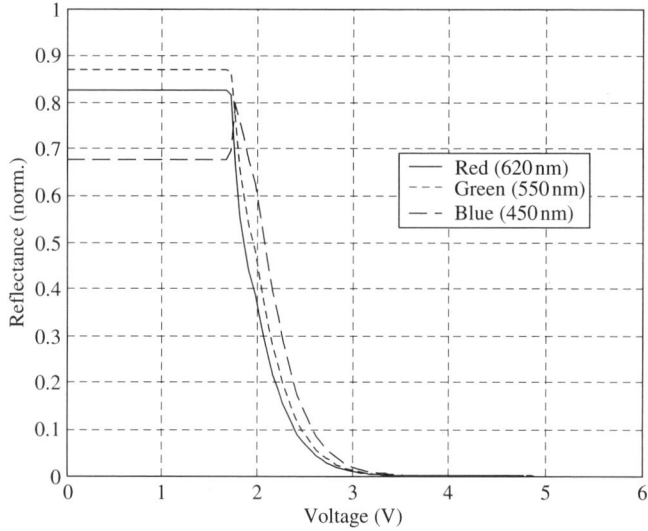

Figure 5.22 R–V response of the 90° MTN normally white LC mode. An optimized $\lambda = 550$ nm, 1.12 μm thick MLC6080 cell is assumed

5.8 FOV of LCDs [Wu S. T., 2001; Bos P. J., 2002]

In the preceding sections, we concentrated on the normal incidence performance of various LC modes used in projection systems. The optical performance at an oblique angle is far more complicated than that at normal incidence. The anisotropic structure encountered along any ray direction is highly dependent upon its specific path. As such, the output polarization depends upon viewing angle, affecting both contrast and brightness. Such effects lead to the well-known gray-scale inversion and contrast loss issues encountered in the early days of direct-view LCD. To visualize this effect, consider a homogeneous LC cell with a uniform tilt angle θ, viewed under crossed polarizers from two different directions (Figure 5.23).

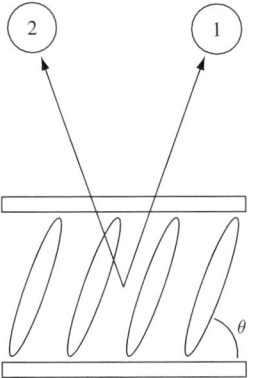

Figure 5.23 FOV issue of LCDs

Since the view direction 1 is along the LC director, the retardation is zero, and so is the resulting transmission between crossed polarizers. However, for view direction 2, the ray experiences an effective phase retardation:

$$\Delta(nd)_{eff} = \left(\frac{n_e n_o}{\sqrt{n_e^2 \cos^2(2\theta) + n_o^2 \sin^2(2\theta)}} - n_o \right) \cdot \frac{d}{\sin\theta} \tag{5.15}$$

that leads to light leakage, T:

$$T = \sin^2\left[\frac{\pi(\Delta nd)_{eff}}{\lambda}\right] \tag{5.16}$$

The influence of the LCD FOV on system performance is quite different for projection systems than that in direct-view circumstances. This is partially because, while the ray angles are considerably smaller, the contrast expectations are considerably higher in projection. Also differences stem from the telecentric illumination condition imposed almost universally in projection systems and shown schematically in Figure 5.24.

The system performance at all points on the image in a telecentric system is then determined by the average of the optical response over a cone angle:

$$Contrast(\lambda) = \frac{\int_0^{2\pi} d\phi \int_0^{\theta_0} I(\theta, \phi, \lambda) T_{Bright}(\theta, \phi, \lambda) d\theta}{\int_0^{2\pi} d\phi \int_0^{\theta_0} I(\theta, \phi, \lambda) T_{Dark}(\theta, \phi, \lambda) d\theta} \tag{5.17}$$

where $I(\theta, \phi)$ is the light intensity profile associated with the cone. In practical terms, lower uniform contrast is observed as the system $f_{/\#}$ decreases. Typical projection systems

126   *LIQUID CRYSTAL DISPLAYS*

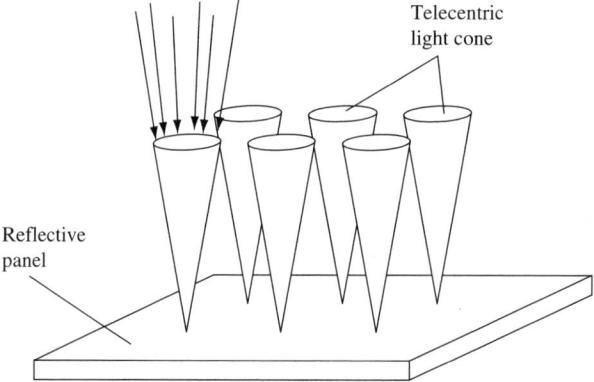

Figure 5.24   Light rays form a cone shape at all points on the LCD in a telecentric projection system

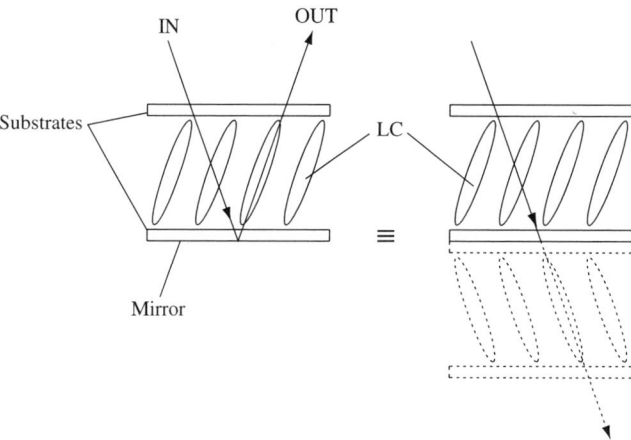

Figure 5.25   FOV in reflective mode equivalent to a two-domain reflection symmetric structure in transmission

operate with illumination half-cone angles $< 14°$, which corresponds to $> f_{/2.0}$. The FOV of reflection LC modes is typically better than that of the equivalent transmissive mode since the doubled optical path creates an effective two-domain reflection symmetric structure (see Figure 5.25).

Recent progress has been made in improving the FOV of direct-view LCDs [Wu S. T., 2001; Bos P. J., 2002] employing techniques that can be roughly categorized into three groups: multidomain, film compensation, and hybrid approaches. A multidomain approach is not suitable for LC panels used in projection systems since structures within the very small pixels produce disclination lines in the multidomain structure, which scatter light and reduce contrast. Conversely, film compensation has been widely adopted to improve the FOV of LC panels in projection systems. Detailed film compensation solutions used within LC projection systems will be addressed in Chapter 7.

References

[Birgeneau R. J., 1978] R. J. Birgeneau and J. D. Lister, Bond orientational order model for smectic B liquid crystals, *J. Phys. Lett.*, 39, pp.399–402, 1978.

[Blinov L. M., 1994] L. M. Blinor and V. G. Chigrinov, *Electrooptic Effects in Liquid Crystal Materials*, p.51, Springer-Verlag, New York, 1994.

[Bone M. F., 1998] M. F. Bone, M. Francis, P. Menard, M. E. Stefanov, and Y. Ji, Novel optical system design for reflective CMOS technology, Proceedings of the 5th Annual Flat Panel Display Strategic and Technical Symposium, p.81, 1998.

[Bos P. J., 2002] P. J. Bos, Field of view issues in LCDs, SID'02 Application Seminar, 2002.

[Chandrasekhar S., 1992] S. Chandrasekhar, *Liquid Crystals* (2nd edition), p.15, Cambridge University Press, Cambridge, 1992.

[Chen J., 2004] J. Chen, M. G. Robinson, and G. D. Sharp, General methodology for LCoS panel compensation, SID'04 Digest, pp.990–993, 2004.

[Coates D., 1990] D. Coates, B. Road, and P. Dorset, *Liquid Crystals: Applications and Uses*, vol. 1, chapter 3, ed. B. Bahadur, World Scientific, London, 1990.

[Cognard J., 1982] J. Cognard, Alignment of nematic liquid crystals and their mixtures, *Mol. Cryst. Liq. Cryst.*, 78, suppl. 1, pp.1–74, 1982.

[de Gennes P. G., 1993] P. G. de Gennes, *The Physics of Liquid Crystals* (2nd edition), Oxford Science Publications, Oxford, 1993.

[Demus D., 1990] D. Demus, *Liquid Crystals: Applications and Uses*, vol. 1, chapter 1, ed. B. Bahadur, World Scientific, London, 1990.

[Frank F. C., 1958] F. C. Frank, On the theory of liquid crystals, *Discuss. Faraday Soc.*, 25, pp.19–28, 1958.

[Friedel G., 1922] G. Friedel, Mesomorphic states of matter, *Ann. Phys.*, 18, pp.273–474, 1922.

[Geary J. M., 1987] J. M. Geary, J. W. Goodby, A. R. Kmetz, and J. S. Patel, The mechanism of polymer alignment of liquid-crystal materials, *J. Appl. Phys.*, 62, pp.4100–4008, 1987.

[Gray G. W., 1984] G. W. Gray and J. W. Goodby, *Smectic Liquid Crystals*, Hayden & Son, Philadelphia, 1984.

[Grinberg J., 1975] J. Grinberg, A. Jacobson, W. Bleha, L. Miller, I. Fraas, D. Bosewell, and G. Myer, A new real-time non-coherent light image converter. The hybrid field effect liquid crystal light valve, *Opt. Eng.*, 14, pp.217–225, 1975.

[Hiroshima K., 1982] K. Hiroshima, Controlled high-tilt-angle nematic alignment compatible with glass frit sealing, *Jpn. J. Appl. Phys.*, 21, pp.L761–763, 1982.

[Hiroshima K., 1984] K. Hiroshima and H. Obi, Controlled low-tilt-angle liquid-crystal orientation on 'omniazimuthally' evaporated SiO films at 60 degree incidence, SID'84 Digest, pp.287–292, 1984.

[Jakeman E., 1972] E. Jakeman and E. P. Raynes, Electro-optic response times in liquid crystals, *Phys Lett.*, 39A, pp.69–70, 1972.

[Kashnow R., 1975] R. A. Kashnow, Quasi-homeotropic twisted nematic liquid crystal device, US Patent 3,914,022, 1975.

[Katayama T., 2001] T. Katayama, H. Natsuhori, T. Moroboshi, M. Yoshimura, and M. Hayakawa, D-ILA™ device for top-end projection display (QXGA), SID'01 Digest, p.976, 2001.

[Kobayashi S., 1990] S. Kobayashi and A. Mochizuki, *Liquid Crystals: Applications and Uses*, vol. 3, chapter 19, ed. B. Bahadur, World Scientific, London, 1990.

[Leadbetter A. J., 1987] A. J. Leadbetter, Structural classification of liquid crystals, in G. W. Gray (ed.), *Thermotropic Liquid Crystals*, John Wiley & Sons, Ltd, Chichester, 1987.

[Lu M., 2000] M. Lu, K. H. Yang, T. Nakasogi, and S. J. Chey, Homeotropic alignment by single oblique evaporation of SiO_2 and its application to high-resolution microdisplays, SID'00 Digest, p.446, 2000.

[Mauguin C., 1948] C. Mauguin, Liquid crystals in convergent light, *C. R. Hebd. Sceances Acad. Sci.*, 151, pp.886–888, 1948.

[Oseen C., 1933] C. W. Oseen, Theory of anisotropic liquids XVIII, *Ark. Math., Astron. Fys.*, 23, 2, pp.1–49, 1933.

[Persham P. S., 1988] P. S. Persham, *Structure of Liquid Crystal Phases*, World Scientific, Singapore, 1988.

[Pindak R., 1981] R. Pindak, D. E. Moncton, S. C. Davey, and J. W. Goodby, X-ray observation of a stacked hexatic liquid-crystal B phase, *Phys. Rev. Lett.*, 46, pp.1153–1158, 1981.

[Schadt M., 1971] M. Schadt and W. Heifrich, Voltage-dependent optical activity of a twisted nematic liquid crystal, *Appl. Phys. Lett.*, 18, p.127, 1971.

[Scheffer T., 1990] T. Scheffer and J. Nehring, *Liquid Crystals: Applications and Uses*, vol. 1, chapter 10, ed. B. Bahadur, World Scientific, London, 1990.

[Sonehara T., 1990a] T. Sonehara, Photo-addressed liquid crystal SLM with twisted nematic ECB (TN-ECB) mode, *Jpn. J. Appl. Phys.*, 29, pp.1231–1234, 1990.

[Sonehara T., 1990b] T. Sonehara and O. Okumura, Photo-addressed liquid crystal SLM with twisted nematic ECB (TN-ECB) mode, SID'00 Digest, p.145, 1990.

[Taylor G. N., 1973] G. N. Taylor, F. J. Kahn, and H. Schonhorn, Surfaced produced alignment of liquid crystals, *Proc. IEEE*, 61, pp.823–826, 1973.

[Tsvetkov V., 1948] V. Tsvetkov and V. Marinin, Dipole moments of molecules of some liquid crystals and electric double refraction of their solutions, *J. Exp. Theor. Phys.*, 18, pp.641–650, 1948.

[Uchida T., 1990] Tatsuo Uchida and Hidehiro Seki, *Liquid Crystals: Applications and Uses*, vol. 3, chapter 5, edited by B. Bahadur, World Scientific, London, 1990.

[Vithana H., 1995] H. Vithana, Y. K. Fung, S. H. Jamal, R. Herke, P. J. Bos, and D. L. Johnson, A well-controlled tilted-homeotropic alignment method and a vertically aligned four-domain LCD fabricated by this technique, SID'95 Digest, pp.873–876, 1995.

[Wu S. T., 1997a] S. T. Wu, C. S. Wu, and C. L. Kuo, Reflective direct-view and projection display using twisted-nematic liquid crystal cells, *Jpn. J. Appl. Phys.*, 36, p.2721, 1997.

[Wu S. T., 1997b] S. T. Wu, C. S. Wu, and K. W. Lin, Chiral-homeotropic liquid crystal cells for high contrast and low voltage displays, *J. Appl. Phys.*, 82, p.4795, 1997.

[Wu S. T., 2001] S. T. Wu and D. K. Yang, *Reflective Liquid Crystal Displays*, John Wiley & Sons, Ltd, Chichester, 2001.

[Yeh P., 1999] Pochi Yeh and Claire Gu, *Optics of Liquid Crystal Displays*, chapter 1, ed. Joseph W. Goodman, John Wiley & Sons, Inc., New York, 1999.

[Zocher H., 1938] H. Zocher and G. Ungar, Structure of nematic layers, *Z. Phys.*, 110, 9–10, pp.529–548, 1938.

6

Retarder Stack Filters

6.1 Introduction

Retarder stack filters (RSFs) [Sharp G. D., 1998; 1999] selectively transform the polarization state of one or more color bands. For example, the red/cyan RSF, shown schematically in Figure 6.1, transforms linear polarized red light to the orthogonal polarization state, while leaving cyan light (blue + green) unchanged. RSFs are fabricated as a laminate of transparent stretched polymer retardation films (typically polycarbonate), each roughly 60 μm in thickness. The number, retardation, and angle of each layer of the RSF are selected for the purpose of precisely controlling polarization at each wavelength. Their unique optical properties make them very suitable for LC projections systems, particularly those based on reflective liquid crystal on silicon (LCOS) technology. RSFs allow a merging of color and polarization in LCOS systems, enabling compact and high-performing projectors (e.g., JVC DLA-SX21SU, DLA-HX1U, DLA-HD2K-SYS, HD-52Z575, HD-61Z575; Canon SX50; Hitachi CP5600).

Since the introduction of RSFs is relatively recent, a complete chapter is devoted to this technology. It will first give an overview of their basic interferometric behavior through a comparison to more conventional interferometers. Their use within projection systems both as color filters and beam splitters will then be summarized and compared to the more conventional dichroic mirrors. Their design will be described in some detail, highlighting both procedure and the underlying principles. To conclude, the chapter will consider subtle symmetry properties of RSF filters that can enhance their system-level performance.

Polarization Engineering for LCD Projection   M. G. Robinson, J. Chen and G. D. Sharp
© 2005 John Wiley & Sons, Ltd

130 RETARDER STACK FILTERS

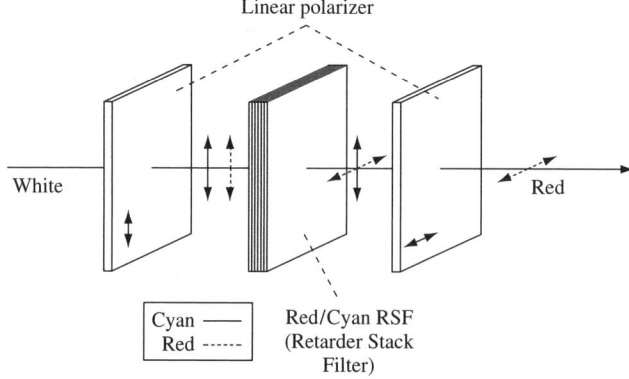

Figure 6.1 Optical performance of a red/cyan RSF

6.2 Principle and Background of RSFs

6.2.1 Single Stage Polarization Interference [Hecht E., 1980, p. 290]

RSFs are an extension of the basic polarization interference concept. The polarizing interferometer (PI) in its simplest form is analogous to other simple interferometers, such as the Mach–Zehnder interferometer (MZI) illustrated in Figure 6.2. Both are two-beam interferometers, though each uses a different mechanism for producing the three necessary steps of: (1) wavefront shearing (to create two beams); (2) introduction of a path-length difference

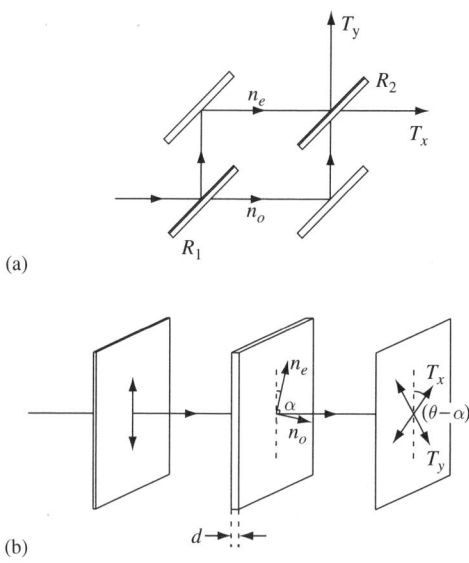

Figure 6.2 Comparison between (a) Mach–Zehnder and (b) polarizing interferometers

(or time delay); and; (3) the subsequent recombination of the two beams in a way that causes some measure of interference.

For reasons of familiarity and ray-tracing simplicity, first consider the MZI. Assume lossless, partially reflecting, splitting/combining mirrors with reflectivity R_1 and R_2, respectively. The splitter reflectivity determines the relative amplitude of the field following each path, while the combiner determines the relative proportion taken from each path to each exit port. For the purpose of comparison with the PI, it is assumed that the total path length between splitter and combiner is d, the refractive index along the upper path is n_e, and that along the lower path is n_o. The electric field exiting each port of the MZI is thus given by:

$$A_x = \sqrt{R_1}\sqrt{1-R_2}e^{-ik_o n_e d} - \sqrt{1-R_1}\sqrt{R_2}e^{-ik_o n_o d}$$
$$A_y = \sqrt{R_1 R_2}e^{-ik_o n_e d} + \sqrt{1-R_1}\sqrt{1-R_2}e^{-ik_o n_o d} \quad (6.1)$$

where $k_o = 2\pi/\lambda$, and λ is the free-space wavelength. Using $T_x = |A_x|^2$, these expressions give the power transmitted into each port as:

$$T_x = R_1(1-R_2) + R_2(1-R_1) - 2\sqrt{R_1 R_2 (1-R_1)(1-R_2)}\cos(k_o \Delta n d) \quad (6.2)$$

where $T_y = (1 - T_x)$ and $\Delta n = (n_e - n_o)$. The last term in the above expression is the interference term, which depends upon optical path-length difference, and oscillates with wavelength.

The PI shows similar behavior, where the amplitudes are instead determined by successive projections onto arbitrarily rotated Cartesian axes. For the simple PI shown in Figure 6.2, we assume a linear input polarization, followed by a uniaxial retarder, with orientation α. The field is first projected onto the retarder coordinates, making it functionally equivalent to the MZI splitter. The electric field is then projected onto a final coordinate system at angle $(\theta - \alpha)$ (usually occupied by a linear polarization analyzer, or PBS), which is equivalent to the MZI combiner.

For an ideal lossless splitter/combiner, we can therefore make the general substitution:

$$\sqrt{R_1} = \cos\alpha$$
$$\sqrt{R_2} = \cos(\theta - \alpha) \quad (6.3)$$

which simplifies the above transmission function to the form:

$$T_x = \sin^2\theta - \sin 2\alpha \sin[2(\theta-\alpha)] - \sin 2\alpha \sin[2(\theta-\alpha)]\cos[k_o \Delta n d] \quad (6.4)$$

It can be shown that this is identical to the transmission function produced by analyzing the PI using the technique of Jones' matrices described in Section 3.3.3.2.

Under the special case, $\alpha = \pi/4$ and $\theta = 0$ or $\pi/2$ (retarder at $\pi/4$ between parallel/crossed polarizers, respectively), the above expressions become:

$$T_x = \cos^2(\pi \Delta n d/\lambda)$$
$$T_y = \sin^2(\pi \Delta n d/\lambda) \quad (6.5)$$

This corresponds to the condition of "maximum fringe visibility" in an interferometer, and is likewise the transmission function of a simple single stage polarization interference filter, i.e., the spectrum of a single retarder between polarizers. These single retarder, polarization interferometer (PI) filter units clearly provide only a rudimentary spectral filtering, but can nevertheless be considered the building block of an RSF.

6.2.2 Multilayer Polarization Interference

To improve spectral discrimination through steeper filter transition slopes and wider pass and stop bands, multiple PI stages or retarder films are necessary. Historically, composite filters were developed that provided spectral outputs that were simply products of the sinusoidal spectra generated by each stage. Lyot and Ohmann [Lyot B., 1933; Yeh P., 1988, chapter 10] independently determined that a multistage filter using a geometrical relationship of retarder thickness $(d, 2d, 4d, \ldots, 2^{N-1}d)$ is sufficient to produce a band-pass profile. The polarization analyzer of one stage forms the input polarizer for the subsequent stage, such that $(N+1)$ polarizers are required. These filters were used to image solar prominences in the early twentieth century, due to their ability to accommodate large acceptance angles with high spectral resolution. Later Solc [Solc I., 1965; Yeh P., 1988, chapter 10] showed examples of polarization interference filters that were free of internal polarizers (Figure 6.3, and are therefore the first predecessors of the RSF.

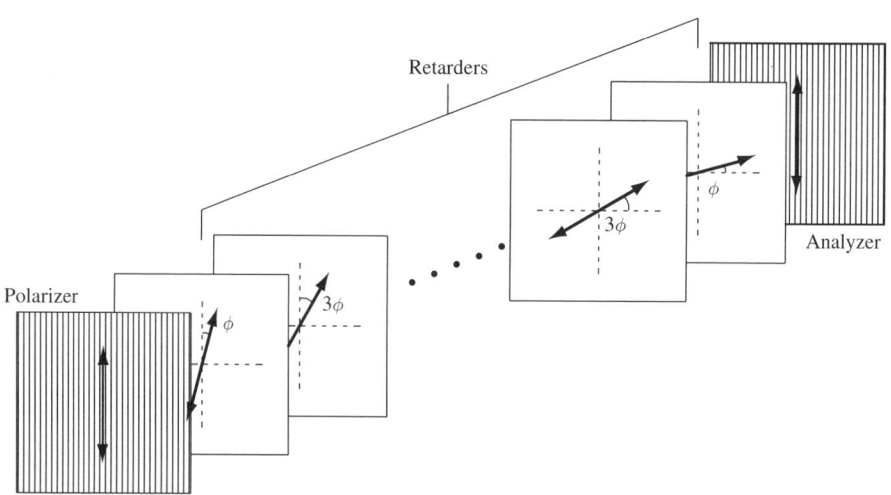

Figure 6.3 Schematic drawing of fan-type Solc filter

The Solc filter demonstrates that multiple retarders, with a single analyzer, can be used to generate a desired spectral profile. By preserving both polarizations, color splitting and combining operations are feasible. A simple two-layer example is used to illustrate the utility of multiple retarders. Figure 6.4 shows a two-layer RSF with identical retarders oriented at

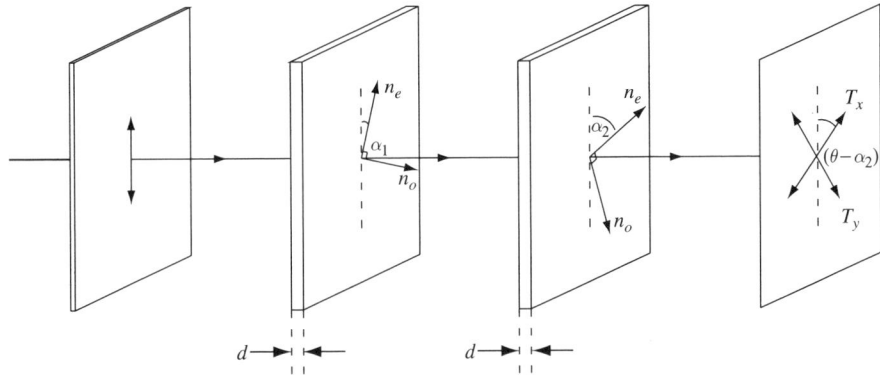

Figure 6.4 Two-stage polarizing interferometer

arbitrary angles α_1 and α_2. Jones' calculus can be used to generate the following analytical form for the output electric field:

$$A_x = A_o e^{-ik_o 2n_e d} + A_1 e^{-ik_o(n_e+n_o)d} + A_2 e^{-ik_o 2n_o d}$$
$$A_y = B_o e^{-ik_o 2n_e d} + B_1 e^{-ik_o(n_e+n_o)d} + B_2 e^{-ik_o 2n_o d} \quad (6.6)$$

where:

$$A_o = \cos\alpha_1 \cos(\alpha_2 - \alpha_1) \cos(\theta - \alpha_2)$$
$$A_1 = -\sin(\alpha_2 - \alpha_1) \sin(\theta - \alpha_2 + \alpha_1)$$
$$A_2 = -\sin\alpha_1 \cos(\alpha_2 - \alpha_1) \sin(\theta - \alpha_2)$$

with similar expressions for the B_is.

The transmission function is then:

$$T_x = (A_o^2 + A_1^2 + A_2^2) + 2A_2(A_o + A_1)\cos(k_o \Delta n d) + 2A_o A_2 \cos(2k_o \Delta n d) \quad (6.7)$$

where, again, $T_y = (1 - T_x)$. This equation demonstrates that the addition of a retarder creates an additional interference term whose relative amplitude can be assigned via the selection of retarder and analyzer orientations. There is clearly a coupling between amplitude and angle, and therefore a non-trivial interdependence, but one can imagine that a solution set exists for specifying a desired transmission profile, implying a potential design methodology. The solution to this inverse problem is the basis of RSF design and will be discussed in Section 6.4.

The net effect of introducing addition layers at prescribed orientation angles is an increase in filter transition slopes and bandwidths. This is best illustrated through example, as shown in Figure 6.5. Each trace in this figure represents the transmission between parallel polarizers of an RSF filter with increasing number of retarders with orientations given by Table 6.1.

134 RETARDER STACK FILTERS

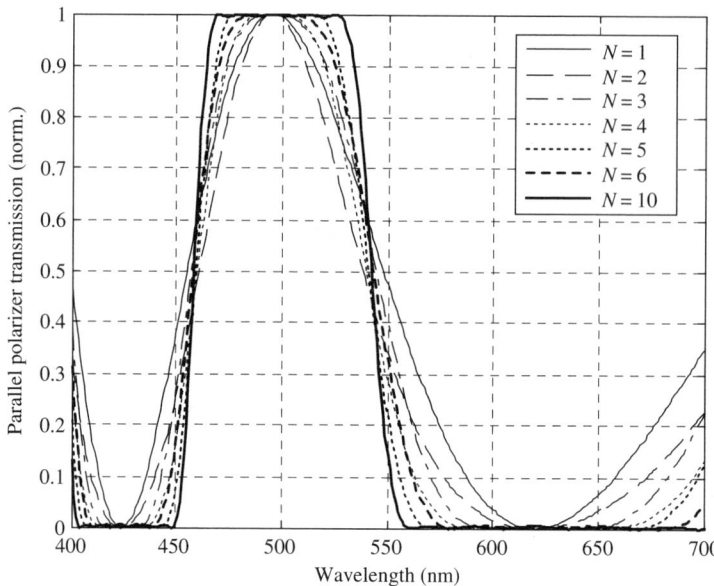

Figure 6.5 Parallel polarizer spectra for RC color RSFs with 1, 2, 3, 4, 5, 6, and 10 retarders each with retardance 1.5 waves at 620 nm

Table 6.1 Retarder angles for the RC RSFs of Figure 6.5. Each retarder has a value of 1.5 waves at 620 nm

N	Orientation in degrees								
1	45								
2	−8.1	−53							
3	10.9	1.8	52						
4	4.5	18.2	2.7	−53.9					
5	−5	0.9	21.9	10.7	−50.1				
6	−5.5	−11.8	1.7	23.1	3.2	−56.6			
10	−4.5	−5.6	1.4	6.3	−4.5	−17.2	−3	25.3	8.6	−53.5

6.3 RSFs in LC Projection Systems

6.3.1 Optical Filters

Conventional optical filtering can be achieved by placing a retarder stack between polarizers, as we have seen. In a projector, the desirable position for the stack is at a location that minimizes the need for additional polarizers (for throughput reasons). Low-contrast operations, such as 10–30% attenuation of one or two primary colors (i.e., color balance or color temperature control), allow the RSF to be placed between the polarization conversion system (PCS) and the clean-up polarizer, where the input polarizing efficiency is adequate. This is also a potential location for yellow notch or yellow plus cyan (double notch) filters that act as trim filters or color enhancement filters. The standard UHP lamp has a strong emission at 578 nm (yellow), which can be neither added to the red nor green primary spectra without severely degrading their color coordinates. As such, most of this light should be removed from the spectrum, ideally before entering the color management assembly. A narrow band in the cyan (roughly 500 nm) can be eliminated for the same argument. One approach is to insert a double notch RSF between the PCS and pre-polarizer, which converts the state of polarization (SOP) of cyan and yellow to the orthogonal polarization. Figure 6.6 shows the spectrum of the standard double notch RSF between ideal parallel polarizers. Note that the double notch is a result of design symmetry and the periodicity of RSFs, versus a cascade of independent notch filters. In the event that the cyan notch is not required, a yellow single notch can also be produced by a retarder stack.

Direct comparison of an RSF to a more conventional dichroic coated filter is difficult as they both inherently perform different optical functions. However, an RSF between polarizers can be compared to a simple dichroic plate in terms of their spectral performance. One key advantage of a simple dichroic is that it does not require either a polarizer or analyzer to operate. In unpolarized or partially polarized systems, this is a distinct throughput advantage since an RSF would require additional polarization components. However, since polarized light is required within LC projection systems, this advantage is not often realized. Advantages of using RSFs over dichroics include their closed form design (see Section 6.4) and their superior angular tolerance. RSFs made from biaxial retardation film (see Section 4.2) perform nearly identically out to $\sim 30°$ incidence angles. In contrast, dichroics show considerable angular intolerance as shown in Figure 4.13. Both these advantages of RSFs make them very attractive compared to dichroics for precise notching filters of the type shown in Figure 6.6.

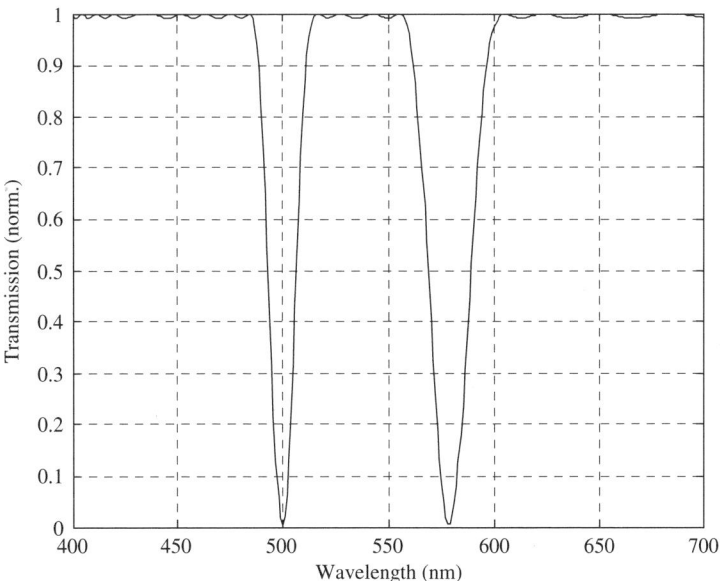

Figure 6.6 Theoretical spectrum of double notch RSF

6.3.2 Color Splitters/Combiners [Sharp G. D., 2002; Chen J., 2004; Robinson M. G., 2004]

An RSF can be used with an achromatic PBS to split and/or combine colored light by virtue of polarization encoding of color, as shown in Figure 6.7.

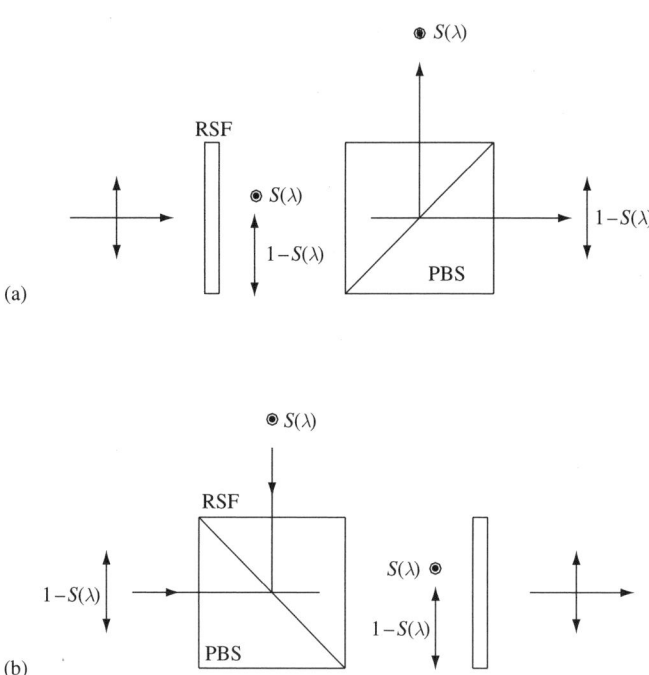

Figure 6.7 RSF/PBS (a) color splitting and (b) recombination

In reflective LCOS projection systems, these two functions can be carried out by a single PBS adjacent to two panels. It performs both color channel separation and recombination together with input/output beam separation. This is then the basic building block of RSF color management systems as will be discussed in detail later in Chapter 8. Combining this dual function building block with the more basic color separating and combining subsystems of Figure 6.7 leads to three-panel architectures, as shown in Figure 6.8.

Many alternative systems can be envisaged through the combination of the RSF/PBS separator and/or combiner subsystems with more conventional dichroic elements such as those shown in Figure 6.9. Despite offering reasonable performance, successful architectures must take into account subtle polarization affects of individual components, as introduced in Chapter 4, to achieve the high contrast demanded of current projection systems. Contrast will be covered in detail in the next chapter. As such, only RSF/PBS systems closely related to those discussed in Chapters 10 and 11 are commmercially feasible at present.

The advantages that RSF/PBS separators/combiners have over dichroic mirrors are related to polarization control and angular performance. Unlike dichroic beam splitting plates that can mix polarization (see Section 4.4.4), PBSs create polarization states that are well defined and

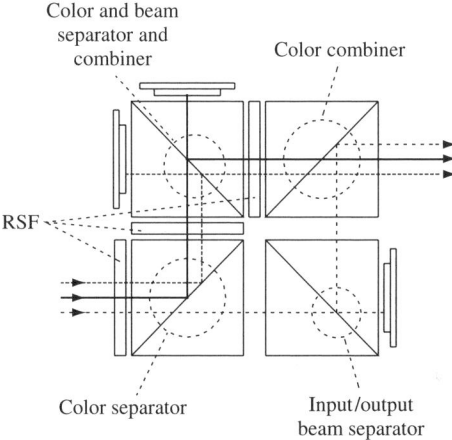

Figure 6.8 Four PBS three-panel LCOS system

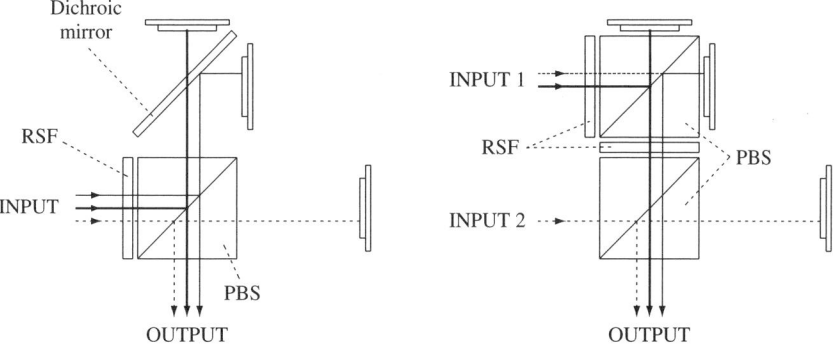

Figure 6.9 Two example three-panel systems using a combination of dichroic mirrors, PBSs, and RSFs. In the second case dichroic mirrors would be used to create the separate input beams

lead to good system contrast without requiring clean-up polarization elements. The angular sensitivity of dichroic beam splitters can introduce chromatic variation within the system that typically requires cleaning up with normal incidence trim filters. Again the angular tolerance of an RSF maintains chromatic integrity over all incidence angles seen in projection systems.

6.4 Design of RSFs

6.4.1 Impulse Response of a Birefringent Network

An RSF can be considered a linear system, and, as such, impulse response analysis can be adopted for its design, as has been widely used for electrical circuits [Aseltine J. A., 1958]. The Fourier transform of a filter's response to an optical impulse is therefore its frequency

138 RETARDER STACK FILTERS

transfer function or filter spectrum. This direct relationship between the time and frequency domains allows for a closed form filter design algorithm since the impulses within the time domain are both finite and discrete.

Consider first the impulse response of the single birefringent plate of Figure 6.10. The normal incidence input optical impulse would have a single electric field direction corresponding in general to linear polarization at an arbitrary angle ϕ from the extraordinary axis of the retarder. This impulse then excites two optical pulses (i.e., the **e-** and **o-**modes) within the material with amplitudes proportional to $(\cos\phi, \sin\phi)$ and electric fields parallel and orthogonal to the optic axis respectively. These propagate at different speeds proportional to the material's refractive indices and emerge separated in time by τ, where:

$$\tau = d(n_e - n_o)/c = d\Delta n/c \qquad (6.8)$$

and d is the retarder thickness, n_e and n_o the extraordinary and ordinary indices, and c the speed of light in vacuum.

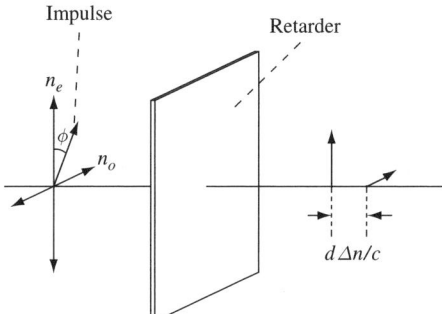

Figure 6.10 Impulse response of a single retarder film

For an impulse incident on an RSF consisting of N arbitrarily oriented retarders, the output pulses number 2^N, since each retarder acts to split any incoming impulse into two, as shown in Figure 6.10. In the case of N identical retarders, however, the output will consist of only $(N+1)$ equally separated pulses. This is because each retarder can only delay impulses by the two discrete periods determined by their common retardance, as shown in Figure 6.11. This finite impulse response (FIR) is analogous to feedforward networks, which are much more amenable to design than infinite impulse response (IIR) feedback networks related to multilayer dichroic coatings. Introducing an output polarizer forces the pulses to interfere in a scalar fashion, making FIR analysis possible. RSFs are therefore designed for linear input and output polarization states.

Since the pulses emanating from a stack consisting of identical thickness retardation films are equally spaced, the filter spectral response (i.e., its Fourier transform) must be periodic. This can be seen by the following example commonly used to describe sampling theory. Suppose a filter has a continuous, aperiodic impulse response $g(t)$; then its corresponding frequency transfer function $G(\omega)$ would also in general be continuous and aperiodic, as shown as Figure 6.12(a). However, suppose a filter has an impulse response which is the same $g(t)$, but sampled at a uniform rate of $1/\tau$ samples/sec; then simple Fourier analysis

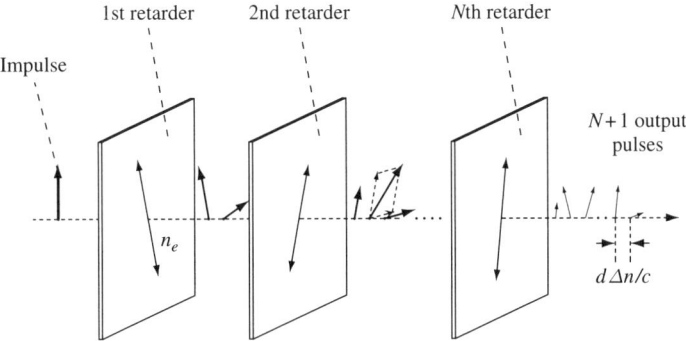

Figure 6.11 Impulse response from N-retarder RSF

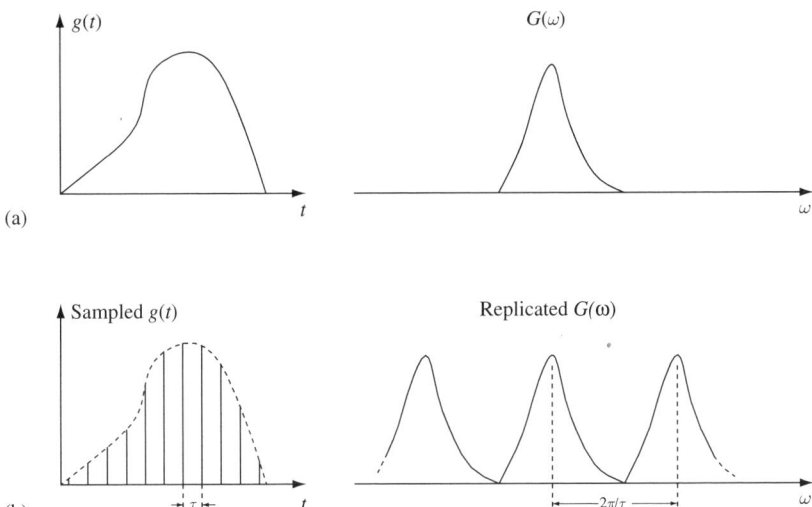

Figure 6.12 Impulse responses and corresponding transfer functions for a network whose impulse response is (a) $g(t)$ and (b) $g(t)$ sampled

yields the periodic transfer function shown in Figure 6.12(b), corresponding to the original $G(\omega)$ replicated with a period $2\pi/\tau$.

Since RSFs made from similar retarders must produce finite sampled impulse responses, their spectral output will always be periodic in frequency (i.e., $\propto 1/\lambda$) and must be designed accordingly. In other words, films with retardance Δnd will have equivalent spectral output at λ_1 and λ_2 if:

$$2\pi d \left(\frac{\Delta n(\lambda_1)}{\lambda_1} - \frac{\Delta n(\lambda_2)}{\lambda_2} \right) = \text{integer} \qquad (6.9)$$

6.4.2 Design Methodology

An RSF design generally starts with a desired response in the wavelength (or frequency) domain. The continuous spectrum is related to a continuous impulse response through Fourier transformation. More importantly, the periodic version of the spectrum is related to the sampled version of the impulse response, as required in FIR filter design.

One method to design the RSF is to define an amplitude transfer function $G(\omega)$, where $G(\omega) = \sqrt{T(\omega)}$, and approximate it by a finite exponential series:

$$G(\omega) \approx C(\omega) = C_0 + C_1 e^{-i\tau\omega} + C_2 e^{-i2\tau\omega} + \cdots + C_N e^{-iN\tau\omega}$$
$$= \sum_{k=0}^{N} C_k e^{-ik\tau\omega} \tag{6.10}$$

with real coefficients C_k. There are multiple solutions for obtaining such a series since it represents a simple Fourier series [Harris S. E., 1964]. The impulse response corresponding to equation (6.10) can then be obtained by taking the inverse Fourier transform, giving:

$$C(t) = C_0 \delta(t) + C_1 \delta(t-\tau) + C_2 \delta(t-2\tau) + \cdots + C_N \delta(t-N\tau)$$
$$= \sum_{k=0}^{k=N} C_k \delta(t-k\tau) \tag{6.11}$$

where N is the number of retarders in the final designed filter.

To determine the retarder angles of an RSF, the impulse response $D(t)$ corresponding to the orthogonal polarization must also be obtained. From energy conservation, the transfer function $D(\omega)$ must satisfy the following relation:

$$D(\omega) \cdot D^*(\omega) = 1 - C(\omega) \cdot C^*(\omega) \tag{6.12}$$

where $D(\omega) = D_0 + D_1 e^{-i\omega\tau} + \cdots + D_N e^{-iN\omega\tau}$ for real coefficients D_k. One method to obtain D_k is that of equivalent roots [Harris S. E., 1964].

Armed with C_k and D_k, the angular recipe for the N-layer RSF can be obtained explicitly using the algorithm of Harris, Amman, and Chang [Harris S. E., 1964].

6.4.3 Impulse Response to RSF Angular Profile Mapping

The algorithm of Harris-Ammann-Chang (HAC) is used to derive the orientation of the individual retarder layers and analyzer from a known FIR. Here we reproduce the algorithm in the original notation [Harris S. E. 1964].

For mathematical convenience, we introduce relative angle notation θ_j between the optic axis orientation of the jth and $(j+1)$th retarder having absolute angles ϕ_j and ϕ_{j+1}

respectively. The input polarizer is always oriented at 0° and the output analyzer at θ_P. Therefore, we define:

$$\begin{aligned}\theta_1 &= \phi_1 \\ \theta_2 &= \phi_2 - \phi_2 \\ &\cdot \\ &\cdot \\ &\cdot \\ \theta_N &= \phi_N - \phi_{N-1} \\ \theta_P &= \phi_P - \phi_N\end{aligned} \tag{6.13}$$

The pulses emitted parallel to the slow and fast axis of the jth retarder are S^j and F^j respectively, corresponding to the e- and o-modes. At the exit of any one retarder, the first isolated pulse oriented parallel to the fast axis has the subscript 0; the next two fast and slow oriented pulses have the subscript 1; and so on. Notice that, in the same way the first pulse to exit is always isolated and oriented along the retarder fast axis, the last pulse is always isolated and oriented along the slow axis. In other words, F_j^j and S_0^j are always zero. Mathematically we can express the output pulses from a series of two retarders as:

$$\begin{aligned}F^2(t) &= F_0^2 \delta(t) + F_1^2 \delta(t-\tau) \\ S^2(t) &= S_1^2 \delta(t-\tau) + S_2^2 \delta(t-2\tau)\end{aligned} \tag{6.14}$$

An RSF with N retarders is shown figuratively in Figure 6.13, where the output from the Nth retarder would be the desired orthogonal C_i and D_i pulse trains. Exiting the final analyzer would, by definition, only be C_i.

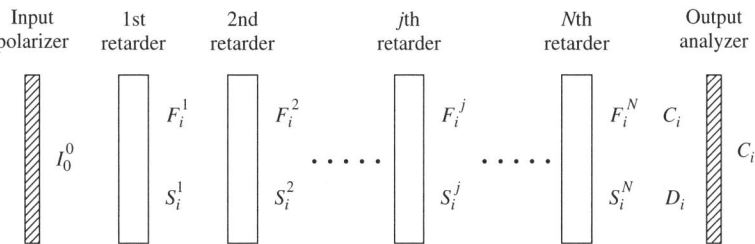

Figure 6.13 Output pulses from individual stages of an RSF with N retarders

The relationship between C_i, D_i, and the impulse response after the last retarder is a simple rotation between its optic axes and that of the analyzer. Mathematically:

$$\begin{pmatrix} F_i^N \\ S_i^N \end{pmatrix} = \begin{pmatrix} \sin\theta_P & -\cos\theta_P \\ \cos\theta_P & \sin\theta_P \end{pmatrix} \begin{pmatrix} C_i \\ D_i \end{pmatrix} \tag{6.15}$$

Since F_N^N and S_0^N are zero:

$$\tan\theta_P = \frac{D_N}{C_N} = -\frac{C_0}{D_0} \tag{6.16}$$

142 RETARDER STACK FILTERS

Similarly, it can be confirmed mathematically that the following relationships hold:
first retarder:

$$\tan \theta_1 = -\frac{F_0^1}{S_1^1} \tag{6.17}$$

$$(F_0^1)^2 + (S_1^1)^2 = 1 \tag{6.18}$$

jth retarder:

$$\tan \theta_j = -\frac{F_{j-1}^j}{S_j^j} \tag{6.19}$$

$$F_0^j F_{j-1}^j + S_1^j S_j^j = 0 \tag{6.20}$$

where the relationships (6.18) and (6.20) are consistent with the conservation of energy.

Using these relations it is possible to determine the impulses a retarder sees from its output. This is simply a formalized procedure of solving for F^{j-1} and S^{j-1} when F^j and S^j are known. In matrix form, the relationships can be summarized as:

$$\begin{pmatrix} F_0^{j-1} \\ F_1^{j-1} \\ F_2^{j-1} \\ \cdot \\ \cdot \\ \cdot \\ F_{j-1}^{j-1} \end{pmatrix} = \frac{1}{[(F_{j-1}^j)^2 + (S_j^j)^2]} \begin{pmatrix} F_0^j & S_1^j \\ F_1^j & S_2^j \\ F_2^j & S_3^j \\ \cdot & \cdot \\ \cdot & \cdot \\ \cdot & \cdot \\ F_{j-1}^j & S_j^j \end{pmatrix} \begin{pmatrix} S_j^j \\ -F_{j-1}^j \end{pmatrix} \tag{6.21}$$

$$\begin{pmatrix} S_0^{j-1} \\ S_1^{j-1} \\ S_2^{j-1} \\ \cdot \\ \cdot \\ \cdot \\ S_{j-1}^{j-1} \end{pmatrix} = \frac{1}{[(F_{j-1}^j)^2 + (S_j^j)^2]} \begin{pmatrix} F_0^j & S_1^j \\ F_1^j & S_2^j \\ F_2^j & S_3^j \\ \cdot & \cdot \\ \cdot & \cdot \\ \cdot & \cdot \\ F_{j-1}^j & S_j^j \end{pmatrix} \begin{pmatrix} F_{j-1}^j \\ S_j^j \end{pmatrix} \tag{6.22}$$

By successive recursion, equations (6.21) and (6.22) create a complete description of pulses entering and leaving each retarder, enforcing the condition of a single input impulse and output pulse train equal to the desired C_i. From the known pulse trains, the relative retarder angles θ_j can be obtained from equation (6.19) and the absolute design angles ϕ_j including the orientation of the output analyzer from equation (6.13). By virtue of this design procedure, the output analyzer is oriented at 90° or 0° (i.e., crossed or parallel polarizers) if the original $C(\omega)$ is 0 or 1 at $\omega = 0$.

6.5 Properties of Retarder Stacks

6.5.1 Unitary Jones' Matrix Representation

In general, the polarization transformation of an RSF for any given optical wavelength (or frequency) can always be written as a unitary Jones' matrix **M**, where:

$$\mathbf{M} = e^{i\phi} \begin{pmatrix} a & b \\ -b^* & a^* \end{pmatrix} \tag{6.23}$$

$|a|^2 + |b|^2 = 1$, and ϕ is the phase common to all polarizations in passing through the structure. The common phase has no impact on the SOP, and can therefore be omitted from analyses concerned only with the SOP. The above statement can be proven through first realizing that an arbitrary retarder of orientation θ_1 and retardance Γ_1 forms a unitary Jones' matrix, as follows:

$$\mathbf{M} = e^{i\phi} \begin{pmatrix} \cos\theta_1 & \sin\theta_1 \\ -\sin\theta_1 & \cos\theta_1 \end{pmatrix} \begin{pmatrix} e^{-i\frac{\Gamma_1}{2}} & 0 \\ 0 & e^{i\frac{\Gamma_1}{2}} \end{pmatrix} \begin{pmatrix} \cos\theta_1 & -\sin\theta_1 \\ \sin\theta_1 & \cos\theta_1 \end{pmatrix} = e^{i\phi} \begin{pmatrix} a_1 & b_1 \\ -b_1^* & a_1^* \end{pmatrix} \tag{6.24}$$

where

$$a_1 = e^{i\frac{\Gamma_1}{2}} \cos^2\theta_1 + e^{-i\frac{\Gamma_1}{2}} \sin^2\theta_1 \quad \text{and} \quad b_1 = -i\sin 2\theta \sin\frac{\Gamma}{2}$$

Any product of unitary matrices is itself unitary, proving that Jones' matrix of an arbitrary RSF is also unitary.

6.5.2 Properties of Symmetric RSF Designs

Based on the above discussion, we can write Jones' matrix for an RSF in unitary form as:

$$\begin{pmatrix} C(\omega) & D(\omega) \\ -D(\omega)^* & C(\omega)^* \end{pmatrix} \tag{6.25}$$

In certain cases, discussed in detail in Chapter 7, specialized relationships are required between the coefficients. One set of solutions involves zero phase delay between diagonal or off-diagonal terms (e.g., $C(\omega) = C^*(\omega)$, which corresponds to rotation invariance of the filter in its non-transformed pass band). Convenient symmetrical representations that exhibit such properties are those consisting of either a pure cosine or sine series. For example, in the case of a cosine series:

$$C(\omega) = \sum_{k=0}^{N} A_k \cos(k\omega\tau) \tag{6.26}$$

which is entirely real, and hence $C(\omega) = C^*(\omega)$.

When designing RSFs for color control, these relationships can be incorporated into the unconverted band (determined by $C(\omega)$), the converted band (determined by $D(\omega)$), or occasionally both. For wavelengths within the pass band of a parallel polarizer filter design with a cosine series frequency response, $C(\omega) \approx 1$ and $D(\omega) \approx 0$. Its Jones' matrix from equation (6.25) is then simply the identity matrix, making the filter functionally isotropic and therefore rotationally invariant in the non-transformed spectral band. This property can be extremely important when correcting geometrical effects in PBSs as discussed in Section 7.6. Furthermore, the impulse response of the $C(\omega)$ cosine series, obtained by taking its Fourier transform, has even symmetry. One set of solutions, discussed in the next section, has physical symmetry in retarder angles with respect to the stack midpoint. This relationship is shown diagrammatically in Figure 6.14, consisting of a reverse in retarder order, with angle reflection and rotation of 90°.

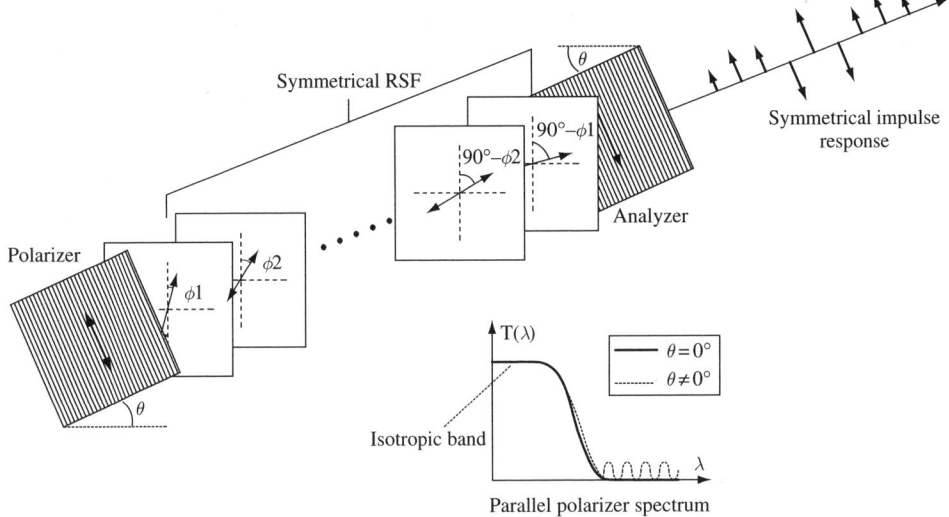

Figure 6.14 Symmetrical properties and rotation invariance of a cosine series, parallel polarizer RSF design

Two other interesting cases can be considered, namely an RSF designed with crossed polarizers using (a) a cosine series and (b) an imaginary sine series. In the latter case $C(0) = 0$ since a sine series cannot have a DC term. For this reason a sine series cannot be used for a parallel polarizer design since this condition is inconsistent with boundary conditions. In these two allowed cases:

(a) The stack has reverse angle symmetry, whereby the retarders beyond the midpoint of the filter are in reverse order and with negative signs (i.e., equivalent to reflected/reverse pass). Its pass band between crossed polarizers is rotationally invariant.

(b) The stack has reverse order symmetry, where the stack appears to have reflection symmetry about its middle, and therefore can be operated in reflection. An off-diagonal sine series designed filter has reflection invariance with respect to its transformed band.

6.5.3 General Properties of Symmetric RSF Designs

This section covers more generally the special case in which a retarder stack can be decomposed into two stacks, possessing certain symmetry properties. Such interactions may occur naturally, as in the case of a double pass arrangement in reflective systems, but may also be synthesized to create a particular response as described in the previous section. Table 6.2 is a convenient reference for the general matrices that will be derived.

Table 6.2 General form of RSF angle symmetry and the corresponding Jones' matrices

Configuration	Stack arrangement	General form of matrix
Forward pass	$\phi_1, \phi_2, \ldots, \phi_N$	$\begin{pmatrix} a & b \\ -b^* & a^* \end{pmatrix}$
Reflected forward pass	$-\phi_1, -\phi_2, \ldots, -\phi_N$	$\begin{pmatrix} a & -b \\ b^* & a^* \end{pmatrix}$
Rotated forward pass	$\phi_1 + \pi/2, \phi_2 + \pi/2, \ldots, \phi_N + \pi/2$	$\begin{pmatrix} a^* & b^* \\ -b & a \end{pmatrix}$
Rotated/reflected forward pass	$\phi_1 - \pi/2, \phi_2 - \pi/2, \ldots, \phi_N - \pi/2$	$\begin{pmatrix} a^* & -b^* \\ -b & a \end{pmatrix}$
Reverse pass	$\phi_N, \ldots, \phi_2, \phi_1$	$\begin{pmatrix} a & -b^* \\ b & a^* \end{pmatrix}$
Reflected reverse pass	$-\phi_N, \ldots, -\phi_2, -\phi_1$	$\begin{pmatrix} a & b^* \\ -b & a^* \end{pmatrix}$
Rotated reverse pass	$\pi/2 + \phi_N, \ldots, \pi/2 + \phi_2, \pi/2 + \phi_1$	$\begin{pmatrix} a^* & -b \\ b^* & a \end{pmatrix}$
Rotated/reflected reverse pass	$\pi/2 - \phi_N, \ldots, \pi/2 - \phi_2, \pi/2 - \phi_1$	$\begin{pmatrix} a^* & b \\ -b^* & a \end{pmatrix}$

Case I: Reverse Order Symmetry

A common, and often desirable, situation is one where light passes twice through an RSF by virtue of a mirror as shown schematically in Figure 6.15.

Mathematically, it is sufficient to analyze the double pass Jones' matrix structure without regard for physical manifestation. In such an arrangement, we only consider the round-trip transformation, allowing arbitrary polarization to be introduced and analyzed. In a true retro-reflecting arrangement, any component that controls the input state of polarization is also encountered in the reverse pass. This may be a linear sheet polarizer acting as both input polarizer and analyzer, effectively constraining the system to operate with parallel polarizers. Alternatively, the input polarizer may be a PBS. In such an arrangement a PBS has three functions it polarizes the input, analyzes the output, and creates distinct paths for the two beams. Using a typical PBS that separates linear states, the configuration is thus constrained to operate with a crossed polarizer input/output relationship.

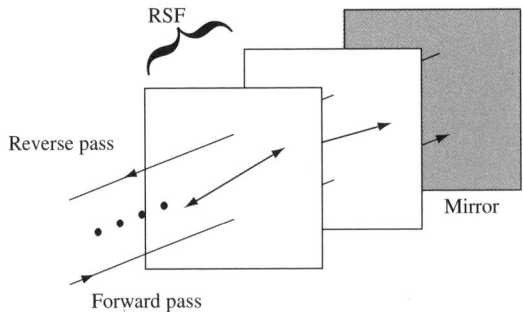

Figure 6.15 Reflective RSF system

The double pass matrix is given as the product of forward and reverse pass matrices of Table 6.1:

$$S = M'M = \begin{pmatrix} a & -b^* \\ b & a^* \end{pmatrix} \begin{pmatrix} a & b \\ -b^* & a^* \end{pmatrix} = \begin{pmatrix} a^2 + b^{*2} & ab - a^*b^* \\ ab - a^*b^* & a^{*2} + b^2 \end{pmatrix} \quad (6.27)$$

Where **M** and **M'** represent the forward pass and reverse pass matrices, respectively. This gives the important result that the off-diagonal components are identical in amplitude. Moreover, using $a = |a|e^{-i\alpha}$ and $b = |b|e^{-i\beta}$ gives $(ab - a^*b^*) = -i2|a||b|\sin(\alpha + \beta)$, showing that the amplitude is in general imaginary. This means that the Fourier series expansion of the off-diagonal(s) can be expressed by a sine series as described in the previous section. If it is desired to generate a design with reverse order symmetry, the impulse response could be specified for the off-diagonal term, with multiple designs generated by determining the roots of the diagonal term. By expressing the off-diagonal term by a sine series, all designs generated will thus possess reverse order symmetry making the sine series expansion an efficient method for identifying reflection mode filters.

If, for instance, a double pass arrangement is used to convert a linear x-input to the orthogonal SOP, the diagonal components must vanish, forcing the constraints $|a| = |b| = 1/\sqrt{2}$ and $(\alpha + \beta) = \pi(1/2 + m)$, where m is an integer. Using the above, the double pass matrix reduces to the form:

$$S = \begin{pmatrix} 0 & -i \\ -i & 0 \end{pmatrix} \quad (6.28)$$

which represents a pure polarization reflection about an axis at $\pi/4$. Physically, the above matrix represents the transformation of an ideal half-wave retarder, with the fast axis oriented along $\pi/4$. This proves a first important point about double pass arrangements: a retarder stack that converts an input x-oriented linear state to the orthogonal state in a round-trip does so using retardation only.

Inserting the above constraint into the forward pass matrix gives the result:

$$M = \frac{1}{\sqrt{2}} \begin{pmatrix} e^{-i\alpha} & -ie^{i\alpha} \\ -ie^{-i\alpha} & e^{i\alpha} \end{pmatrix} = \frac{1}{\sqrt{2}} \begin{pmatrix} 1 & -i \\ -i & 1 \end{pmatrix} \begin{pmatrix} e^{-i\alpha} & 0 \\ 0 & e^{i\alpha} \end{pmatrix} \quad (6.29)$$

where the latter decomposition represents a linear retarder with retardation 2α, oriented parallel to the input polarization, followed by a pure quarter-wave retarder with orientation $\pi/4$. The above matrix illustrates a second point: a circular state must exist at the mirror (or midpoint) in order to achieve full polarization conversion in a double pass. From a design standpoint, an effective double pass converter can thus be designed by constraining the single pass to produce a very precise circular state. Note that the above is not the matrix of an ideal quarter-wave retarder, which would further require linear eigenpolarizations, but is more appropriately termed a circular polarizer. A pure quarter-wave retarder is thus a further constrained subset of the circular polarizer family, where $\alpha = 0$.

It can be shown that the reverse pass stack can be written as a quarter-wave retarder with orientation $\pi/4$, followed by a linear retarder with retardation 2α and orientation 0. To the extent that a stack possessing the above symmetry converts a range of wavelengths to the orthogonal linear state, it can be regarded as a half-wave retarder over that range of wavelengths. In addition to having stable retardation, the structure can furthermore be considered to have wavelength stable eigenpolarizations over that range of wavelengths. By definition, the round-trip matrix can thus be diagonalized via a $\pi/4$ rotation. A compound element with stable behavior over an extended band is called an achromatic half-wave retarder. Based on the above, an achromatic half-wave retarder can be designed by pairing an achromatic circular polarizer with the reverse order stack. However, it should not be construed that reverse order symmetry is a necessary constraint for realizing achromatic half-wave behavior.

Now consider an additional constraint, in which the double pass Jones' matrix is required to induce no change in the state of polarization of an x-oriented linear state over a prescribed range of wavelengths. Such is the case with spectral edge filters, where one portion of the spectrum is converted to the orthogonal SOP, while another retains the input SOP. The combination of the former example with the present thus gives the overall behavior of reflection mode stacks used for the purpose of filtering wavelengths.

In this case, the magnitude of the off-diagonal terms must vanish in double pass, giving three possible solutions: $|a| = 0$, $|b| = 0$, and/or $\alpha + \beta = m\pi$. The degenerate cases force the SOP at the mirror (or midpoint) to be either x-oriented, or y-oriented. The third, more general case can be inserted into the double pass matrix to give:

$$\mathbf{S} = \begin{pmatrix} e^{-i2\alpha} & 0 \\ 0 & e^{i2\alpha} \end{pmatrix} \tag{6.30}$$

This result is consistent with the above, in which retardation is doubled in the second pass. Inserting the constraint into the forward pass matrix gives:

$$\mathbf{M} = \begin{pmatrix} \cos\eta\, e^{-i\alpha} & -\sin\eta\, e^{i\alpha} \\ -\sin\eta\, e^{-i\alpha} & \cos\eta\, e^{i\alpha} \end{pmatrix} = \begin{pmatrix} \cos\eta & \sin\eta \\ -\sin\eta & \cos\eta \end{pmatrix} \begin{pmatrix} e^{-i\alpha} & 0 \\ 0 & e^{i\alpha} \end{pmatrix} \tag{6.31}$$

where, without loss of generality, we have taken $|a| = \cos\eta$, $|b| = \sin\eta$. The above decomposition is of course that of a linear retarder with stable optic axis along 0 and arbitrary retardation 2α, followed by a pure rotator with arbitrary rotation η. It is left to the reader to verify that the reverse pass stack is that of a pure rotator with rotation $-\eta$, followed by a linear retarder also with stable optic axis along 0 and retardation 2α. The above matrix shows that any pure linear state with arbitrary orientation η at the midpoint will be returned to the original x-orientation in the second pass. This is a property of reciprocal systems, which can be contrasted with Faraday devices that accumulate rotation in a double pass. From a design standpoint, double pass arrangements that preserve the input polarization in a round-trip can thus be designed by constraining the polarization to be linear at the mirror over a prescribed range of wavelengths. In summary, a filter with reverse order symmetry is thus a pure linear half-wave retarder with $\pi/4$-orientation in the converted band, and a pure linear retarder with 0-orientation, followed by a rotator in the unconverted band. The retardation of the last depends upon the impulse response and the design selection resulting from the finite number of roots generated.

The above double pass analysis can be applied to any true-reflection arrangement containing lossless reciprocal elements, and is thus quite general. What follows is a treatment of symmetric stacks that are more specific to optical filtering. That is, certain symmetric stacks having properties that are attractive when one band of wavelengths has polarization converted relative to another band. Examples pertain specifically to a common case in which one band retains the original linear polarization, while the second band is converted, by whatever means, to the orthogonal linear state.

Case II: Reverse Order with Reflection

A network consisting of a general stack, followed by the reverse order reflected version, has the following matrix:

$$\mathbf{S} = \mathbf{M'M} = \begin{pmatrix} a^2 - b^{*2} & ab + a^*b^* \\ -[ab + a^*b^*] & a^{*2} - b^2 \end{pmatrix} \quad (6.32)$$

Physically, the above symmetry (and that of subsequent cases) is easily created with traveling wave filters by simply adhering to the prescription for the reverse order stack. In true reflection mode, however, generating the exact Jones' matrix of equation (6.32) is not straightforward. An achromatic quarter-wave retarder, inserted between the stack and the mirror, gives a Jones' matrix that contains terms with the same amplitude. However, the composite matrix further contains a residual achromatic half-wave retardation.

As in the reverse order case, the off-diagonal terms of this matrix are identical in magnitude. However, in this case there is a π phase shift between off-diagonal components created by the reflection of stack angles. Another consequence of the reflection is that the off-diagonals are in general pure real, $(ab + a^*b^*) = 2|a||b|\cos(\alpha + \beta)$, and thus can be expanded in a Fourier cosine series. If it is desirable to generate a design with such symmetry, the impulse response would again be specified for the off-diagonal term, where multiple designs are similarly generated by determining the roots of the diagonal term. It can further be shown that by expressing the off-diagonal term by a cosine series, all designs generated will possess reverse order reflection symmetry.

Again consider the case in which a stack with reverse order reflection symmetry transforms an x-oriented linear polarization to the orthogonal state. Again forcing unity-amplitude off-diagonal components gives the requirements $|a| = |b| = 1/\sqrt{2}$ and $(\alpha + \beta) = m\pi$, where, m is an integer. The resulting composite Jones' matrix is given by:

$$\mathbf{S} = \begin{pmatrix} 0 & 1 \\ -1 & 0 \end{pmatrix} \quad (6.33)$$

corresponding to a pure $\pi/2$ rotator. The effect of the sign change in the reverse order stack is thus to change the conversion mechanism from pure retardation to pure rotation.

Substituting the above constraints into the matrix for the forward stack gives the transformation:

$$\mathbf{M} = \frac{1}{\sqrt{2}} \begin{pmatrix} e^{-i\alpha} & e^{i\alpha} \\ e^{-i\alpha} & e^{i\alpha} \end{pmatrix} = \frac{1}{\sqrt{2}} \begin{pmatrix} 1 & 1 \\ -1 & 1 \end{pmatrix} \begin{pmatrix} e^{-i\alpha} & 0 \\ 0 & e^{i\alpha} \end{pmatrix} \quad (6.34)$$

The above shows, as before, that an arbitrary input retardation is permitted, provided that the optic axis is stable along the input direction. This retarder is followed by a pure $\pi/4$ rotation, which is doubled by the second stack. As the $\pi/2$ rotation between the external retarders renders them effectively crossed, the composite matrix in general contains zero retardation in the converted portion of the spectrum.

The addition of reflection to the stack allows us to make a series of inverse observations relative to the previous case: A linear state must exist at the midpoint in order to achieve full polarization conversion in a double pass. From a design standpoint, an effective double pass converter can thus be designed by constraining the single pass to produce a very precise $\pi/4$-oriented linear state. Note that the above is not the matrix of a pure rotator, which would further require $\alpha = 0$.

It is a straightforward exercise to show that the reverse pass stack can be written as a pure $\pi/4$ rotator, followed by a linear retarder with retardation 2α and orientation 0. It can clearly be seen therefore that, where rotation is doubled, retardation is nullified in the reverse pass. To the extent that a stack possessing the above symmetry converts a range of wavelengths to the orthogonal linear state, it can be regarded as a pure rotator over that range of wavelengths. As such, stacks with such symmetry can be considered to have no axis. They therefore exhibit optical activity over the converted band, possessing circular eigenpolarizations. By definition, the round-trip matrix can thus be diagonalized via a conversion from a linear to a circular basis set. A compound element with circular eigenpolarizations that are stable over an extended band is called a pure achromatic rotator. Based on the above, such a pure achromatic rotator can be designed by pairing a stack that generates a $\pi/4$-oriented linear polarization, with the reverse order reflected version of the stack. However, it should not be construed that all stacks exhibiting pure achromatic rotation are required to have such symmetry.

As before, now consider a second portion of the spectrum, in which the input state of polarization is unchanged by the stack (an eigenpolarization of the composite structure). In order to maintain the input state of polarization, we again force the off-diagonal components to zero. Inserting the constraint $(\alpha + \beta) = (m + 1/2)\pi$ gives the first matrix as:

$$\mathbf{M} = \begin{pmatrix} \cos\gamma & -i\sin\gamma \\ -i\sin\gamma & \cos\gamma \end{pmatrix} \begin{pmatrix} e^{-i\alpha} & 0 \\ 0 & e^{i\alpha} \end{pmatrix} \quad (6.35)$$

which represents a linear retarder with 0-orientation and arbitrary retardation 2α, followed by a second linear retarder with $\pi/4$-orientation and arbitrary retardation 2γ. It is left as an exercise to verify that the reverse order reflected stack consists of a linear retarder with $-\pi/4$-orientation and arbitrary retardation 2γ, followed by a linear retarder with 0-orientation and arbitrary retardation 2α. The composite stack, given by equation (6.32) shows that the central retarder is eliminated by the second stack.

As previously stated, the half-wave wavelength is in general converted via pure rotation if an even number of identical retarders is used. Moreover, the above symmetry insures that the entire converted band is changed to the orthogonal state using pure rotation. If the entire stack is rotated between crossed polarizers, the transmission spectrum is invariant in the converted band, though oscillation in general occurs in the full-wave band due to compound retardation.

Case III: Reverse Order Reflection/Rotation

Using Table 6.1, a network consisting of a general stack followed by a reverse order reflected version with an additional $\pi/2$ rotation has the following matrix:

$$\mathbf{S} = \begin{pmatrix} |a|^2 - |b|^2 & 2a^*b \\ -2ab^* & |a|^2 - |b|^2 \end{pmatrix} \tag{6.36}$$

In this case, the diagonal components are pure real and, as such, there is in general zero relative phase, or retardation. As in the previous cases, such a structure can be used to convert a linear input to the orthogonal state of polarization. Imposing the condition $|a| = |b| = 1/\sqrt{2}$ gives the composite matrix as:

$$\mathbf{S} = \begin{pmatrix} 0 & 1 \\ -1 & 0 \end{pmatrix} \begin{pmatrix} e^{-i(\alpha-\beta)} & 0 \\ 0 & e^{i(\alpha-\beta)} \end{pmatrix} \tag{6.37}$$

As in the previous case, the mechanism for conversion is rotation. In this case, it is accompanied by a linear retardation, much like the unconverted band of the previous cases. The retardation depends on the phase difference between diagonal and off-diagonal terms, which depends upon the impulse response and the specific design selected.

Inserting the constraint into the input stack matrix gives the result:

$$\mathbf{M} = \frac{1}{\sqrt{2}} \begin{pmatrix} e^{-i\alpha} & e^{-i\beta} \\ -e^{i\beta} & e^{i\alpha} \end{pmatrix} \tag{6.38}$$

This matrix can be decomposed into a sequence of three matrices, corresponding to a coordinate transformation in complex Cartesian space, as described by Cayley and Klein [Arfken G., 1985]:

$$\mathbf{M} = \begin{pmatrix} e^{-i(\frac{\alpha+\beta}{2})} & 0 \\ 0 & e^{i(\frac{\alpha+\beta}{2})} \end{pmatrix} \frac{1}{\sqrt{2}} \begin{pmatrix} 1 & 1 \\ -1 & 1 \end{pmatrix} \begin{pmatrix} e^{-i(\frac{\alpha-\beta}{2})} & 0 \\ 0 & e^{i(\frac{\alpha-\beta}{2})} \end{pmatrix} \tag{6.39}$$

This is very illustrative, in that it represents the transformation as a $\pi/4$ rotator flanked by linear retarders. It is straightforward to show that the transformation of the output stack is of similar form, though the matrices are in reverse order, and the linear retarders are oriented at $\pi/2$. Because the input (difference) retardation is external to the rotator, the effect is cumulative. The output (sum) retarders are effectively crossed, and therefore eliminated from the composite matrix.

From a design consideration, a stack with such symmetry would likely be obtained by specifying the impulse response of the diagonal term. A series that is specified in the power domain has real terms, and, as such, zero retardation between diagonal terms.

Once again, filter applications call upon the stack to convert the SOP of one band, while leaving that of another band unconverted. By examination of the composite matrix, the reverse order–reflection–rotation case exhibits the more interesting behavior in the unconverted band. Only two solutions exist: $|a| = 0$ or $|b| = 0$. In other words, the input stack must behave as a linear retarder, while the second stack is that of an equivalent crossed retarder. Naturally, the result is the identity matrix.

Because the response is proportional to the identity matrix, the filter can be rotated between crossed polarizers with, in general, zero transmission of the unconverted band.

Case IV: Reverse Order Rotation

The final case to consider is the degenerate case of reverse order rotated symmetry. In this case, the center elements are crossed, eliminating any net change in the SOP, as are the next adjacent elements, and so on. Because each element is crossed with its counterpart in the symmetric stack, the composite matrix is in general the identity matrix. This case is of no particular use at normal incidence, unless for example active switching elements are inserted between the stacks. The benefit of the arrangement is that LC devices can be made to vanish in a particular voltage state, allowing the entire structure to vanish at all wavelengths. At another voltage state, the LC device manipulates the SOP between the stacks, modifying the composite matrix to perform a desired filter function.

References

[Arfken G., 1985] G. Arfken, *Mathemateial for Physicists* (3rd edition), pp.211–212, Academic Press, Orlando, FL, 1985.

[Aseltine J. A., 1958] J. A. Aseltine, *Transform method in linear system analysis*, McGraw-Hill, New York, 1958.

[Chen J., 2004] J. Chen, M. G. Robinson, and G. D. Sharp, LCOS projection color management using retarder stack technology, ASID'04, p.134, 2004.

[Harris S. E., 1964] S. E. Harris, E. O. Ammann, and I. C. Chang, Optical network synthesis using birefringent crystals. I. synthesis of lossless networks of equal-length crystals, *J. Opt. Soc. Am.*, 54, p.1267, 1964.

[Hecht E., 1980] E. Hecht and A. Zajac, *Optics*, Addison-Wesley World Student Series Edition, 5th printing, Addison-Wesley, Reading, MA, 1980.

[Lyot B., 1933] B. Lyot, Optical apparatus with wide field using interference of polarized light, *C. R. Acad. Sci., Paris*, 197, p.1593, 1933.

[Robinson M. G., 2004] M. G. Robinson, J. Chen, and G. D. Sharp, Optimization of three-panel liquid crystal on silicon video projection system, IDW'04, p.1683, 2004.

[Sharp G. D., 1998] G. D. Sharp, 'Color polarizer for polarizing an additive color spectrum along a list axis and its complement along a second axis', US Patent 5,751,384, 1998.

[Sharp G. D., 1999] G, D. Sharp and J. B. Birge, Retarder stack technology for color manipulation, SID'99 Digest, p.1072, 1999.

[Sharp G. D., 2002] G. D. Sharp, M. G. Robinson, J. Chen, and J. Birge, LCOS projection management using retarder stack technology, *Displays*, 23, p.139, 2002.

[Solc I., 1965] I. Solc, Birefringent chain filters, *J. Opt. Soc. Am.*, 55, p. 621, 1965.

[Yeh P., 1988] P. Yeh, *Optical Waves in Layered Media*, John Wiley & Sons, Ine., New York, 1988.

7
System Contrast

7.1 Introduction

Contrast is considered the most important performance specification of a projection system, as it ultimately influences the number of true gray levels and the color fidelity. In this chapter, we will only focus on methods to improve the system contrast for reflective LCOS systems, leaving contrast improvements for transmissive LCD systems for Chapter 9. One of the major challenges for LCOS-based projectors is to achieve the contrast promoted by DLP-based systems. These challenges are uniquely associated with polarization management in reflective displays. As projectors operate at finite $f_{/\#}$s, precise polarization control of each ray must be maintained through the optical train in order to meet system goals. This is particularly so in the space directly adjacent to the panel.

To tackle system contrast, it is convenient to split polarization effects into those that affect on-axis, normal incidence rays (often referred to as "head-on") and those associated with off-axis rays. The small depolarization effects that reduce contrast can in general be considered perturbations, and therefore allow the on- and off-axes effects to be considered independently. "Head-on" compensation schemes for LC modes will first be considered, followed by off-axis schemes, which include:

- FOV of clean-up polarizers
- off-axis properties of a homeotropic LC cell
- geometrical effects with MacNeille cube PBSs
- LC effects with fast LCOS panels, specifically the 45° TN mode
- contrast loss due to surface reflections
- skew ray effects in retarder stack polarization filters.

Polarization Engineering for LCD Projection   M. G. Robinson, J. Chen and G. D. Sharp
© 2005 John Wiley & Sons, Ltd

154 SYSTEM CONTRAST

A general three-step procedure used to optimize the system contrast is:

1. Optimize head-on contrast.
2. Maximize the contrast for off-axis rays without affecting head-on performance.
3. Reduce reflections from the compensating components that otherwise reduce contrast.

Any one system will have a contrast determined by the performance of several factors and the extent to which they are compensated. At the end of the chapter, a general expression will be given that incorporates the various factors that contribute to the system contrast. This expression allows specific system contrasts to be estimated, and will be used for this purpose in later chapters.

7.2 On-axis Contrast

7.2.1 Head-on Contrast of LC Mode [Chen J., 2004a]

Certain LC modes, described in Chapter 5, require a compensator to improve the head-on contrast of the LC panel. The specific compensation scheme depends upon the specific LC mode. For 90° TN and VA TN mode, little if any compensation is needed for head-on contrast improvement. Conversely, non 90° twisted LC modes, such as the 45° TN and 63.6° MTN modes, have head-on contrasts of only ~40:1 without compensation. The methodology developed here to determine the head-on compensation scheme is universal and can be implemented with any LC modes operating in either transmission or reflection.

In an LCOS system, the optimum in-plane compensation scheme should ensure that the state of polarization (SOP) of the reflected beam is identical to the SOP of the incident beam in a round-trip. Based on the general property of the Jones' matrix (see Case I, Section 6.5.3), this imposes the requirement that the SOP of light at the mirror must be linear. The design methodology for head-on contrast is as follows:

1. An arbitrary linear SOP at the mirror interface is chosen (indicated as position **M** on the equator of the Poincaré sphere shown in Figure 7.1).

2. The SOP (**I** in Figure 7.1) of the light exiting the LCOS panel is calculated from the initial SOP **M** at the mirror and the known LC OFF-state director profile.

3. The retardance ($\Delta nd \equiv \Gamma$) and orientation (ϕ) of a single birefringent compensator required to convert the **I** into the original input SOP, **O**(0, 0), is calculated using:

$$\phi = \frac{1}{2} \tan^{-1}\left(\frac{\cos(2\xi)\cos(2\theta) - 1}{\cos(2\xi)\sin(2\theta)}\right) \quad (7.1)$$

$$\Gamma = \frac{\lambda}{4\pi} \sin^{-1}\left(\frac{|\mathbf{OB} \times \mathbf{BI}|}{|\mathbf{OB}| \cdot |\mathbf{BI}|}\right) \quad (7.2)$$

where (θ, ξ) is the SOP **I** on the Poincaré sphere in the figure, and **B** is its associated compensating retarder orientation at the equator.

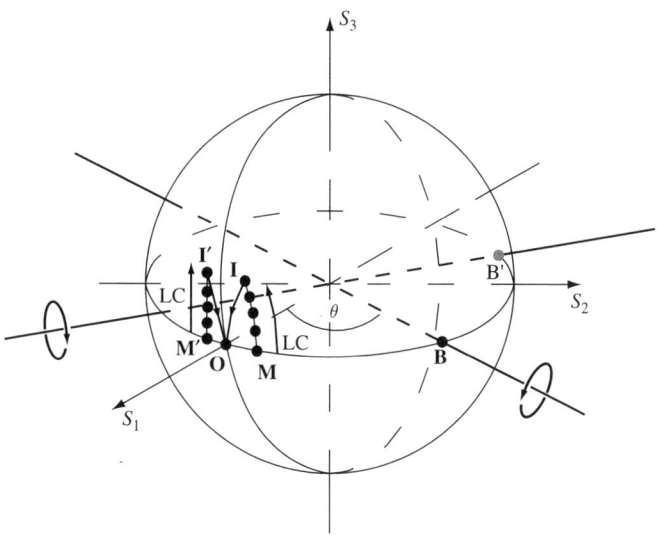

Figure 7.1 Poincaré sphere illustration of the "head-on" LCOS design methodology. The preconditioned input SOP **I** is at the entrance and exit of the panel, and **O** is the original SOP of incident light. There are two sets of solutions, (**M**,**I**, **O**) and (**M′**, **I′**, **O**), which have near-reflective symmetry about 135° (or 45°)

By sweeping through all possible linear SOPs at the mirror interface, a solution set can be obtained as shown in Figure 7.2. This shows that there are two solutions for any value of retardance, related in angle by a reflection about 45° (135°).

The orientation and retardation of a compensator required for a normally white 45° MTN mode is calculated and shown in Figure 7.2. The rubbing direction of the top substrate is 90° to avoid depolarization from anisotropic reflections as discussed later in Section 7.4. The polarization state **I** is obtained using Berreman 4 × 4 algebra based on the energized LC state ($V_{rms} = 5.0$ V). Figure 7.3 shows the reflectance versus voltage (R–V) performance at three different wavelengths (450 nm, 550 nm, and 620 nm) employing a single retarder as the compensator (17 nm at 135°). The contrast improvement is clearly demonstrated. It is interesting to note that the brightness is also improved through compensation, which can be further optimized by adjusting the compensation value and the applied voltage while maintaining good contrast.

The existence of two orientations for any given retardance value would be expected if the transformation of polarization imparted by the LC on the different linear mirror interface SOPs were similar. The results of Figure 7.2 strongly suggest similar LC transformations, which in turn implies the situation is one of small, perturbative, independent polarization transformations.

If perturbative transformations are assumed, it can be shown that the associated Jones' matrices can be permuted. Mathematically, this can be demonstrated by considering two Jones' transformation matrices representing small polarization transformations ($\alpha, \beta, \gamma, \delta \ll 1$):

$$\mathbf{A} \approx \begin{pmatrix} 1+\alpha & -\beta^* \\ \beta & 1+\alpha^* \end{pmatrix} \quad \text{and} \quad \mathbf{B} \approx \begin{pmatrix} 1+\gamma & -\delta^* \\ \delta & 1+\gamma^* \end{pmatrix} \quad (7.3)$$

156 SYSTEM CONTRAST

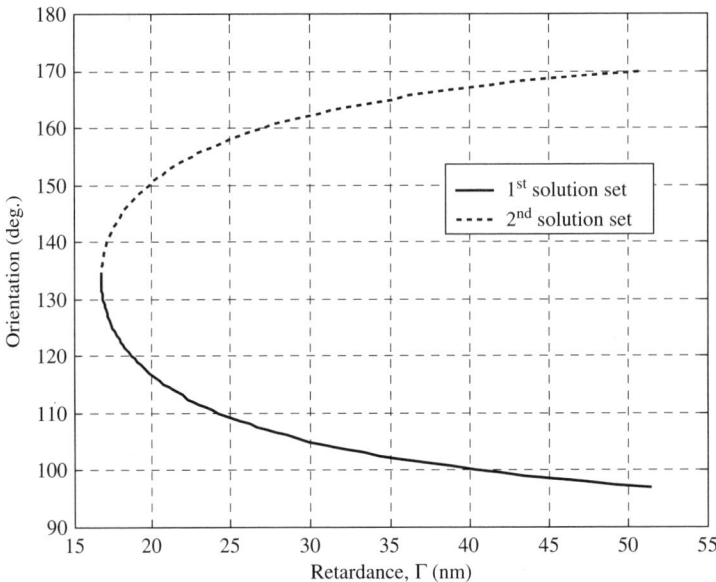

Figure 7.2 Orientation and magnitude of a single compensator for "head-on" contrast of a 45° TN LCOS panel. The LC MLC6792 is assumed with 90° buffing at the glass substrate. $V_{off} = 5$ V. There are two sets of solutions for any one retardation value having reflective symmetry about 135°

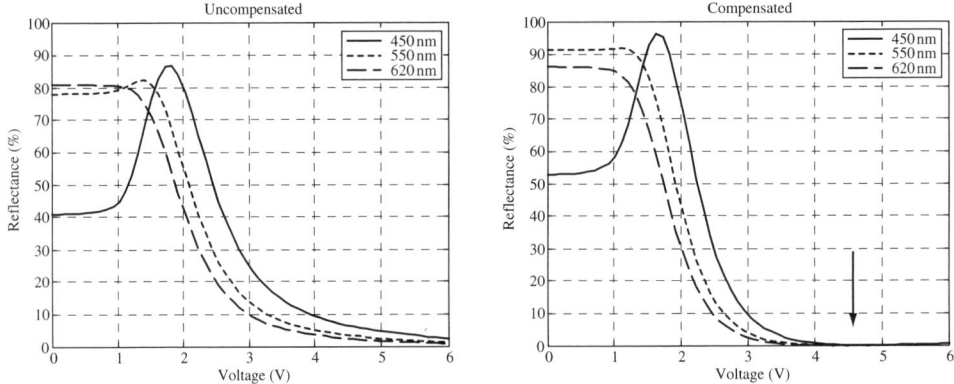

Figure 7.3 $R-V$ performance of the 45° TN display without and with compensation. The specific compensation solution chosen lies on the curve of Figure 7.2 (17 nm at 135°)

and comparing the products **AB** with **BA**. If products between the small variables are assumed negligible (e.g., $\beta\delta \ll 1$), then **AB** = **BA**. From an order of magnitude standpoint, the matrix products agree to better than 1% if the polarization transformations are equivalent to the maximum achievable by retarders of <15 nm.

Since a typical LCOS panel in its OFF-state has <15 nm, the transformation order can be reversed, making it only necessary to calculate the transformation of the LC on the linear input SOP. From this transformed SOP (θ_1, ξ_1), it is then possible to determine the complete solution set from the analytic expressions describing the mapping of the SOP from the LC output onto a linear SOP at the mirror. This yields the following relationship between the retardance, Γ, and angle, θ, of any one solution:

$$\Gamma(\theta) = \frac{\lambda}{2\pi} \cos^{-1}\left(\frac{\sin 2(\theta_1 - \theta)}{\sqrt{\tan^2 2\xi_1 + \sin^2 2(\theta_1 - \theta)}}\right) \quad (7.4)$$

where λ is the wavelength of light at which the compensation scheme is optimum.

This general result relies on knowledge of the transformation of the LC, and hence requires the accurate LC director profile. If two solutions are known, however, then the polarization state can be inferred by solving for the two unknowns in equation (7.4). In practice, this requires that two compensators of different retardance values are rotated adjacent to the panel and the angles determined at the transmission minimum. These two solutions can be used to determine θ_1 and ξ_1 and therefore compute the complete solution set without any additional knowledge of the LC properties.

By way of example, we can take the rigorously calculated solution set of Figure 7.2 and choose two arbitrary solutions on the curve, say (25 nm, 158°) and (40 nm, 167°), which yield an LC transformed SOP of $(\theta_1, \xi_1) = (-1.44°, 5.47°)$. Placing these values into the expression above leads to the solution set agreeing almost exactly with the more rigorous solution shown in Figure 7.2.

The solution yielding the best contrast in a final system is dependent upon the interaction of components, and in particular interface reflections.

In the transmissive system, discussed in Chapter 9, the optimum in-plane compensation scheme occurs when the exiting SOP is identical to the incident SOP, assuming a normally white mode. The design methodology here is as follows:

1. Set the SOP (on the equator of the Poincaré sphere) to be identical to the SOP of incident light (**O** position in Figure 7.1).

2. Calculate the SOP **I** in Figure 7.1 for light exiting the panel based on the LC profile.

3. Using equations (7.1) and (7.2), determine the magnitude Δnd and orientation ϕ of a single layer compensator that converts **I** to **O**.

In this case, there is only one solution.

7.2.2 Normal Incidence Pre- and Post-polarizers

In a typical LCOS projection system, a PBS is used to separate the input from the reflected beam. For the conventional MacNeille-type cube PBS, embedded dielectric coatings are designed to be highly reflecting for **s**- and transmitting for **p**-polarized light. The former can be readily achieved, whereas the suppression of **p**-reflectivity over a finite angular range is challenging (see Chapter 4). Typical MacNeille PBSs reflect 3–10% of **p**-polarized light at $f_{/2.8}$. As shown in Figure 7.4, leakage from <100% **p**-transmittance (T_p) is equal to $T_p.R_p \sim 5\%$, where $R_p = (1 - T_p)$ for a PBS with no absorption. This yields an unacceptable

158 SYSTEM CONTRAST

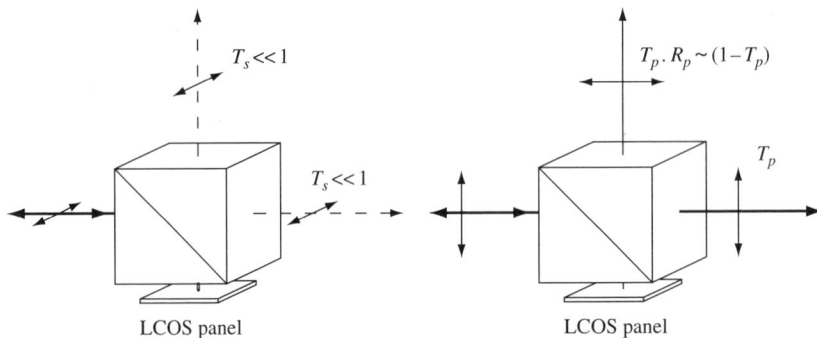

Figure 7.4 The leakage of **s**- and **p**-transmittance in a reflective system using a PBS

system contrast of 20:1 even if a perfect LCOS panel is assumed. To achieve acceptable system contrast levels, a supplemental (clean-up) polarizer must be included in the reflection pass to block the reflected **p**-light.

Clean-up polarizers are a necessity, as all conventional PBS technologies, including wire grid PBSs, reflect unacceptable amounts of **p**-polarization.

Assuming infinite contrast panels, and that leakages contributing to contrast loss are small and independent, the following mathematical expressions describe the contrasts for the systems shown in Figure 7.5:

$$\frac{1}{CR_A} = \frac{T_p.R_p}{C_{P1}} + \frac{T_s}{C_{P2}} \qquad (7.5)$$

$$\frac{1}{CR_B} = \frac{T_s}{C_{P1}} + \frac{T_p.R_p}{C_{P2}} \qquad (7.6)$$

where C_{P1} and C_{P2} are the contrast for the pre- and post-clean-up sheet polarizers respectively (~1000:1 in each case). From equations (7.5) and (7.6), the head-on contrast can

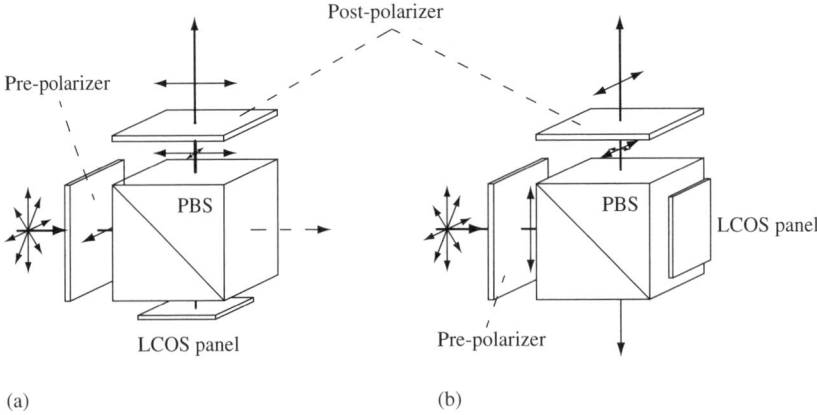

Figure 7.5 Pre- and post-clean-up polarizers to enhance PBS/LCOS system contrast

be ~20 000:1 in the case where two polarizers are used and ~1000:1 if only one is used to clean up the reflected beam. Better PBS coatings and higher contrast polarizers often allow the single polarizer approach to achieve acceptable levels of system contrast. This is currently the standard approach taken with MacNeille-based systems.

7.3 Off-axis Effects

7.3.1 Homeotropic Liquid Crystals

The off-axis behavior of a homeotropic LC layer is an important characteristic, as it frequently determines the maximum achievable contrast. A perfect homeotropic LC layer has no head-on retardation, and therefore zero modulation of normal incidence light. Off-axis light, however, experiences retardation. For a ray propagating at a small angle θ with respect to the homeotropic axis, i.e., $\sin^2\theta \ll n_o^2, n_e^2$, a phase retardation is introduced between the two orthogonally polarized **e**- and **o**-modes of the system, which is given by [Yeh P., 1999, chapter 4]:

$$\Gamma(\theta) = (n_e - n_o)d\frac{(n_o + n_e)}{2n_o n_e^2}\sin^2\theta \qquad (7.7)$$

with the **e**- and **o**-polarization axes tracking the azimuthal incidence plane orientation.

Equation (7.7) predicts light leakage at azimuthal angles where both **e**- and **o**-modes are launched (maximal along $\phi = 45°/135°$ and zero along $\phi = 0°/90°$, where the crossed polarizer and analyzer are aligned along $\phi = 0°/90°$). This "cross-like" FOV pattern, shown below in Figure 7.7(a), can be inferred geometrically from the index ellipsoid construct shown in Figure 7.6. The leakage between the spokes of the "cross" acts to reduce contrast in finite $f_{/\#}$ systems.

The off-axis birefringence of a homeotropic LC cell can be effectively compensated by a negative c-plate with the identical retardation value. Choosing a c-plate where

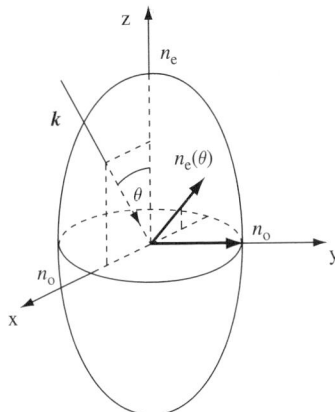

Figure 7.6 Index ellipsoid showing the **e**- and **o**-mode indices and optic axes of a homeotropic LC for an incident ray in the xz plane

160 SYSTEM CONTRAST

n_o (c-plate) = n_e (LC) and n_e (c-plate) = n_o (LC), the net retardation of such a combination is almost zero, or:

$$\Gamma(\theta) = \Gamma_1(\theta) - \Gamma_2(\theta) = \frac{(n_e + n_o) d \sin^2\theta}{2 n_o^2 n_e^2} (n_e - n_o)^2 \approx 0 \tag{7.8}$$

where $\Gamma_1(\theta)$ and $\Gamma_2(\theta)$ are the angular dependent retardations for a homeotropic LC cell and negative c-plate respectively.

An alternative compensation scheme uses two orthogonal uniaxial a-plates. The retardation of a single a-plate at oblique angles (θ, ϕ) is equal to [Yeh P., 1999, chapter 4]:

$$\Gamma_{a\text{-}plate}(\theta, \phi) = (n_e - n_o) d \left[1 + \frac{\sin^2\theta}{2 n_o^2} \left(\frac{n_o}{n_e} - \frac{n_o + n_e}{n_e} \cos^2\phi \right) \right] \tag{7.9}$$

where θ is the polar angle of incidence (which is assumed small, i.e., $\sin\theta \approx \theta$) and ϕ is the azimuth angle. For orthogonal a-plates, the combined retardation under the assumption that eigenpolarization axes are parallel equals:

$$\Gamma(\theta, \phi) = \Gamma_{a\text{-}plate}(\theta, \phi) - \Gamma_{a\text{-}plate}(\theta, 90-\phi) = -\frac{(n_e^2 - n_o^2) d \sin^2\theta \cos(2\phi)}{2 n_o^2 n_e} \tag{7.10}$$

If the plane of incidence contains either of the a-plate optical axes, the eigenpolarization vectors are exactly parallel, and the net retardation becomes:

$$\Gamma(\theta) = -\frac{(n_e^2 - n_o^2) d \sin^2\theta}{2 n_o^2 n_e} \tag{7.11}$$

This can be equated to the magnitude of retardation from the homeotropic LC cell, but with opposite sign since in general $n_o \sim n_e$. In this manner, we can use crossed a-plate compensation as an alternative to a negative c-plate to compensate the ±45° azimuth. In all other planes, the optical axes of the a-plates do not remain orthogonal, producing rotation of the incident polarization axes. As such, compensating the ±45° azimuth compromises the contrast in the 0°/90° azimuth. This tends to rotate the orientation of the high-contrast "cross" as is clearly seen in Figures 7.7(a), (b), and (c).

The polarization rotation caused by crossed a-plates in the $\phi = 0°$ and 90° planes is of opposite sense for forward and return passes. This means they are self-compensating when used in reflection, producing a good FOV (see Figure 7.8) and very high contrast. In principle, this scheme can be used for the transmissive VA mode when considered an unfolded version of the reflective case, i.e., (λ/4, 45°), (λ/4, −45°), LC(λ/2), (λ/4, −45°), (λ/4, 45°). Moving both sets of plates to a single side is also effective, leading to a (λ/4, 45°), (λ/2, −45°), (λ/4, 45°), LC(λ/2) compensation scheme.

7.3.2 Off-axis Property of Sheet Polarizers

The general property of sheet polarizer has been illustrated in Section 4.3.1. Off-axis behavior of sheet polarizers, and the methods to correct for leakages at oblique angles, will be described in this section. First we consider the case of crossed conventional o-type sheet polarizers, with the assumption that the head-on leakage for ideal crossed polarizers is

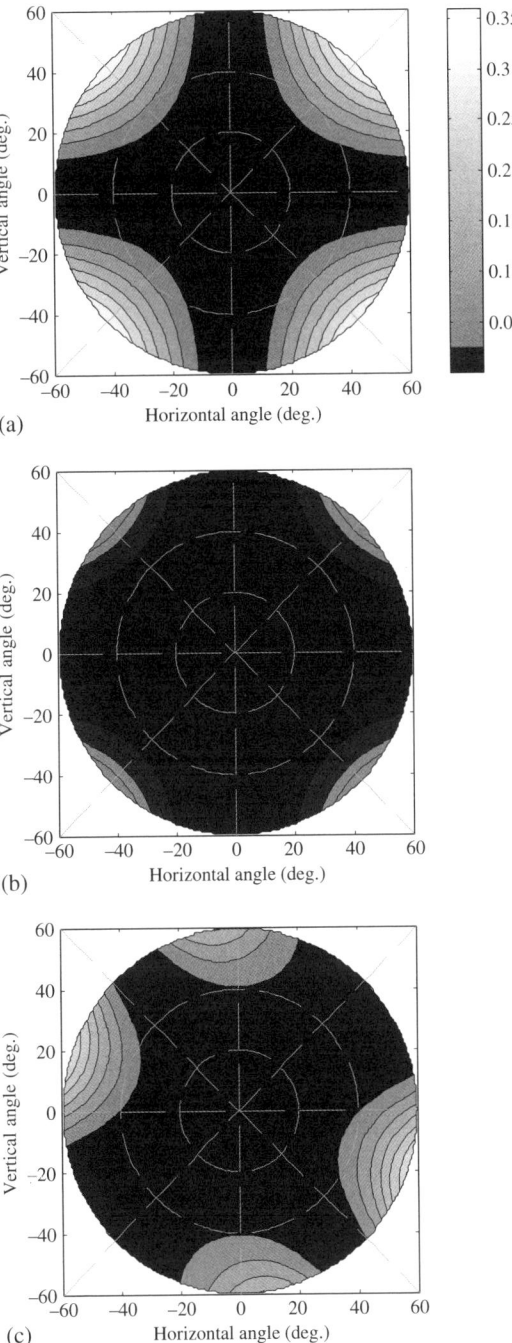

Figure 7.7 (a) FOV for $\lambda = 550$ nm of a homeotropic LC cell with retardance 275 nm; (b) FOV of the same homeotropic LC cell with negative c-plate (-275 nm) compensation; (c) FOV of the same homeotropic LC cell with crossed 275 nm a-plates (45°, 135°)

162 SYSTEM CONTRAST

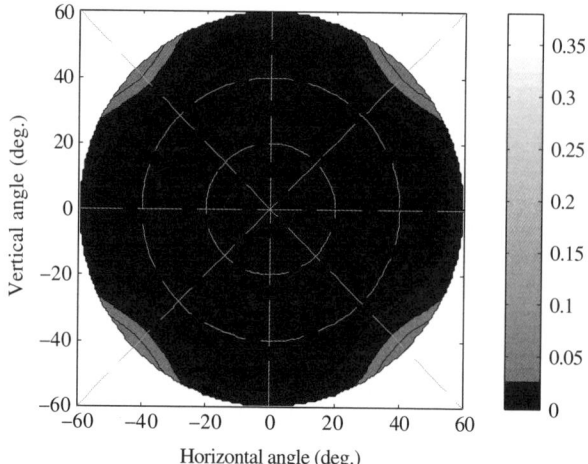

Figure 7.8 Reflection FOV of a 137.5 nm[1] homeotropic LC cell compensated with crossed 137.5 nm a-plates (45°, 135°)

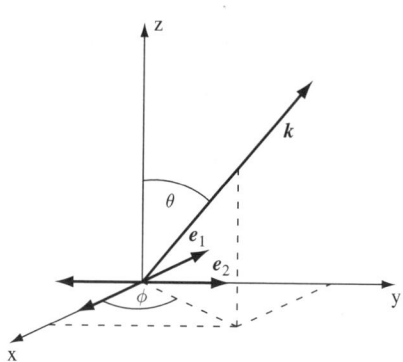

Figure 7.9 Coordinate system used to analyze two crossed **o**-type polarizers

zero; o-type polarizers transmit light with the electric vector orthogonal to an absorption axis aligned with its optical axis. The vector configuration is shown in Figure 7.9.

The internal angle of propagation is defined using conventional polar coordinates (θ, ϕ):

$$\boldsymbol{k} = k(\sin\theta\cos\phi, \sin\theta\sin\phi, \cos\theta) \quad (7.12)$$

The polarization state of the transmitted light after the first polarizer, with absorption axis $\boldsymbol{e}_1$, is then given by:

$$\boldsymbol{T}_1 = \frac{\boldsymbol{k} \times \boldsymbol{e}_1}{|\boldsymbol{k} \times \boldsymbol{e}_1|} \quad (7.13)$$

---

[1] Equivalent to 275 nm since the light passes through it twice in reflection.

and that exiting the second polarizer, with absorption axis e_2, by:

$$T_2 = \frac{k \times e_2}{|k \times e_2|} \quad (7.14)$$

If e_1 and e_2 are orthogonal, then normal incidence light is not transmitted. However, in the general case T_1 is not orthogonal to T_2 leading to finite transmission or leakage. The leakage of unpolarized light through crossed o-type polarizers oriented at $\phi = 0°$ and $90°$ is given by [Yeh P., 1999, chapter 3]:

$$T = \tfrac{1}{2}|T_1 \cdot T_2|^2 = \frac{\sin^4 \theta \sin^2 \phi \cos^2 \phi}{4(1 - \sin^2 \theta \sin^2 \phi)(1 - \sin^2 \theta \cos^2 \phi)} \quad (7.15)$$

This expression describes the FOV plot shown in Figure 7.10, where it is clear that the leakage occurs in the bisecting ($\phi = 45°$, $135°$) planes[2].

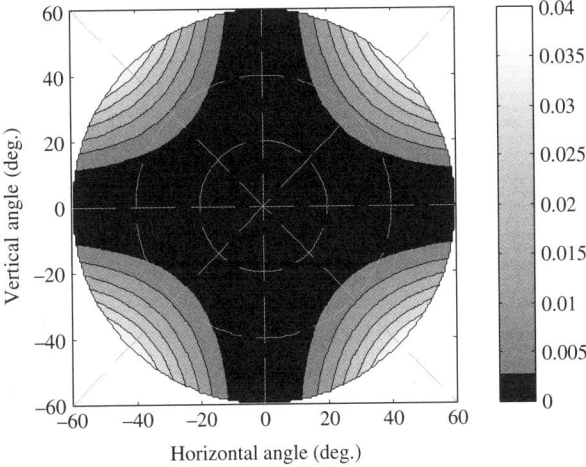

Figure 7.10 FOV of crossed o-type polarizers

The leakage at oblique angles for crossed o-type polarizers can be explained by visualizing the polarization transformations of specific rays on the Poincaré sphere (Figure 7.11). For normal incidence rays, the linear polarization state imparted by the first polarizer (**I**) is orthogonal to that of the second crossed polarizer (**O**), i.e., they are diametrically located on the equator. As the polar angle of a ray increases in the plane bisecting the polarizer axes, these polarization states are transformed through geometrical projection into **I'** and **O'**, which are no longer orthogonal. It can be proven that this rotation is equal to

$$\frac{\pi}{4} - \frac{1}{2}\cos^{-1}\left(\frac{\sin^2 \theta}{\cos^2 \theta - 1}\right)$$

---

[2] Note that the polar angle is the incidence angle in the media, which is determined by Snell's law ($\sin^{-1}(\sin \theta / n)$) where θ is the incidence angle in air.

164 SYSTEM CONTRAST

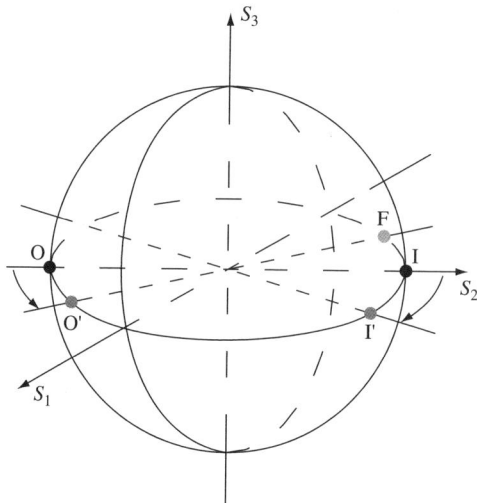

Figure 7.11 Visualization of off-axis leakage for crossed o-type polarizers

A desirable compensation scheme transforms the polarization state **I'** into **F**, which is orthogonal to the analyzer state **O'**, prior to the analyzing polarizer. However, the compensator must maintain head-on performance, requiring that it make zero contribution at normal incidence. Schemes with the optic axis parallel/perpendicular to the polarizer, or c-plate compensators, are such examples. At least two compensation schemes have been developed, which achieve this [Chen J., 1998]. One approach is to use a biaxial ($N_Z = 0.5$) half-wave retarder, whose working principle is shown in Figure 7.12.

Note that, whereas a biaxial half-wave plate converts the polarization **I'** to **F** as desired, a uniaxial half wave does not due to optical axis instability.

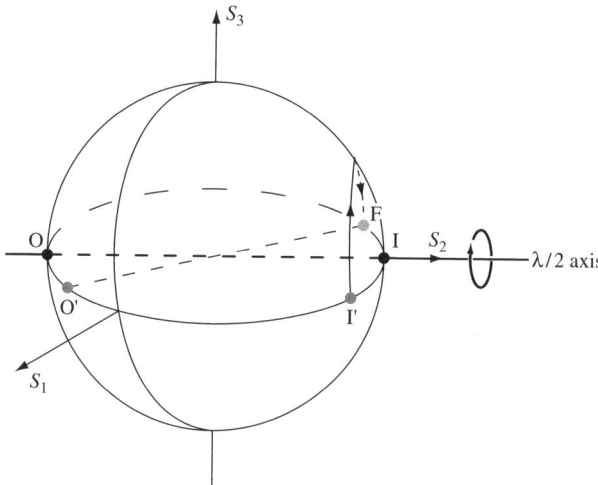

Figure 7.12 Half-wave biaxial film compensation of off-axis leakage from crossed polarizers

Of the many other possible solutions, a second uses two uniaxial materials, the first of which is a λ/4 a-plate (137 nm) with the second a positive c-plate (80 nm). Its principle is shown in Figure 7.13, which delivers the FOV performance shown in Figure 7.14. The optimized values for the compensation films alter when the retardation of the TAC substrate is taken into account [Chen J., 1998]. Compared to the non-compensated case, the leakage at oblique incidence can be reduced by an order of magnitude with compensation.

The off-axis leakage of two crossed **e**-type polarizers is similar to that of two **o**-type polarizers, since for any given ray, projected absorption axes are always orthogonal to the

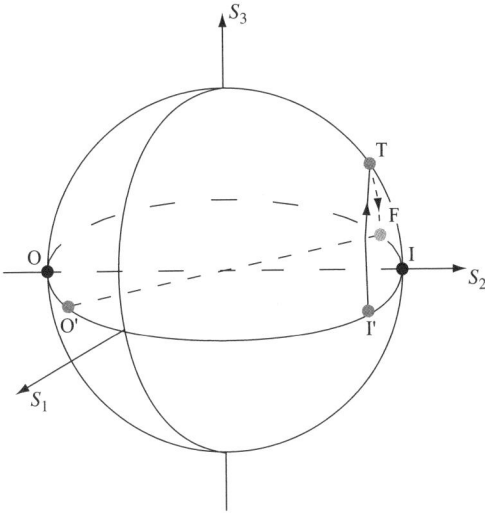

Figure 7.13 Compensation scheme using the combination of a- and c-plates. A λ/4 a-plate (137 nm) moves the polarization state from **I′** to **T**, then a positive c-plate (80 nm) moves **T** to **F**

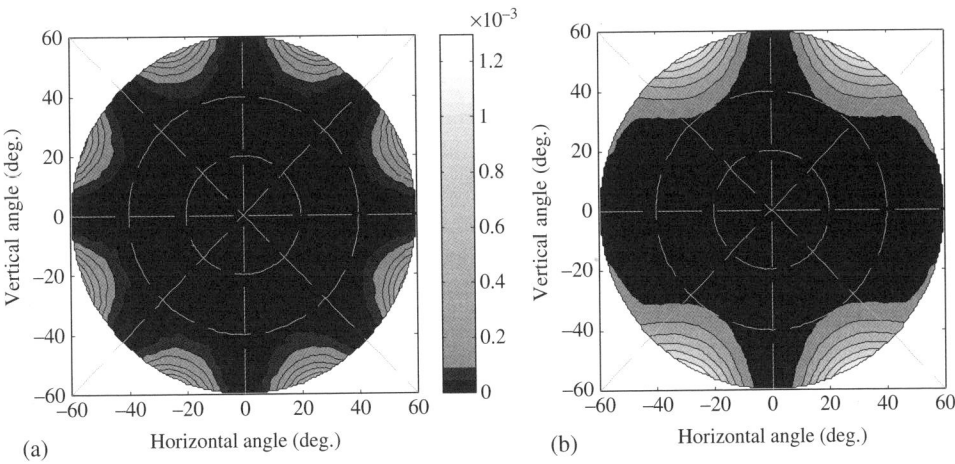

Figure 7.14 (a) Crossed polarizer FOV with half-wave, biaxial ($Nz = 0.5$) compensation film; (b) crossed polarizer FOV compensated with a 137 nm a-plate and an 80 nm c-plate

166 SYSTEM CONTRAST

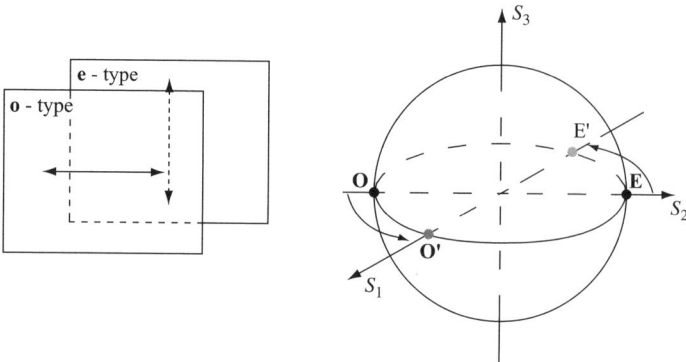

Figure 7.15 Off-axis compensation effect with crossed **o**- and **e**-type polarizers

projected transmission axes. Therefore, the same compensation schemes used for **o**-type can be applied to **e**-type polarizers.

The off-axis leakage for crossed polarizers can be dramatically reduced if **o**- and **e**-type polarizers are used together, as illustrated in Figure 7.15.

In the figure the **O'** and **E'** linearly polarized transmission states associated with crossed **o**- and **e**-type polarizers rotate to the same extent about the equator of the Poincaré sphere for all rays and thus remain orthogonal. Crossed in this context implies parallel optic axes. If the extraordinary n_e and ordinary n_o refractive indices of both **o**- and **e**-type polarizers are equivalent, then the off-axis leakage through crossed polarizers is completely eliminated [Lazarev P., 2001].

The compensation techniques are applicable to all polarizing elements that behave in a similar manner to **e**- and **o**-type sheet polarizers such as the multilayer birefringent and wire grid reflective polarizers introduced in Chapter 4.

7.3.3 Geometrical PBS Compensation

For PBSs, and in particular the MacNeille type, the polarization axes for skew rays are rotated geometrically with respect to the system coordinate axes as described in Section 4.5. There are two situations where this can create leakage within projection systems. The first is when PBSs are used in sequence as polarizer and analyzer, and the second is when a PBS is used with a sheet polarizer.

7.3.3.1 Paired MacNeille PBSs

For two MacNeille PBSs in series the beam splitting surfaces can be either parallel or perpendicular. Pupil mapping of the **p**-polarization of the first PBS and the analyzing **s**-polarization of the second PBS are shown in Figure 7.16 for both cases. In the first case, where the surfaces are parallel, the optic axes remain orthogonal and therefore there is no leakage. However, in the second case, the axes rotations are opposite and leakage occurs for rays out of the angular plane (as defined in Figure 4.18) even for perfect (i.e $T_p = 100\%$, $T_s = 0\%$) PBSs.

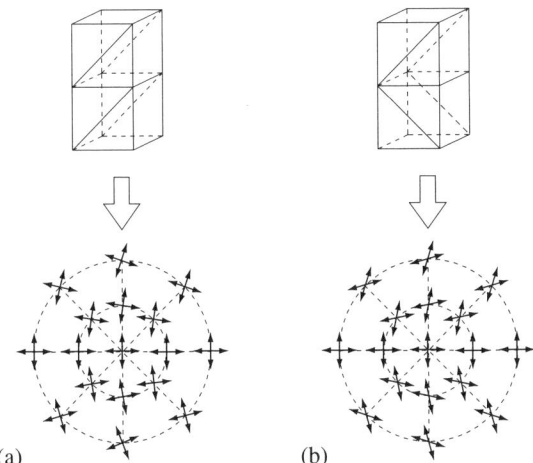

Figure 7.16 Polarization pupil maps showing **p**-axes of the first PBS and **s**-axes of the second, for two paired PBSs arrangements

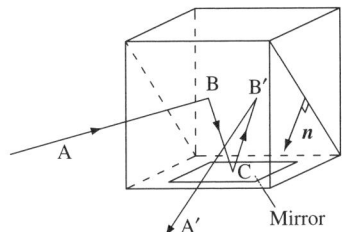

Figure 7.17 Skew ray path in a reflective LCOS system leading to compounded polarization axes rotation

In LCOS projection systems this is relevant when a PBS is used adjacent to a panel as both polarizer and analyzer, as shown in Figure 7.17. Unfolded, this is equivalent to the situation shown in Figure 7.16(b).

The FOV plot of the leakage from the system shown in Figure 7.17 can be calculated, and is shown in Figure 7.18. Considering rays in the geometric plane, the leakage is equal to $\sin^2(2\sin\theta/n) \propto \theta^2$, where n is the refractive index of the PBS, which implies that the system contrast is proportional to NA^{-2} (or $\sim f_\#^2$) [Rosenbluth A. E., 1998]. Assuming the refractive index of an ideal PBS is equal to 1.5 and that the light distribution in the cone angle is uniform, the average leakage for an $f_{/2.5}$ cone angle is $\sim 2\%$, equivalent to $\sim 50{:}1$ contrast.

To compensate for this unacceptable contrast, a QWP can be inserted with axis aligned to the normal incidence polarization axes. A QWP acts as an HWP in double pass, whose Jones' matrix can be written as $\begin{pmatrix} 1 & 0 \\ 0 & -1 \end{pmatrix}$. For polarization states, this represents a reflection transformation about the system polarization axes, and corresponds to the desired compensating transformation.

The simplest half-wave component is a uniaxial plate. Considering this first, there are two possible orientations of the uniaxial QWP relative to the PBS: parallel ($\phi = 0°$) or

168 SYSTEM CONTRAST

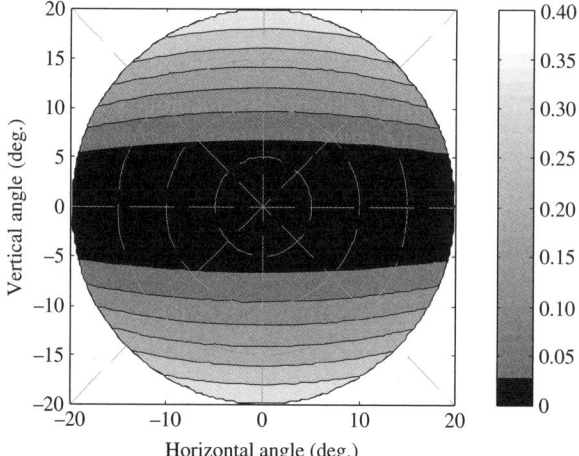

Figure 7.18 FOV of ideal MacNeille PBS/mirror reflective system ($n_{PBS} = 1.5$)

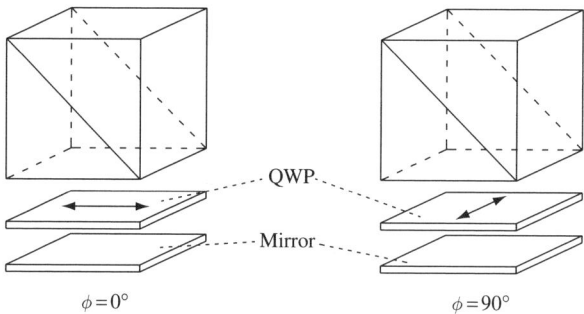

Figure 7.19 Two possible orientations for a QWP for cube PBS/mirror skew ray compensation

perpendicular ($\phi = 90°$) to the PBS **p**-axis (see Figure 7.19). For each configuration the simulated ($\lambda = 550$ nm) FOV leakage plots are shown in Figure 7.20. From these plots it is evident that there is a slightly preferable orientation of a QWP for the PBS/mirror system, whereby the QWP optic axis is parallel to the normal incidence **p**-polarization axis.

The preference in QWP orientation can be explained by examining the FOV of a uniaxial retarder. Here retardation decreases with increasing incidence angle $\theta (\propto \cos\theta)$ in the plane containing the optic axis, and increases to a lesser degree in the orthogonal plane ($\propto 1/\cos\theta$). It is therefore preferable to have the uniaxial optic axis where the retardation changes least in the geometric plane (i.e., $\phi = 0°$ in Figure 7.19). In the case when a biaxial QWP is used, no orientation preference is seen and its overall performance is an average between the two uniaxial orientation options due to increased leakage in the $\phi = 45°/135°$ planes. In a practical situation, there are other factors that can alter the preferential orientation, such as optical bandwidth, dispersion mismatch, interface reflection, and non-ideal PBS performance.

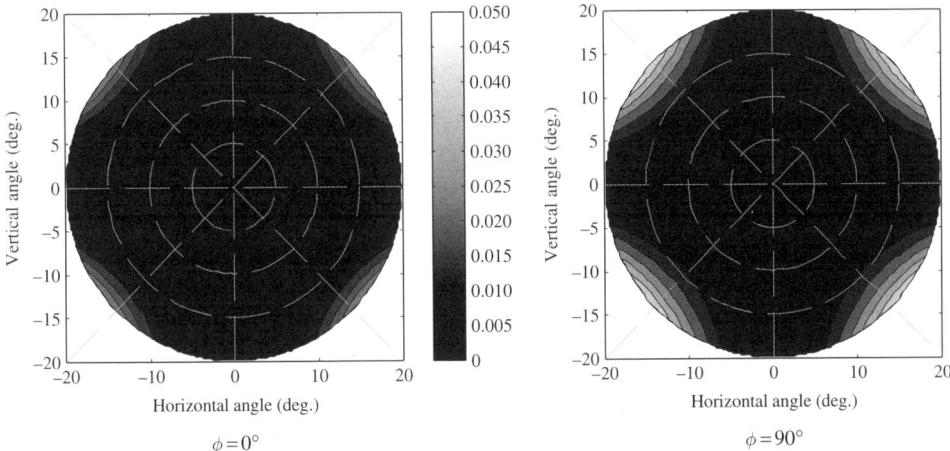

Figure 7.20 FOV of ideal PBS/mirror compensated by a uniaxial QWP with two orientations ($\lambda = 550$ nm, no interface reflections taken into account)

Assuming a perfectly coated $n = 1.85$ cube with 534 nm $< \lambda <$ 572 nm filtered uniform UHP $f_{/2.8}$ illumination and no surface reflections, the calculated photopic contrast values with a dispersionless 137.5 nm QWP at 0° is ~40 000:1. At 90° the contrast is ~10 000:1.

7.3.3.2 MacNeille PBS/Sheet Polarizer Combination [Robinson M. G., 2003]

In the case where a cube PBS is followed by a sheet polarizer (or vice versa), all rays are analyzed with essentially the same polarization axis, leading to leakage for off-axis skew rays from the pupil plot mismatch shown in Figure 7.21.

Desired compensation would transform between these pupil maps requiring on- and off-axis rays to be treated differently. A solution is therefore not achievable using a simple a-plate film retarder.

The optimum solution must comply with both the following mapping requirements:

1. All rays in the $\phi = 0°$ plane do not require correction.

2. The PBS pupil map has an odd (reflection) symmetry about the x-axis.

One solution is a uniaxial retarder with an out-of-plane tilt, i.e., an o-plate. To comply with the above mapping properties, such a plate would have the optic axis:

- parallel or perpendicular to the on-axis polarization direction
- project with odd symmetry about the x-axis for skew rays.

To exhibit these properties in a single retarder, the optic axis is of the form:

$$e = (\cos \alpha, 0, \sin \alpha) \tag{7.16}$$

where α is the o-plate tilt angle.

170 SYSTEM CONTRAST

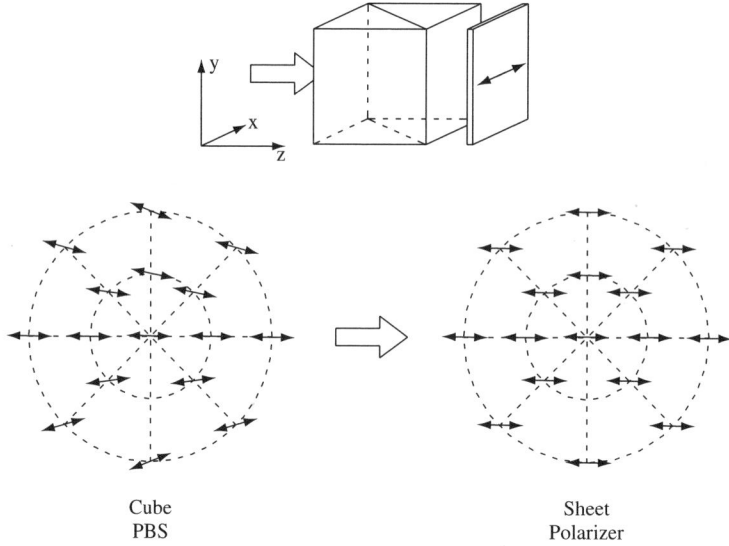

Figure 7.21 Polarization pupil mapping of a PBS and a sheet polarizer

To determine a suitable compensating transformation, the polarization states in the geometric plane of the PBS need to be reflected about an axis that is midway between the original and the desired polarization axes. A half-wave retardation will perform the desired reflection transformation assuming the geometrical projection of its optic axis for the off-axis rays has the correct orientation. To determine whether a suitable tilt angle exists that projects a bisecting optic axis, it is necessary to calculate the projected axes of an o-plate whose director is of the form given in equation (7.16).

The propagation vector of a ray in the $\phi = 90°$ geometric plane can be written:

$$k = (0, \sin\theta, \cos\theta) \tag{7.17}$$

The uniaxial optic axis as seen by the ray in the (xy) plane is then derived from the vector cross-product $(k \times e)/|k \times e|$, and the result projected onto the 2D (xy) plane to give $(\sin\theta \sin\alpha, \cos\theta \cos\alpha)$.

Finally, the angle that the optic axis projection forms with the y-axis is given by:

$$\rho = \cos^{-1}\left[\frac{(1,0) \cdot (\sin\theta \sin\alpha, \cos\theta \cos\alpha)}{|\sin\theta \sin\alpha, \cos\theta \cos\alpha|}\right] \tag{7.18}$$

To correctly reflect the polarization state exiting the PBS in this plane, this projection angle ρ should be $\beta/2$, where β is the geometrical polarization rotation angle defined by equation (4.14). For the small polar angles seen in projection systems, $\beta \sim \theta$, allowing the expression to be expanded and solved for α. Solutions for α as a function θ are shown in Figure 7.22. The solution assumes similar refractive indices for an o-plate and PBS. The tilt angle can be scaled according to Snell's law in situations where indices are mismatched.

As can be seen, there is a near independence of α on θ, which implies that a simple uniaxial material with its optic axis at $\sim 26.25°$ essentially compensates all rays if it imparts a

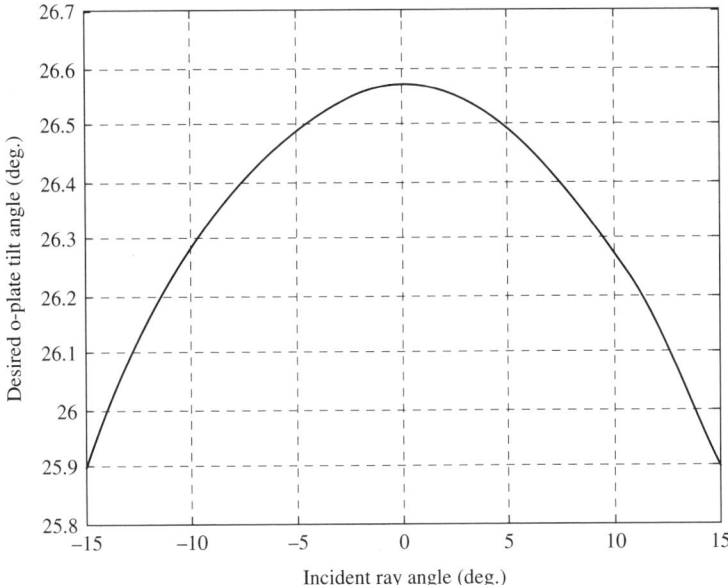

Figure 7.22 The o-plate tilt angle α that corrects polarization mismatch for rays with incident polar angle θ in the geometric plane of a cube PBS/sheet polarizer system

half-wave retardance. This condition is achieved for small polar angles typical of projection systems if the following relation holds:

$$\left(\frac{n_e n_o}{\sqrt{n_o^2 \cos^2 \alpha + n_e^2 \sin^2 \alpha}} - n_o \right) d = \frac{\lambda}{2} \tag{7.19}$$

The amplitude leakage of a ray passing through a perfect o-type $\phi = 0°$ oriented sheet polarizer followed by a perfect (i.e., $T_p = 100\%$, $T_s = 0\%$) MacNeille PBS can be calculated analytically. Assuming a propagation vector k, a PBS with coating normal n, and an o-type polarizer with optic axis e, the **p**-leakage component is:

$$p_r = \frac{k \times e}{|k \times e|} \cdot \frac{k \times (k \times n)}{|k \times (k \times n)|} \tag{7.20}$$

with k defined within a material of index equivalent to both polarizer and PBS. If the indices differ between PBS and sheet polarizer, a Snell's law transformation is necessary. The intensity FOV plot of this leakage is shown in Figure 7.23(a). The total leakage can be obtained by summing over a finite ray set making up the illumination cone. At $f_{/2.8}$, with a high-index ($n = 1.85$) cube PBS, the total wavelength-independent **p**-leakage is $\sim 0.25\%$. Simulation indicates that with a 35 µm thick, $\alpha = 21.3°$ quartz o-plate between the polarizer and PBS, the leakage reduces to $<0.0025\%$ at $\lambda = 550$ nm. Over a 100 nm band, the leakage remains less than 0.0075%. The associated leakage FOV plot is shown in Figure 7.23(b).

172   SYSTEM CONTRAST

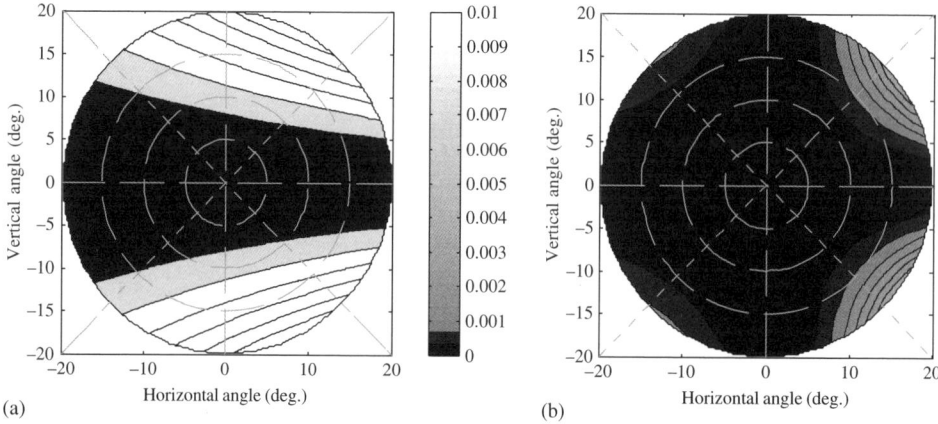

Figure 7.23 FOV plots for PBS/sheet polarizer system (a) without and (b) with o-plate compensation; $n = 1.5$ for PBS, o-plate, and o-type polarizer. Uniform o-plate with $\alpha = 26°$, $\Delta nd = 330$ nm. Both luminance FOV plots use 534–572 nm filtered UHP illumination

Similar compensation can be achieved with splayed birefringent structures such as liquid crystal polymers (LCPs) [Seiberle H., 2003]. From a practical, commercial standpoint, splayed LCP solutions are attractive, offering low cost, mass manufacturability, and robustness to the harsh environments seen in LCOS video projection systems. Average tilts of greater than 25° are possible with typical splayed director profiles, and layer thicknesses of up to 1 μm allow half-wave retarders to be manufactured. Using splayed structures there are two possible orientations, shown in Figure 7.24, which exhibit the correct compensating symmetry.

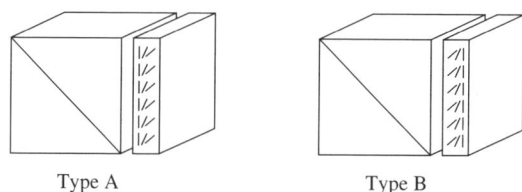

Figure 7.24 Two splayed LCP orientations for single PBS compensation

The performance of both splayed LCP orientations in Figure 7.23 have been simulated and experimentally verified using the set-up shown in Figure 7.25.

The accompanying simulation in this case uses 4×4 Berreman calculus, which includes scaling and offsets consistent with the experimental system angle-dependent transmission and finite contrast. Experimental and simulation results are shown in Figure 7.26. Both type "A" and "B" orientations of the same plate are presented. Both results demonstrate successful compensation, as an uncompensated ray at incidence angle θ has a leakage of $\sim (\sin\theta/n)^2$, or $\sim 1\%$ at 10°, which is $\sim 5 \times$ the compensated case. The LCP plate in this experiment is chosen to have a retardance that highlights the effect of orientation with respect to the PBS.

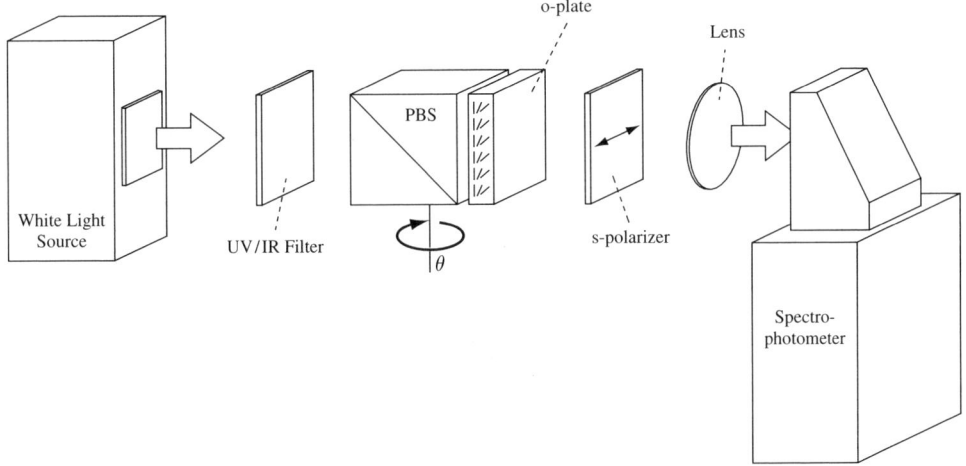

Figure 7.25 Experimental verification set-up for the LCP o-plate compensator

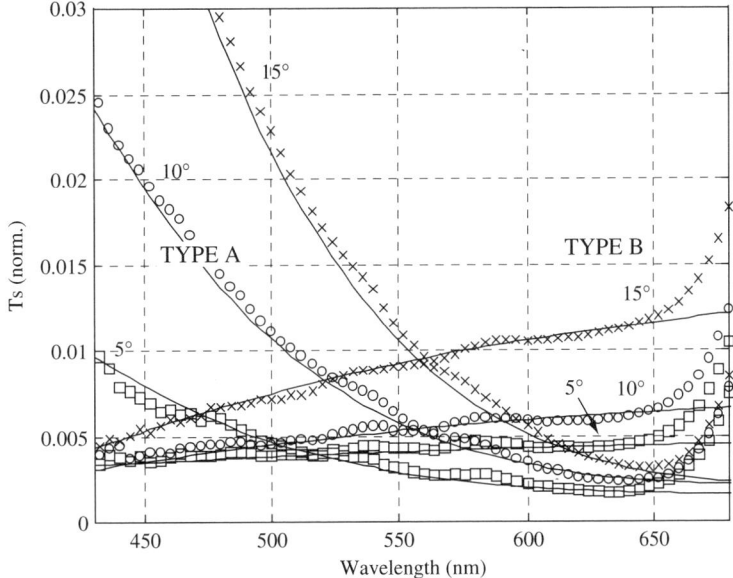

Figure 7.26 Experimental and simulation results for an LCP compensation scheme ($n_{PBS} = 1.65$, $n_o = 1.55$, $n_e = 1.68$, $\alpha_{air} = 0°$, $\alpha_{sub} = 50°$)

An actual LCP compensator optimized for visible wavelengths would be of type "A", but would have a lower retardance value.

Based on simulation, the performances of an optimized homogeneous uniaxial o-plate and an LCP version are similar, but the different material structures and dispersions lead to subtle differences. In general the uniaxial plate is more achromatic.

Placing an o-plate compensator after a cube PBS transforms the pupil map at the output to one equivalent to a sheet polarizer, thereby allowing control of polarization between successive PBS stages. The reflective PBS/o-plate/mirror system when unfolded is equivalent to two casacded stages, thereby allowing an o-plate to be used in place of a QWP with certain advantages. These include greater achromaticity and a greater angular alignment tolerance. These properties stem from the back-to-back half-wave reflection transformation that compensates for errors in retardance and angle. Figure 7.27 compares the average leakage as a function of wavelength at $f_{/2.5}$ for the PBS/QWP/mirror and PBS/o-plate/mirror situations and clearly shows the increased achromaticity of the o-plate option. The substantially increased achromaticity allows the o-plate to be considered in situations where more than one color band illuminates a single panel, as is the case for temporally modulated, full-color reflective projection systems (see Chapter 11). The QWP compensation scheme is, however, favored for narrowband operation due to a lower null value.

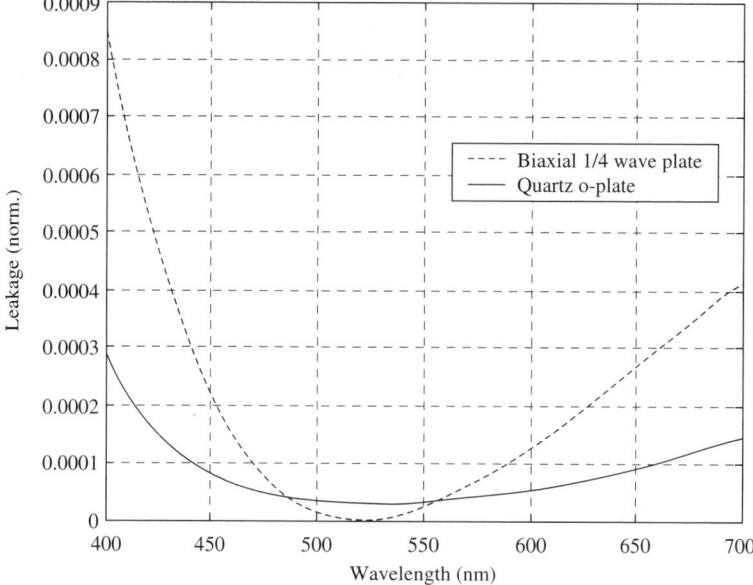

Figure 7.27 Total leakage from a cube PBS and mirror reflective system assuming $f_{/2.5}$ uniform illumination. Both conventional QWP and o-plate compensation performance are shown

The o-plate angular alignment tolerance is shown in Figure 7.28. Here total leakage for varying illumination cone angles is shown as a function of wavelength with 1° alignment error for both o-plate and QWP compensator.

While MacNeille PBS compensation is essential in reflective LC projection systems, alternative PBS components do not suffer to the same degree from geometrical effects. Both the multilayer birefringent cube and the wire grid plate PBSs behave essentially as o- and e-type sheet polarizers and analyzers, and effectively require no compensation (see Section 7.3.2).

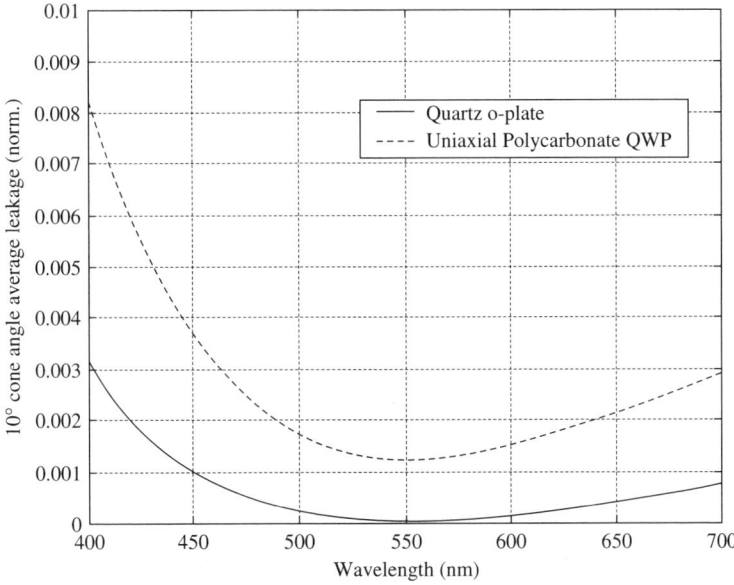

Figure 7.28 Cone average leakage for different compensation schemes with 1° misalignment

7.4 PBS/LCOS Compensation

The generic configurations for reflective LCOS modulating systems are shown in Figure 7.29 for the common MacNeille and wire grid PBSs. The compensation schemes depend on the specific LCOS mode, but can be conveniently split into VA-like (i.e., homeotropic aligned OFF-state) and more general TN modes.

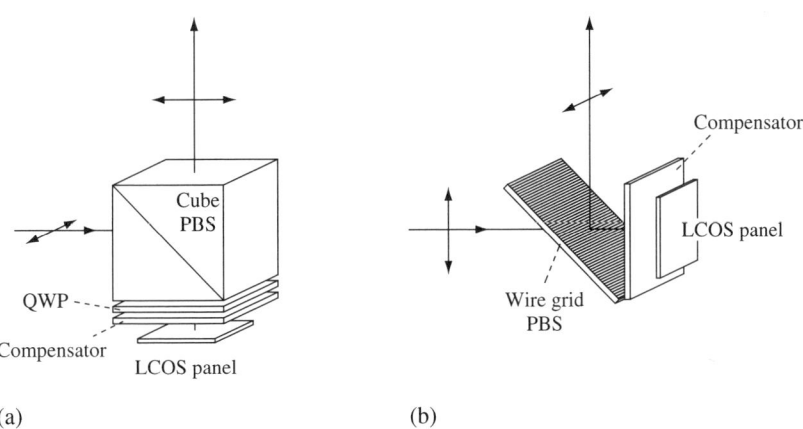

Figure 7.29 The two established reflective LCOS systems using: (a) MacNeille PBS, (b) wire grid plate (WGP) PBS

7.4.1 VA LCOS Mode

PBS/LCOS compensation schemes are more complicated than those for the PBS/mirror system, and depend greatly on the specific LCOS mode. The VA mode will be considered first, as this currently yields the highest contrast in cube PBS-based systems, and is also a good first approximation to most LCOS panels in their OFF-state. Conventional compensation of VA panels with a MacNeille cube PBS utilizes a single uniaxial QWP aligned parallel or perpendicular to the normal incidence polarization axes of the cube. This approach is taken since on-axis compensation is unnecessary. Simulating the perfect PBS FOV for each QWP orientation (see Figure 7.30), an obvious preference is seen. When compared to the

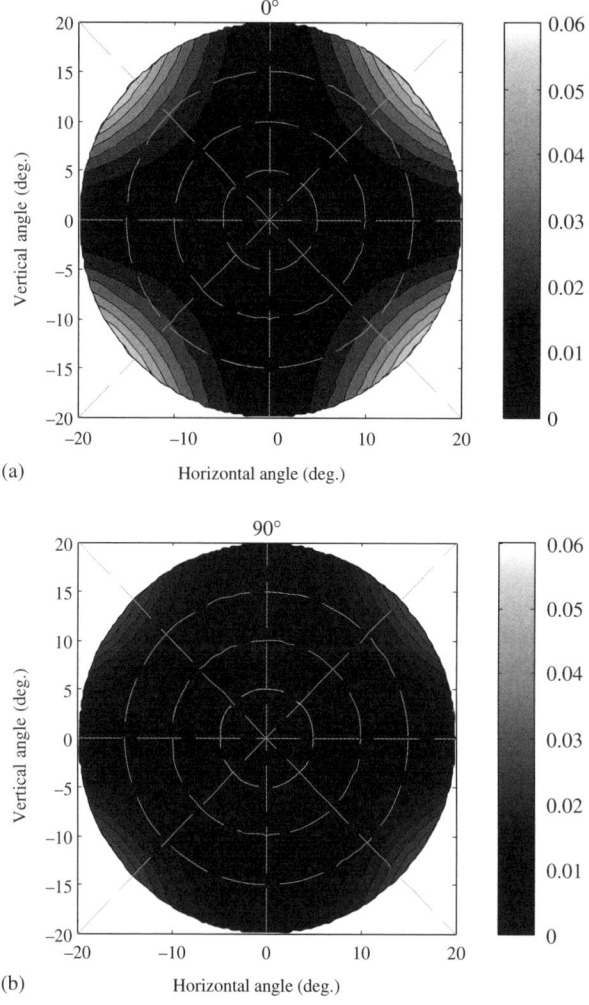

Figure 7.30 FOV of PBS/VA (LCOS) compensated by a uniaxial 137.5 nm QWP at $\lambda = 550$ nm, whose optic axes are parallel to the normal incidence (a) **p**- ($\phi = 0°$) and (b) **s**-axes ($\phi = 90°$) respectively

PBS/mirror case of Figure 7.19, the preferred orientation is switched. This preference can be explained in terms of maintaining the net compensating retardance as a function of polar angle in the geometric plane. A homeotropic LC layer tends to reduce in retardance as a function of polar angle, which is matched by the increase in retardance of a $\phi = 90°$ oriented uniaxial QWP.

Assuming a perfect $n = 1.85$ cube with 534 nm $< \lambda <$ 572 nm filtered uniform UHP $f_{/2.8}$ illumination and no surface reflections, the calculated photopic contrast values with a 250 nm VA LC and dispersion-less 137.5 nm QWP at 0° is ~2200:1, whereas at 90° the contrast is ~12 000:1.

Though certain panels approximate an ideal VA OFF-state alignment, in general there is some residual retardance as a result of finite pretilt at the LC boundaries. In the case where this residual retardance is small (<10 nm), compensation can be achieved by a slight rotation of the QWP skew ray compensator. Note that this represents a valid on-axis solution. The simulated photopic contrast of a green (534 nm $< \lambda <$ 572 nm) UHP $f_{/2.8}$ illuminated system and the subsequent optimum QWP orientation angle is shown in Figure 7.31 as a function of the residual retardance of a 275 nm LCOS panel.

In practical situations, where PBSs reflect a finite amount of **p**-polarized light, the phase of the reflected light affects the QWP compensation. This can manifest itself as a reversal of the optimum orientation of the QWP with respect to the PBS axes (i.e., 0° yields the best contrast). Since this is very dependent on specific PBS coatings it is necessary to determine the optimum orientation experimentally.

Replacing the QWP with an o-plate as above can also be considered for panels with ideal LC VA mode alignment. Unfortunately, unlike the uniaxial QWP, the out-of-plane retardance of the LC in the panel OFF-state is not compensated and leads to low contrast. However,

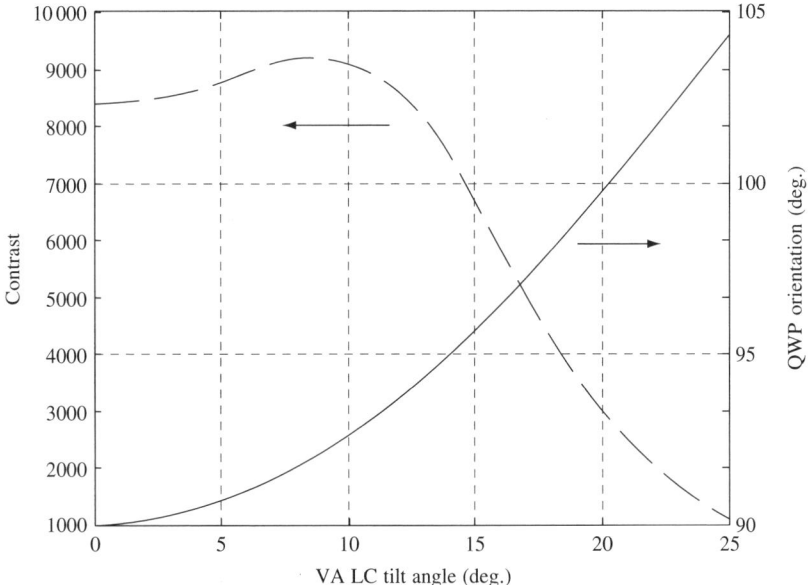

Figure 7.31 System contrast of a VA LCOS panel with an ideal MacNeille PBS and associated optimum QWP orientation angle as a function of 275 nm VA pretilt

all merits of the o-plate mentioned in Section 7.3.3.2 can be restored if a further negative c-plate or crossed a-plate compensator is used to offset the LC out-of-plane birefringence.

The self-compensating nature of a WGP PBS requires LC-only compensation. For VA panels this requires primarily an out-of-plane negative c-plate compensator with matched retardance, or near-equivalent crossed a-plates (see Section 7.3.1). For a panel with finite pretilt, an additional small a-plate retarder is required, which can be either incorporated into an out-of-plane compensator in the form of a biaxial c-plate or created through mismatched crossed a-plates. More recently, on-axis compensation by rotating the wire grid PBS about its surface normal has been demonstrated utilizing its subtle birefringent properties (Section 4.5.3) [Kurtz A. F., 2004]. The same group has also developed more complex compensation schemes using a combination of a- and c-plates with VA LCOS panels, and clean-up wire grid polarizers [Mi X. D., 2003].

Using perfect crossed e- and o-type polarizers to mimic the empirical wire grid telecentric modulating system model, the photopic contrast of an uncompensated 250 nm VA LCOS panel is $\sim$3000:1 assuming 534 nm $< \lambda <$ 572 nm filtered uniform UHP $f_{/2.5}$ illumination and no surface reflections. With either crossed a-plate or c-plate compensation, contrast in excess of 50 000:1 is predicted, but in practice is more typically limited to $\sim$20 000:1 through scattering.

7.4.2 General LCOS Mode [Chen J., 2004]

In the case where in-plane retardance is large (e.g., non-90° TN LC modes), a more general compensation scheme is needed that addresses separately out-of-plane and residual retardance contributions. This approach can be applied to WGP PBS or conventional MacNeille-based systems, as shown in Figure 7.29.

Considering first a WGP PBS system, the desired head-on compensation can be determined as described in Section 7.2.1. A small uniaxial a-plate compensating only the head-on retardance can deliver contrast up to 3000:1, which can sometimes be sufficient. Simulated contrast of a typical $f_{/2.5}$ telecentric wire grid system is shown in Figure 7.32 for the equivalent 45° TN mode used to derive the solution set of Figure 7.2. Figure 7.33 shows the simulated system contrast versus orientation for two specific a-plate retardation values (17 nm, 26.2 nm). Clearly the "C" curve of Figure 7.2 is preserved even after taking into account FOV effects.

From Figure 7.32 it can be seen that larger retardance values at angles closer to the LC front-surface optic axis (180°) are preferred. However, these larger retardance solutions are intolerant of high incidence angles. For practical retarder components that can exhibit a non-uniformity of optic axis over the part, especially at elevated temperatures, it is better to choose a lower retardation solution. Below the lower retardance limit of $\sim$17 nm, there is no angular orientation that yields high contrast. For this reason, a suitable practical compensation solution for the case in point would be an $\sim$18–19 nm retarder at $\sim$145° (taking into account a maximum component alignment and retardance tolerance of $\pm$5° and $\pm$1 nm respectively).

Further improvement to this basic compensation scheme is to include some negative out-of-plane retardance to compensate for the majority of the LCs OFF-state birefringence. This can be accomplished efficiently by combining an in-plane residual compensator (a-plate) with a negative c-plate in the form of a single biaxial film. Choosing the tolerant 17 nm (at $\sim$135°) in-plane retardation solution, the biaxiality of compensator film can be varied and

PBS/LCOS COMPENSATION 179

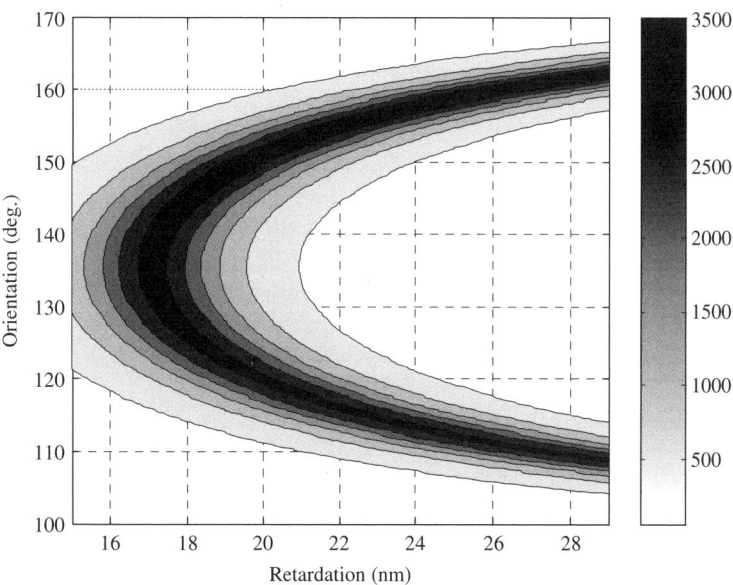

Figure 7.32 Contrast versus orientation and retardation for a 45° TN LCOS mode in a WGP PBS modulating system illuminated at $f_{/2.5}$

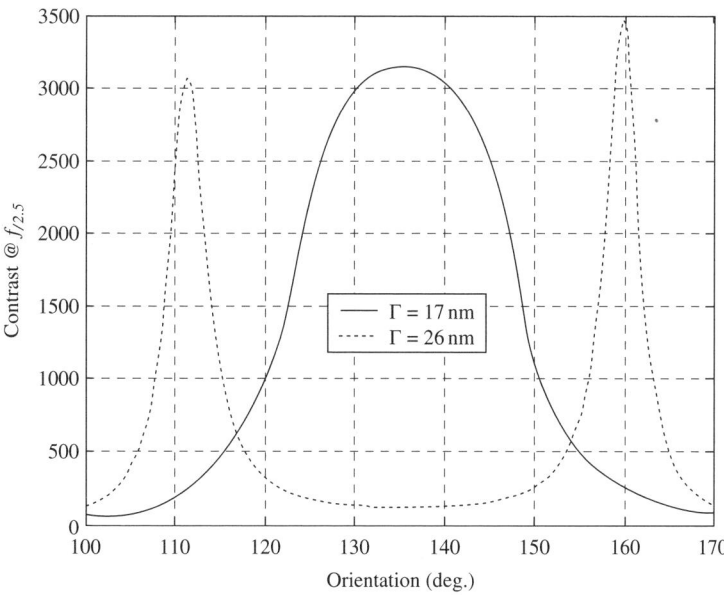

Figure 7.33 Simulated system contrast of a WGP PBS modulating system as a function of uniaxial a-plate orientation

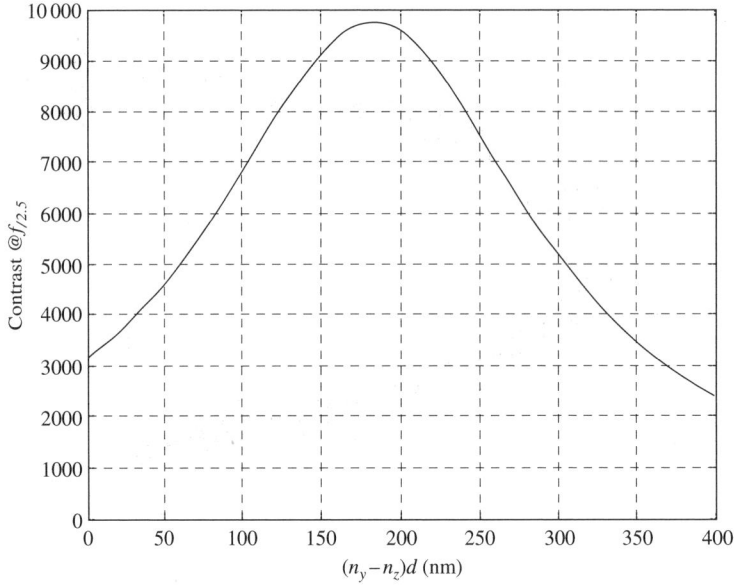

Figure 7.34 Average contrast of a 45° LCOS panel at $f_{/2.5}$ versus Nz retardation ($= (n_y - n_z)d$) for a WGP PBS when a biaxial compensation film is used; $(n_x - n_y)d = 17$ nm

the contrast calculated for $f_{/2.5}$, 534–572 nm filtered UHP illumination. The results are shown in Figure 7.34, which clearly shows a maximum contrast when the out-of-plane retardation, $(n_x - n_z)d$, is approximately ~ -200 nm. This result is perhaps not surprising as this value matches the $+250$ nm retardance of the LC in the panel OFF-state when twice the in-plane retardance is subtracted. This is justified since the measured in-plane retardance corresponds to the proportion of LC that is not vertically oriented and is not aligned with the input polarization. A second in-plane contribution of equal magnitude is aligned with the input polarization.

In the case of MacNeille PBS modulating systems, compensation of the TN LCOS mode has the additional complexity of having to compensate for the PBS geometry. Based on the previous compensation schemes, a two-film approach should perform well, whereby a small uniaxial a-plate compensator corrects for head-on contrast and a uniaxial QWP corrects for both PBS geometry and the LC out-of-plane retardance [Chen J., 2004a]. Although in general both retarders can be oriented independently, aligning the QWP to the PBS's s-polarization axis yields more achromatic solutions. Figure 7.35 shows the simulated contrast for an $f_{/2.8}$ illuminated MacNeille PBS modulating system. An ideal $n = 1.5$ cube PBS is assumed (i.e., $R_s = 0\%$, $T_p = 100\%$) having no surface reflections.

Where there are multiple solutions for compensating WGP systems, there is a single optimum solution for MacNeille based systems. This is further highlighted in Figure 7.36, which shows simulated system contrast versus orientation of two arbitrary compensation values (17 nm, 26 nm). Once again the lower retardation solution has better tolerance with respect to orientation.

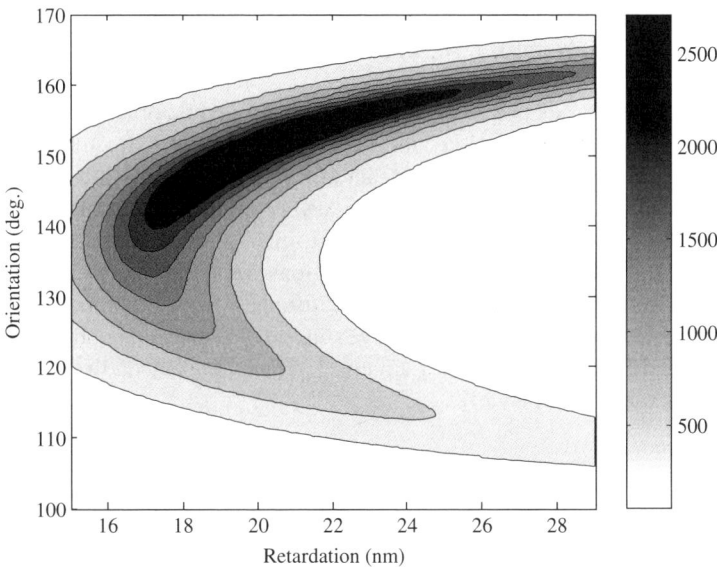

Figure 7.35 Contrast versus a-plate orientation and retardation for 45° TN LCOS mode with a perfect MacNeille PBS and 90° oriented QWP. Uses 534–572 nm filtered UHP illumination at $f_{/2.8}$

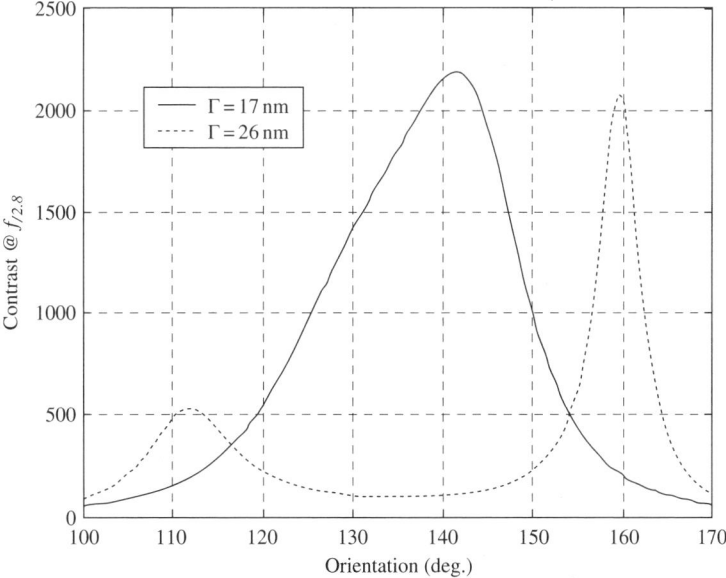

Figure 7.36 Simulated 45° TN LCOS system contrast versus rotation angle of a-plate (17 nm and 26 nm) with a perfect MacNeille PBS and 90° oriented QWP. Uses 534–572 nm filtered UHP illumination at $f_{/2.8}$

182 SYSTEM CONTRAST

7.4.3 Influence of the Reflections from Interfaces on System Contrast
[Chen J., 2004a; Shimizu J. A., 2004]

Many components in a projection system are free standing while others are optically bonded together. In each case reflections can occur due to index mismatch at surfaces. Unlike transmissive systems where the reflections from interfaces result mainly in a loss of brightness, these reflections in an LCOS system can have severe impacts on contrast (sequential and ANSI). The impact from the reflection of light from any one interface strongly depends on its location, which leads to different specifications for AR coatings and index-matched bonding of the various optical components. The strongest impact comes from components between the PBS and panels, such as the QWPs used to compensate cube PBSs. Assuming the most general PBS panel system of Figure 7.37, the impact of the various reflection surfaces can be examined separately.

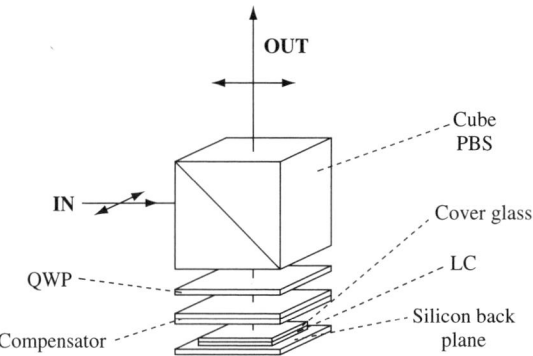

Figure 7.37 Schematic showing a general PBS/panel modulating system

(a) QWP
As already described in Section 7.3.3, a cube PBS/panel system without QWP geometrical compensation is limited to ~50:1 contrast. With the QWP, the same system can achieve ~10 000:1. This component therefore forms a division since reflections from the interfaces before the QWP are not compensated for PBS geometry, while those after it are. The total leakage from the reflections of a QWP can be approximated by:

$$\frac{1}{50} \times R_f + \frac{1}{10\,000} \times R_b \qquad (7.21)$$

where R_f and R_b are the reflection coefficients of the front and back of a QWP respectively. Clearly the AR specifications for the surfaces before the QWP, including AR coating for PBS, are much more important than those after. This can lead to preferential reflection specification of parts. For example, commercially available QWPs typically consist of many layers, such as a polycarbonate (PC) QWP retarder layer, ($n = 1.58$), a high-durability pressure-sensitive adhesive (PSA) layer ($n < 1.48$), and a glass substrate ($n \sim 1.5$). The index mismatch between the PSA and the PC results in a large reflection making it preferential to have the PC facing the PBS (see Figure 7.38).

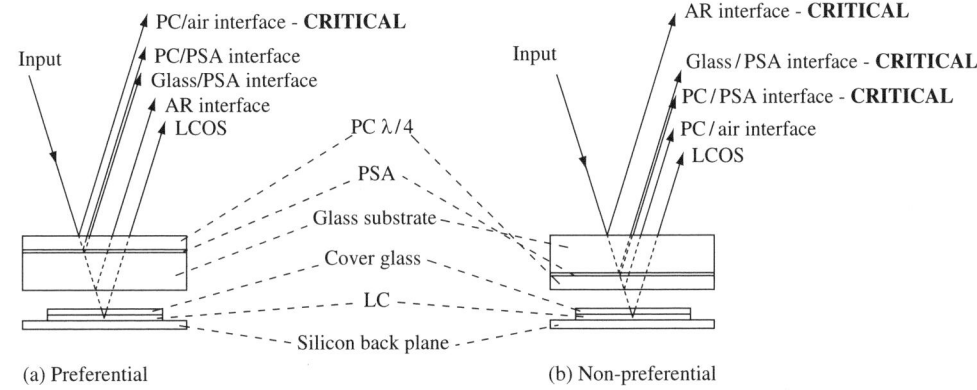

Figure 7.38 Preferential arrangement of a laminated PC QWP

(b) Compensator

For certain LC modes such as the 45° twisted mode, a fractional wave plate compensator is necessary to achieve acceptable head-on contrast (see Section 7.2.1). Usually, a compensator is sandwiched between two glass substrates. Depending on location, the reflections from the various interfaces can be grouped into three categories as shown in Figure 7.39. Since a compensator is placed behind the QWP, the contrast leakage from reflection group 1 is equal to $1/10\,000 \times R_f$ for a cube PBS with QWP, or $1/20\,000 \times R_f$ for a WGP PBS.

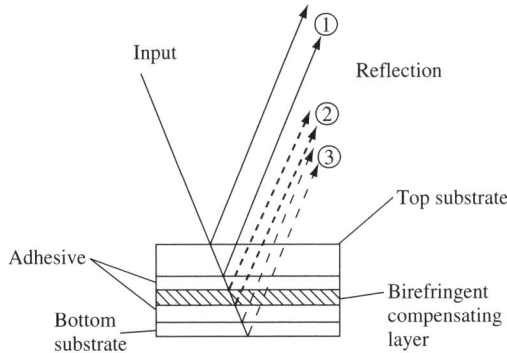

Figure 7.39 Cross-sectional diagram of a typical LCOS compensator, showing incident and reflected beams

The function of a compensator is to offset residual birefringence, and as such the leakage of the compensator under crossed polarizers is equal to that of the uncompensated LCOS panel. The contrast leakage from group 3 is therefore equal to the reflectance divided by the head-on contrast of the uncompensated LCOS panel, which is about 40:1 for either of the 45° and 63.7° TN LC modes when 5 V is applied, i.e, $1/40 \times R_b$.

For the second group emanating from the interfaces between the birefringent material and the bounding media, the situation is more complicated as reflections at these interfaces act to mix polarization. The refractive indices of a compensator are denoted by n_e and n_o, with a refractive index n_g for the bounding glass media. Assuming the optic axis is at an angle θ to the incident

184   SYSTEM CONTRAST

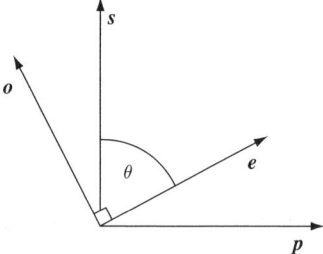

Figure 7.40 Vector diagram illustrating compensator axes orientations. The input light is s-polarized and the compensator axis is oriented at θ

polarization, as shown in Figure 7.40, polarization mixing occurs from the difference between the amplitude reflection of extraordinary and ordinary waves. For the front surface only, this results in a light leakage equal to [Yeh P., 1999, p. 313]:

$$I_{leakage} = \frac{\Delta n^2}{16\bar{n}^2} \sin^2(2\theta) \tag{7.22}$$

where $\bar{n}$ is the average index $(n_e + n_o)/2$.

Reflection from the back side can be treated in a similar way, but has a phase difference with respect to the front-surface reflection resulting in interference. Depending on material parameters, the total reflection can be enhanced or diminished. In the worst case, the total leakage is four times that of equation (7.22). Assuming $n_e = 1.7$, $n_o = 1.5$, and $\theta = 45°$, the contrast limit due to reflection alone from a single compensator/bounding material interface would be about 1000:1, which is unacceptable. Clearly, a compensator with low Δn would reduce this effect. However, in the case of highly birefringent materials such as LCPs, acceptable contrast can only be achieved by choosing material parameters that result in destructive interference.

(c) LCOS Panel

Contrast loss due to the isotropic reflections from the front-cover glass of the LCOS panel and all other interfaces up to the one bounding the LC can be treated the same as that from the back of a compensator. Depending on the LCOS mode, it is equal to the uncompensated $1/40 \times R_f$ leakage for either the 45° or 63.7° TN LCOS mode as for the group 3 compensator reflections of Figure 7.39. In the absence of a compensator, the contrast loss due to these reflections would be $1/10000 \times R_f$ in the cube PBS case, and $1/20000 \times R_f$ with a wire grid PBS. Reflection from the anisotropic interface between the bounding layer and the LC can be treated the same as a reflection of a birefringent retarder interface (i.e., group 2 in Figure 7.39). From equation (7.22), this strongly suggests that the input LC director should be aligned to the incoming polarization direction to achieve high contrast, even if the brightness must be sacrificed to meet this requirement [Shimizu J. A., 2004].

Table 7.1 summarizes the leakages due to isotropic reflections only. Allowing the sum of the reflections to create a background contrast limit of 2000:1, and allocating the same reflection budget to all five critical interfaces, the reflection at these surfaces must be <0.45%. This is a demanding though achievable reflection requirement and must be considered seriously in the design of projection systems.

PBS/LCOS COMPENSATION

Table 7.1 Fraction of off-state projected light emanating from isotropic component surface reflections

Surface	Fraction transmitted
PBS back	0.02
QWP front	0.02
QWP back	0.0001
Retarder front	0.0001
Retarder back	0.025
Cover-glass front	0.025
Index-matched ITO	0.025

To minimize the leakage from the mixed polarization reflections present with high-birefringence compensators, the orientation θ, retardance Γ ($= \Delta n d$), and birefringence Δn should be chosen to destructively interfere with front and back reflections. Considering again the 45° TN LCOS mode with a wire grid polarizer PBS by way of example, we can first obtain the orientation θ and retardation Γ set that follow the high-contrast "C" locus of Figure 7.32, which assumes no reflections. Second, the contrast contribution from all reflections as a function of the compensator retardance Γ, and birefringence Δn, can be calculated. Shown in Figure 7.41, the obvious stripes indicating good contrast are expected from interference effects.

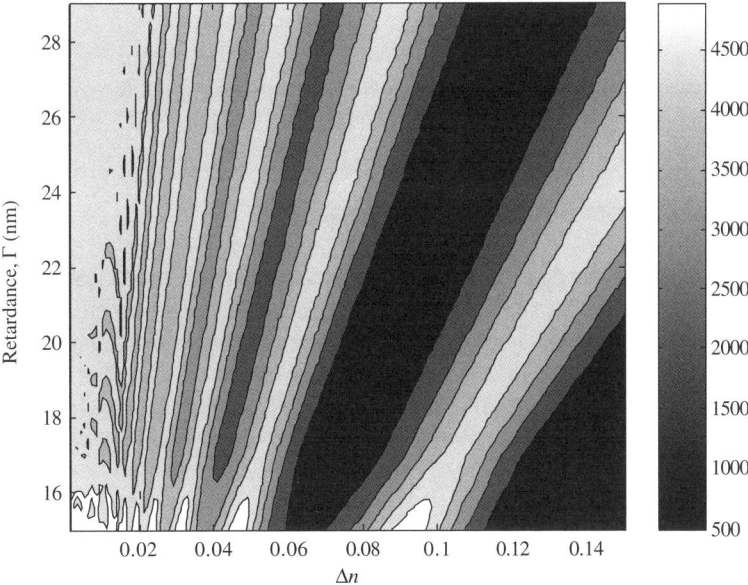

Figure 7.41 Contrast limit of a 45° TN LCOS mode with wire grid PBS due to the interface reflections assuming a finite Δn. All isotropic surfaces are assumed to have 0.25% reflection. The average Δn of the compensator is 1.621

186 SYSTEM CONTRAST

The best combination of the three parameters θ, Γ, and Δn can be determined by plotting the two contributions to the contrast as a function of retarder value, whose orientations are consistent with the maximum contrast "C" locus of Figure 7.32. The net reciprocal contrast is then the sum of the individual reciprocal contrast terms in the usual way. Assuming a material has a fixed birefringence $\Delta n = 0.1$, the total contrast can be plotted as shown in Figure 7.42, which yields the optimum compensation with retardance $\Gamma = 17$ nm (oriented at $\sim 142°$).

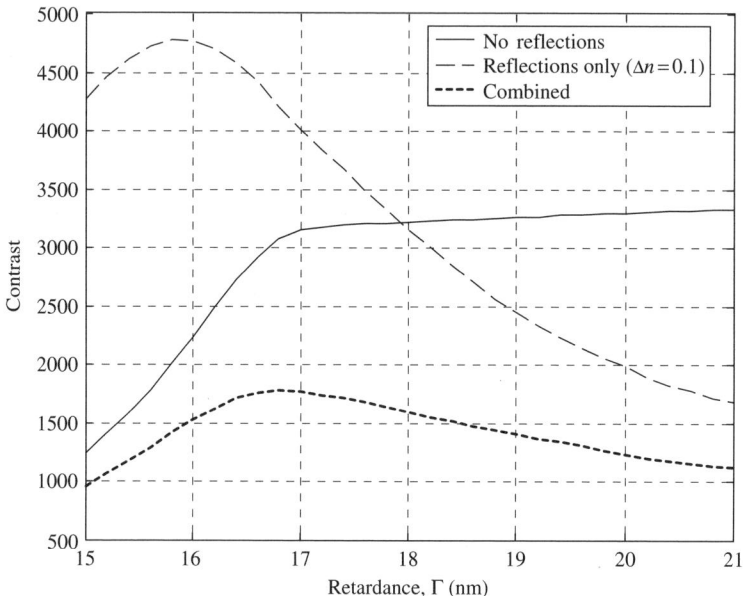

Figure 7.42 Final system contrast versus compensator retardation assuming $\Delta n = 0.1$. Optimum value for retardation is 17 nm at 142° determined from Figure 7.32

7.5 ANSI Contrast Enhancement

ANSI contrast described in Section 2.2.6 is a key system metric and is particularly difficult to control in reflective LC technologies, where reflections can light up otherwise dark areas of the display. Reducing component count and having good index matching in the projection path of a system is important in achieving high ANSI contrast. For this reason, reflective wire grid polarizers cannot in general be used as exit clean-up polarizers.

Unwanted back reflection from interfaces within a multi-element projection lens can be fully or partially eliminated by an exit 45° oriented QWP since these reflections occur after the final analyzer of the system [Robinson M. G., 2004]. It works in the same way as an optical isolator and is situated directly after the analyzer. The working principle is shown in Figure 7.43. Linear polarized light incident on the QWP is transformed to circular polarization, which, when reflected by subsequent components (such as a projection lens element), passes back through the QWP and is converted to the orthogonal linear polarization. The analyzer then absorbs or redirects this light preventing its illuminating dark panel regions and reducing ANSI contrast.

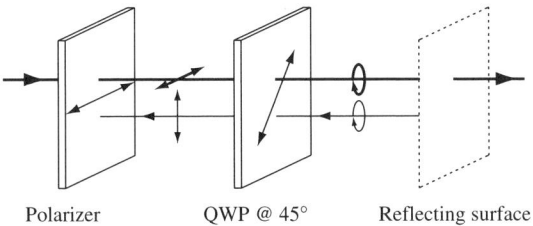

Figure 7.43 Working principle of isolating QWP for ANSI contrast

If a PBS is used as the exit analyzer (such as in the CQ3 configuration described in Chapter 10), the isolating QWP can work with a retarder stack filter to enhance the ANSI contrast. In this case, the reflected green light is dumped to an unused port of the PBS adjacent the blue panel as shown in Figure 7.44.

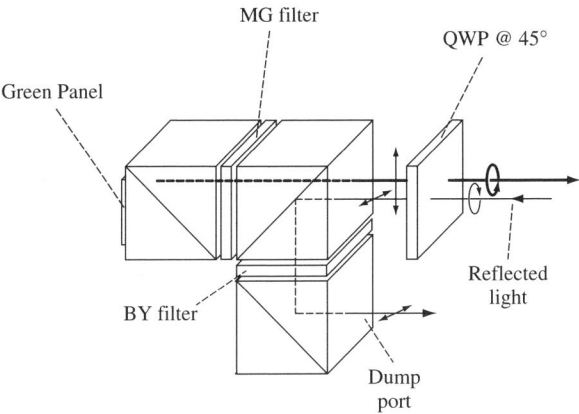

Figure 7.44 Working principle of an isolating QWP compensator for ANSI contrast enhancement in systems with a PBS analyzer

7.6 Skew Ray Compensated Retarder Stack Filters [Sharp G. D., 2002; Chen J., 2004b]

Retarder stack filters (RSFs) introduced in Chapter 6 are generally designed for normal incidence operation between crossed or parallel polarizers. However, in projection systems RSFs are used in combination with cube MacNeille PBSs to separate and/or recombine color bands. As introduced in Section 4.5, skew rays experience rotated polarization axes with MacNeille PBSs, which in general cause an increase in filter leakage and system performance degradation. Two common RSF PBSs system configurations, one having parallel and the other having orthogonal dichroic coated planes, are shown in Figure 7.45.

The ideal RSF compensates for the relationship between ray direction and (effective) input/output polarizer orientation. To do this the filters must have rotation invariance in the first case (i.e., its spectrum is independent of filter orientation), and reflection invariance in the second (i.e., similar spectra are obtained when polarizer and analyzer are rotated with

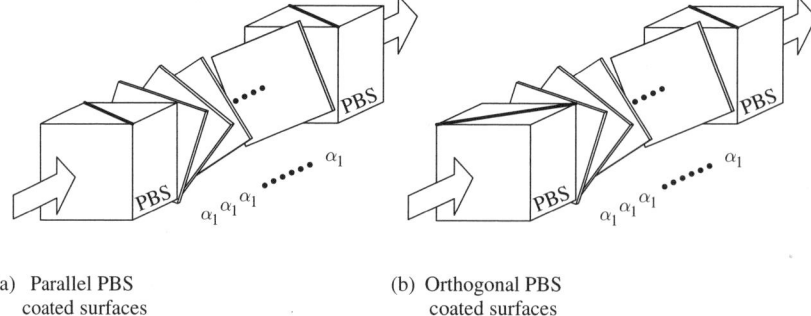

(a) Parallel PBS coated surfaces

(b) Orthogonal PBS coated surfaces

Figure 7.45 Two common RSF PBS configurations used in projection systems

opposite angles). The two cases are illustrated in Figure 7.46, which also highlights the equivalence between a reflection-invariant filter design and one that is rotation invariant with an integrated 0° oriented HWP. By carefully designing the filters these properties can be incorporated as the following examples show.

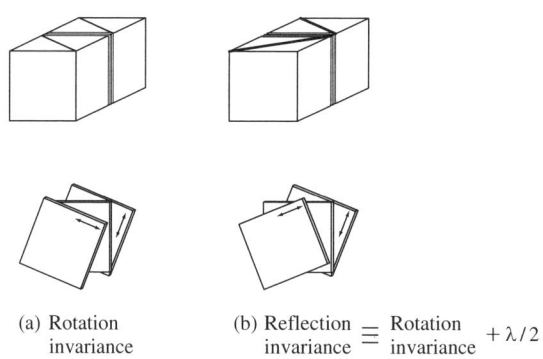

(a) Rotation invariance

(b) Reflection invariance $\equiv$ Rotation invariance $+ \lambda/2$

Figure 7.46 RSF desired property for PBS compensation

Example 1 Conventional Design

Figure 7.47 shows crossed polarizer spectra of a conventional red/blue (*RB*) RSF, where the solid and dashed lines correspond to 0° and 45° orientation of the optic axis with respect to the polarizers. Sandwiched between parallel-aligned $n = 1.5$ cube PBSs the filter performance is poor, with 2% leakage at $f_{/2.5}$ as shown in Figure 7.48. The 2% leakage represents a 20-fold increase over the original 0.1% on-axis design.

Example 2 Rotation-invariant Design

Figure 7.49 shows crossed polarizer spectra of a rotation-invariant *RB* RSF, where again the solid and dashed lines correspond to 0° and 45° orientation of its optic axis with respect to the polarizers. Their similarity is due to rotation invariance. In this case, when sandwiched between parallel-aligned $n = 1.5$ cube PBSs, the filter performance is good, with 0.2% leakage at $f_{/2.5}$ (see Figure 7.50). Adding a HWP to this design makes it reflection invariant over a band determined by the HWP, which results in comparable 0.2% leakage in the orthogonal PBS case.

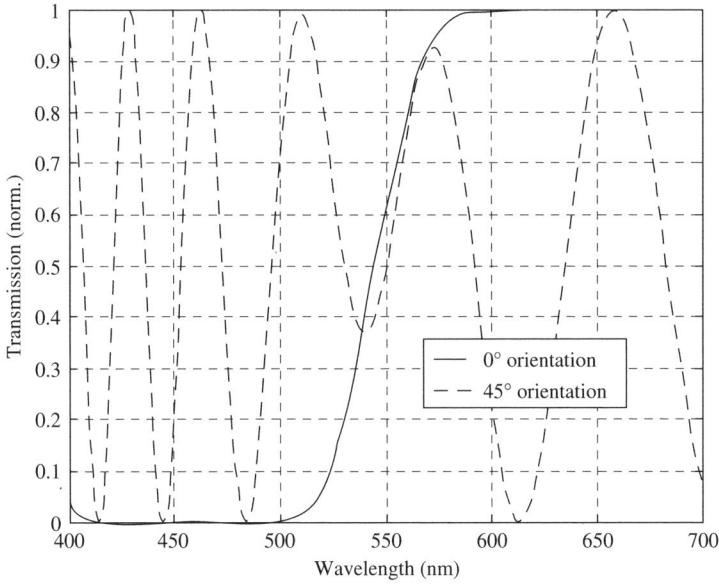

Figure 7.47 Spectra of a conventional *RB* RSF under crossed polarizers

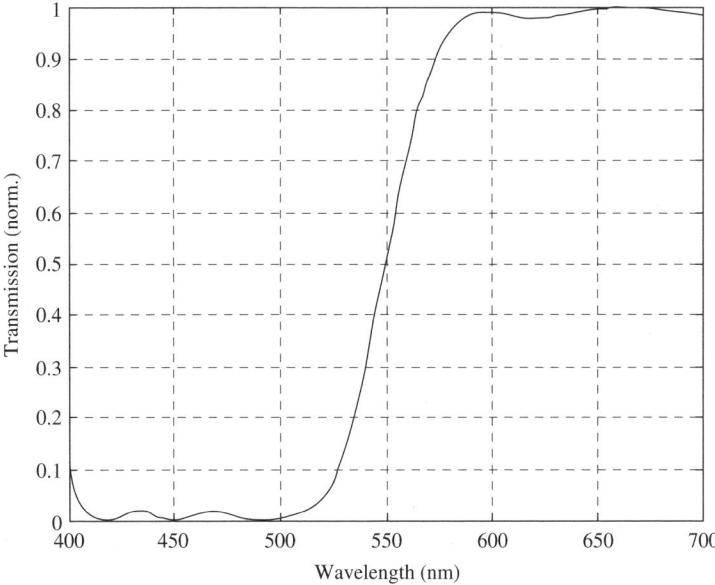

Figure 7.48 $f_{/2.5}$ cone average spectrum of the *RB* filter of Figure 7.47 sandwiched between two parallel PBSs

190 SYSTEM CONTRAST

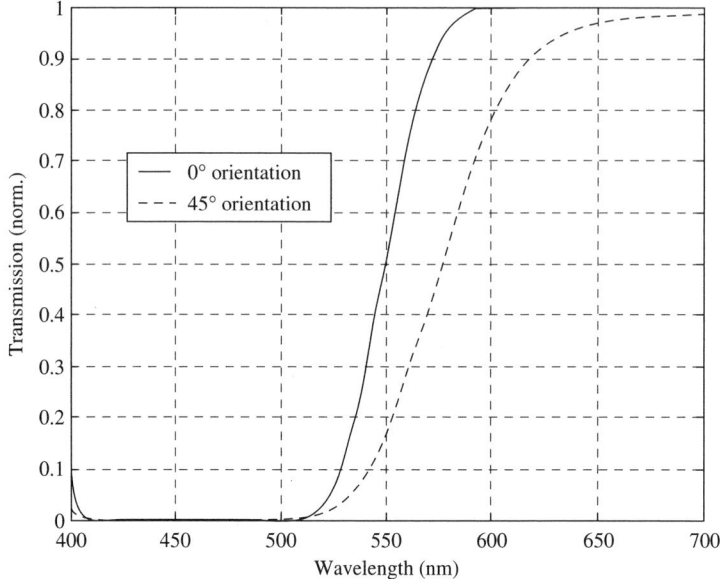

Figure 7.49 Spectra of a rotation-invariant *RB* RSF under crossed polarizers

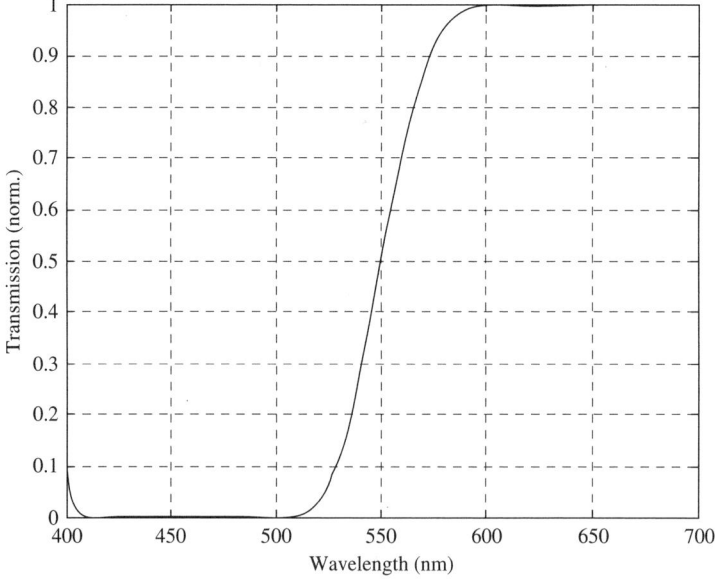

Figure 7.50 $f_{/2.5}$ cone average spectrum of the rotation-invariant *RB* filter of Figure 7.49 sandwiched between two parallel PBSs

ALTERNATIVE PROJECTION SYSTEMS

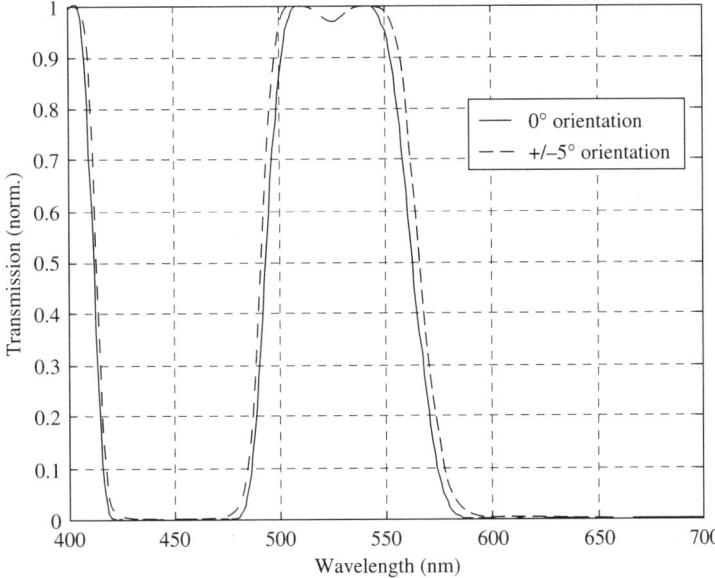

Figure 7.51 Spectra of an MG filter with reflection invariance in the red

Example 3 Reflection-invariant Design

An advantage of using a reflection-invariant RSF is that no HWP is required for compensated operation between orthogonal PBSs. Figure 7.51 shows the spectra of a reflection-invariant magenta/green (*MG*) filter (used in the CQ3 architecture of Chapter 10) between parallel polarizers (0°, 0°) (dashed line) and for polarizers rotated by ±5° (solid line). Clearly, its critical spectrum in the red has the required reflective property desired for skew ray compensation and low system leakage.

7.7 Alternative Projection Systems

The chapter so far has been concerned solely with compensation of telecentric projection systems, as these represent the standard approach to LC video projection. It is, however, not exclusive since it is possible to build systems that can be off axis and/or off telecentric, assuming angle-tolerant polarization components are used. Wire grid PBSs and sheet polarizers are two common elements that allow these alternative approaches. The former lend themselves to off-telecentric systems, while the latter are used off-axis as they require physical separation of the input and output beams.

7.7.1 Off-telecentric Wire Grid PBS System

A typical off-telecentric system is shown in Figure 7.52. Characteristic of this approach is the field lens at the panel. The distance between this lens and the aperture stop is its focal length,

192 SYSTEM CONTRAST

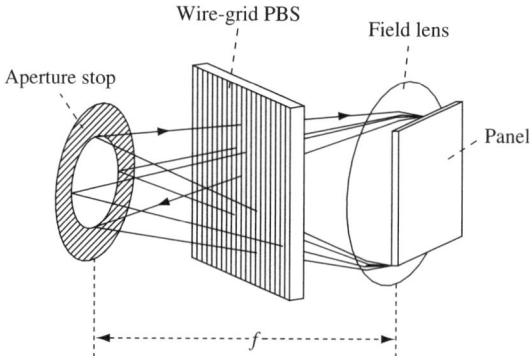

Figure 7.52 Schematic of a typical off-telecentric wire grid modulating system

ensuring telecentricity at the panel. In cases where the aperture stop is small, the system when unfolded has two wire grid PBSs with planes parallel, unlike the perpendicular unfolded telecentric case. The effect of this is to reduce contrast significantly since the wire grid self-compensation seen in conventional telecentric systems no longer applies. Essentially, the c-plate contribution is additive. This is exacerbated at extreme panel positions since the performance is no longer an average of rays centered about the optical axis. In the case of an uncompensated off-telecentric system, the contrast at the center can exceed 10 000:1, whereas at the extreme positions on the panel where the principal rays propagate with polar coordinates (15°, 90°/270°) through the wire grid PBS, the contrast can be <100:1. To restore compensation, the polarization states of reflected light from the panel can be transformed with a 0° or 90° oriented QWP (i.e., an HWP in reflection) into those seen in the telecentric case. In practice, inserting a simple uniaxial QWP in an LED-illuminated off-telecentric system ($f \sim 70$ mm) increased the contrast at the field edge by a factor of four, from $\sim$100:1 to $\sim$400:1.

7.7.2 Off-axis System [Bone M. F., 1998]

An off-axis system is one whose illuminating and reflecting light is incident on the panel at an angle (see Figure 7.53).

By illuminating the panel at an angle, the input and output beams become physically separated and independent polarizers and analyzers can be used. From a polarization standpoint, the polarization pupil maps of the polarizer and analyzer are also offset in angle as shown in Figure 7.54. If **o**-type polarizers and analyzers are used there is a mismatch in the effective absorption axes for skew rays that results in leakage and a loss in contrast. This is greater than the leakage calculated for two crossed **o**-type sheet polarizers in Section 7.3.2 as the beams are offset from normal incidence. The limiting contrast can be calculated from the system parameters, and is typically $\sim$2500:1 [Bone M., 2001]. Compensation is clearly necessary to meet current standards. By using the sheet polarizer compensation techniques of Section 7.13 (pairing an **o**- with an **e**-type polarizer; 0° oriented HWP etc.), system contrasts should exceed the desired >10 000:1 range.

ALTERNATIVE PROJECTION SYSTEMS

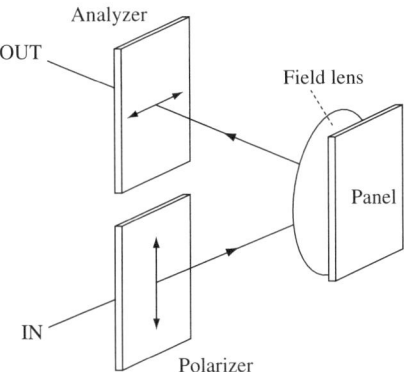

Figure 7.53 Off-axis system schematic

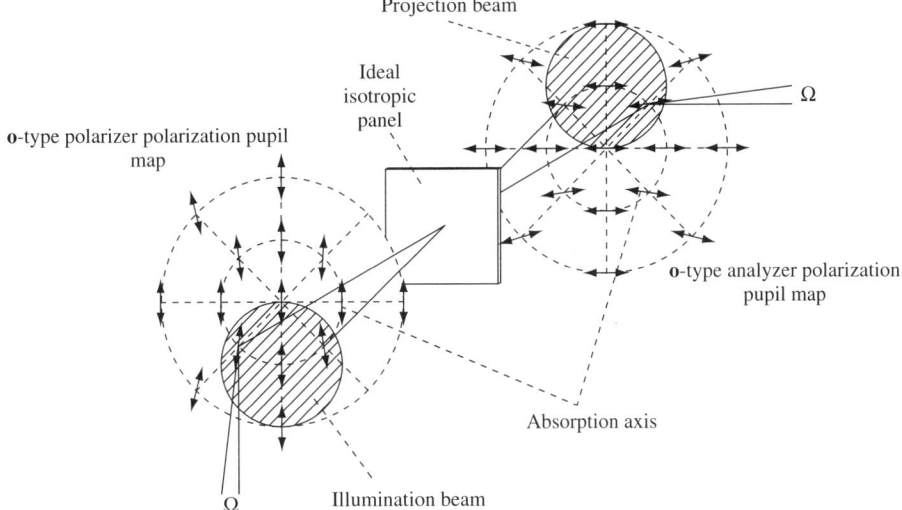

Figure 7.54 Ray mapping in an off-axis system between polarization pupil maps for **o**-type polarizer and analyzer. The values of the effective polarization axes rotation Ω for skew rays are opposite, resulting in contrast loss

Once the sheet polarizer and analyzer are geometrically compensated, the system contrast then depends on panel compensation. Due to the separation of the input and output beams, compensators can be placed in three locations. Placing a compensator adjacent the panel allows the compensation schemes covered in Section 7.4 to be used, which effectively cancels out the on- and off-axis panel retardation. Placing a compensator in the output projected beam is not attractive from an imaging quality standpoint. There is, however, an incentive to remove a compensator from adjacent the panel to avoid high cosmetic specification and situate it in the illumination path. The geometry of the system makes compensation in the illumination path complex and requires further exploration.

7.8 Overall System Contrast

Many aspects affecting system contrast have been discussed somewhat in isolation. For any one system, contrast is determined by many constituent components, their mutual interaction, and any compensation that is used. In general, these factors are not easily recognized and isolated, especially in the more complex reflective systems. In an attempt to simplify this, an approximate expression for system contrast can be written in the form of equation (7.23), with the approximation stemming from the reasonable assumption that individual system leakage terms are small and independent:

$$\frac{1}{C_{sys}} \approx \frac{1}{C_{ideal\_PBS}} + \frac{1}{C_{ideal\_panel-comp}} + \frac{1}{C_s} \quad (7.23)$$

C_{ideal_PBS} is the contrast attainable with ideal PBSs. The leakage in this case would emanate from the imperfect compensated panel and any ARs between compensating elements. The leakage due to the AR coatings can be calculated readily as described in Section 7.4. In most cases where the compensating elements are index matched and have low birefringence this term becomes independent of the LC mode. Therefore it can be written:

$$\frac{1}{C_{ideal\_PBS}} = \frac{1}{C_{AR}} + \frac{1}{C_{compensation}} \quad (7.24)$$

The compensation term is then the calculated leakage over a finite illumination cone of a panel with appropriate PBS and compensation (see for example Figure 7.31).

The second term of equation (7.23) represents the leakage if panel compensation was perfect. It can be split into two terms:

$$\frac{1}{C_{ideal\_panel-conp}} = \frac{1}{C_{coating-only}} + \frac{1}{C_{pre-cond}} \quad (7.25)$$

The first term relates to PBS coatings assuming the clean-up polarizers are geometrically matched to the PBS. In the common case where a single panel is illuminated in reflection and analyzed in transmission by a PBS, this leakage is the cone average of the on-axis equation (7.5):

$$\frac{1}{C_{coating-only}} = \frac{\langle T_p R_p \rangle}{C_{P1}} + \frac{\langle T_s \rangle}{C_{P2}} \quad (7.26)$$

The second term is more subtle, and relates to the extent to which the clean-up polarizers are not geometrically matched to the PBS as introduced in Section 7.3.3.2. It can be written explicitly by equation (7.27) for the common case of an o-type input "clean-up" polarizer with a MacNeille PBS:

$$\frac{1}{C_{pre-cond}} = \left\langle \left(\overline{[k \times (k \times n)]} \cdot \overline{(k \times e)} \right)^2 R_p \right\rangle \quad (7.27)$$

where the bars above a vector product expression imply the vector is normalized, and the angular brackets imply an average over the illumination. The expression is simply the average reflected **p**-component of a ray as defined by the PBS with normal **n** when pre-polarized with an **o**-type polarizer of optic axis **e**.

The final term $1/C_s$ of equation (7.23) is a general term introduced to cover various light scattering phenomena that are common to panels and systems. Although this term can be further split into quantifiable contributions such as diffraction from pixel boundaries, it is beyond the scope of this book to consider these higher order effects.

With the expression of equation (7.23) it is possible to estimate system contrast for system comparison as we will do later, but also useful in isolating key contributors to low-contrast performance for future improvement.

References

[Bone M., 2001] M. F. Bone, Method to improve contrast in an off-axis projection engine, SID'01 Digest, p.1180, 2001.

[Bone M. F., 1998] M. F. Bone, M. Francis, P. Menard, M. E. Stefanov, and Y. Ji, Novel optical system design for reflective CMOS technology, proceedings of the 5th Annual Flat Panel Display Strategic and Technical Symposium, p.81, 1998.

[Chen J., 1998] J. Chen, K. H. Kim, J. J. Lyu, J. H. Souk, J. R. Kelly, and P. J. Bos, Optimum film compensation modes for TN and VA LCDs, SID'98 Digest, p.315, 1998.

[Chen J., 2004a] J. Chen, M. G. Robinson, and G. D. Sharp, General methodology for LCOS panel compensation, SID'04 Digest, p.990, 2004.

[Chen J., 2004b] J. Chen, M. G. Robinson, J. R. Birge, and G. D. Sharp, Compensated color management systems and methods, US Patent 6,816,309 B2, 2004.

[Kurtz A. F., 2004] A. F. Kurtz, B. D. Silverstein, and J. M. Cobb, Digital cinema projection with R-LCOS displays, SID'04 Digest, p.166, 2004.

[Lazarev P., 2001] P. Lazarev and M. Paukshto, Low leakage off-angle in E polarizers, *J. SID*, 9, pp.101–105, 2001.

[Mi X. D., 2003] Xiang-Dong Mi, Andrew F. Kurtz, and David Kessler, Display apparatus using a wire grid polarizing beam splitter with compensator, US Patent Application, US2003/0128320 A1, 2003.

[Robinson M. G., 2003] M. G. Robinson, J. Chen, and G. D. Sharp, Wide field of view compensation scheme for cube polarizing beam splitters, SID'03 Digest, p.874, 2003.

[Robinson M. G., 2004] M. G. Robinson, J. Chen, and G. D. Sharp, Optimization of three-panel liquid crystal on silicon video projection system, IDW'04, p.1683, 2004.

[Rosenbluth A. E., 1998] A. E. Rosenbluth, D. B. Dove, F. E. Doany, R. N. Singh, K. H. Yang, and M. Lu, *IBM J. Res. Dev.*, 42, p.356, 1998.

[Seiberle H., 2003] H. Seiberle, C. Benecke, and T. Bachels, Photo-aligned anisotropic optical thin films, SID'03 Digest, p.1162, 2003.

[Sharp G. D., 2002] G. D. Sharp, J. Chen, M. G. Robinson, and J. R. Birge, Skew-ray compensated retarder-stack filters for LCOS projection, SID'02 Digest, p.1338, 2002.

[Shimizu J. A., 2004] J. A. Shimizu, P. Janssen, S. Yakovenko, and D. Anderson, Contrast improvements for scrolling color LCOS, SID'04 Digest, p.145, 2004.

[Yeh P., 1999] P. Yeh and C. Gu, *Optics of Liquid Crystal Displays*, Wiley-Interscience, New York, 1999.

8

Color Management

8.1 Introduction

LC microdisplay panels modulate light intensity but cannot discriminate between colors. In order to get full-color images from a white light source it is therefore necessary to separate light into at least the red, green, and blue (RGB) primary colors to allow color generation through their independent modulation. Combining and imaging the spatially modulated colored beams then forms the desired full-color image. Separating the primary colors can be by either spatial or temporal means, or a combination of both, and constitutes a significant part of the overall system design. By way of introduction, this chapter will discuss how color is defined within a white light illuminated projection system and will present color band definitions that can be employed within specific spatial and temporal color separation techniques. Conventional spatial color separation and recombination using dichroic mirrors will then be considered, followed by more recent innovations employing polarization filters and polarizing beam splitters. Temporal methods using both rotating dichroic filters and polarization LC switches will then be presented, culminating with brief introductions to less generic approaches. Throughout, the polarization aspects of the specific color management schemes will be emphasized where applicable.

8.2 System Color Band Determination

In a conventional three-color video projection system, white light is filtered into three individual R, G, and B spectra, with the most saturated colors obtained with the narrowest pass band filters. The more saturated the primary colors, the larger the color gamut as discussed in Chapter 2, and consequently the more colors that can be represented by the display. Narrow-band filtering, however, leads to low system brightness. The trade-off between brightness

---

Polarization Engineering for LCD Projection M. G. Robinson, J. Chen and G. D. Sharp
© 2005 John Wiley & Sons, Ltd

198   COLOR MANAGEMENT

and color gamut is inherent in the design of the system. To avoid low throughput, filters are typically chosen to obtain the minimum color gamut desired by the system typically derived from current TV standards [Süsstrunk S., 1999; Hunt R., 2004]. Since ultra-high pressure (UHP) mercury lamps are universally accepted currently as the only viable projection sources for commercial RPTVs, compatible filter cut-off wavelengths can be calculated using the color calculation methods presented in Chapter 2. Figures 8.1(a) and (b) show diagrammatically the effect of long- and short-pass primary color filters on the primary RGB color coordinates.

Primary color points that are able to represent all colors within the color gamut must lie in the regions defined by the lines joining the standard points as shown in Figure 8.2.

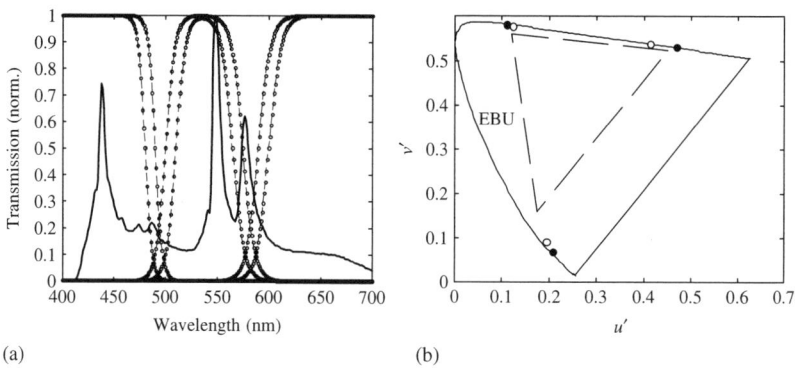

Figure 8.1 Effect of primary filtering on a UHP source spectrum: (a) color filter profiles superimposed on the source spectrum, (b) resultant color points

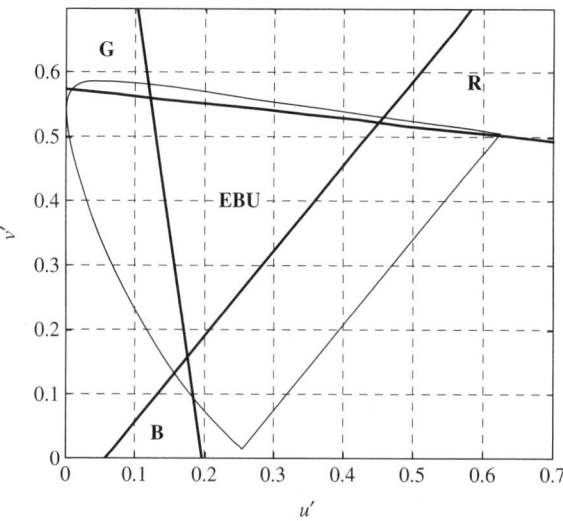

Figure 8.2 R, G, and B regions for acceptable primary color points that allow full representation of the color gamut (EBU in this case)

The optimum primary spectra are those that have color coordinates that lie in the allowed regions of Figure 8.2 and yield the maximum lumen output when combined to form an acceptable white. The following procedure can be used:

1. Choose convenient filter profiles with linear transition slopes that define the blue, green, and red color bands, with typical 15, 20, and 25 nm 10–90% values for the respective filters.
2. Scan the defining 50% cut-off wavelengths, creating for each cut-off set the resulting on-screen spectra by multiplying the filter profiles with the unfiltered source.
3. Calculate the primary color points and corrected white lumens at a desired white color temperature.
4. Obtain the maximum lumens for an allowable set of primary color points based on a desired color gamut.

Note that a 420 nm, 10 nm 10–90% UV cut filter is assumed together with a uniform 1000:1 system spectral contrast, in order to better approximate an actual system. These modifications have minor impact on the result.

Using this method, a UHP illuminated projection system adhering to the EBU standard of Figure 8.2 yields the maximum output lumens at white temperatures <12 000 K with filter cut-off values given in Table 8.1.

Table 8.1 Optimum filter cut-off values to achieve the EBU standard gamut with UHP illuminated projection

Red cut-off:	595 nm
Blue cut-off:	503 nm
Green short cut-off:	535 nm
Green long cut-off:	565 nm

In this specific case, the allowed blue 50% cut-off value is restricted between 501 and 508 nm, which limits the solution space considerably. Furthermore, in most red-starved UHP systems, the red cut-off should be as short as possible, effectively fixing it at its shortest 595 nm allowable limit. These transitions are therefore effectively fixed. The two defining green transitions are, however, less restricted with regard to maximizing lumen output. Figure 8.3 shows the system lumen output (normalized to the unfiltered lamp spectrum) as a function of the two green cut-off values.

As is shown in Figure 8.3, despite the 535/565 cut-off values producing maximum output, near-equivalent solutions exist anywhere on the boundary line connecting ~(535, 565) and ~(480, 575). It is interesting to note that similar output (if not slightly more) can be achieved by filtering out cyan and yellow light!

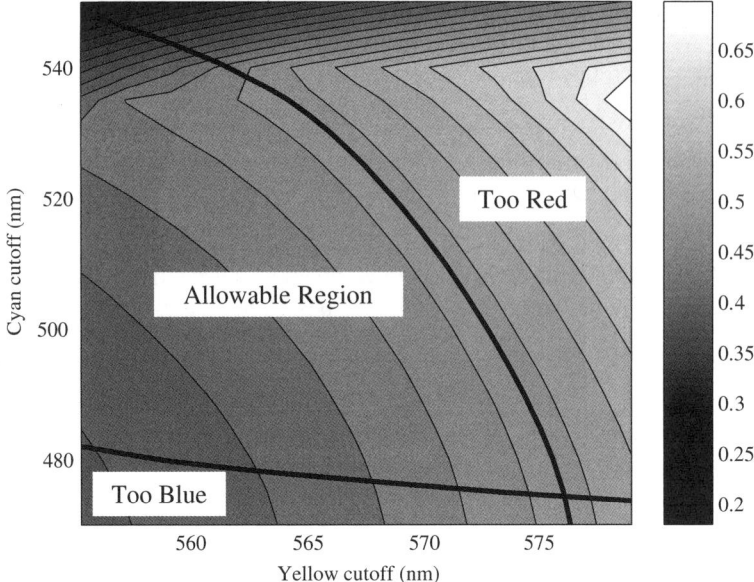

Figure 8.3 Corrected 8000 K white lumen output for cyan and yellow green filter transitions; 503 and 595 nm are taken for the blue and red cut-off values respectively

In practical cases, filter transition tolerances must be included in the final filter specification. The following sensitivity to filter transition values can be used to determine final filter values:

1. The red filter cut-off affects primarily the u' coordinate of the projected red $\left(\frac{\Delta u'}{\Delta \lambda_c} \approx 0.006\,\text{nm}^{-1}\right)$.

2. The yellow cut-off of the green spectrum affects the projected green u' coordinate $\left(\frac{\Delta u'}{\Delta \lambda_c} \approx 0.002\,\text{nm}^{-1}\right)$.

3. The cyan cut-off of the green band affects the green v' coordinate $\left(\frac{\Delta v'}{\Delta \lambda_c} \approx 0.001\,\text{nm}^{-1}\right)$.

4. The cyan cut-off of the blue mainly affects its v' coordinate $\left(\frac{\Delta v'}{\Delta \lambda_c} \approx 0.004\,\text{nm}^{-1}\right)$.

Any theoretical analysis carried out in the remaining chapters will take the filter transition values of Table 8.1 unless otherwise stated.

A corrected white can be obtained from any set of primary spectra, yielding different output lumens. For projected primary spectra created by filtering the UHP source of Figure 8.1 by filters with cut-offs given in Table 8.1, the luminous output can be calculated as a function of corrected white color temperature (see Figure 8.4). It is assumed that primary color modulation is independent, as for three-panel projection systems. Also shown in Figure 8.4 are the corrected lumens for modified primary spectra: (a) a 20% lower blue, (b) a 600 nm red cut-off instead of 595 nm, and (c) 505, 570 nm green cut-off values. The attenuated blue situation is often encountered in RPTV systems where polarizers, folding

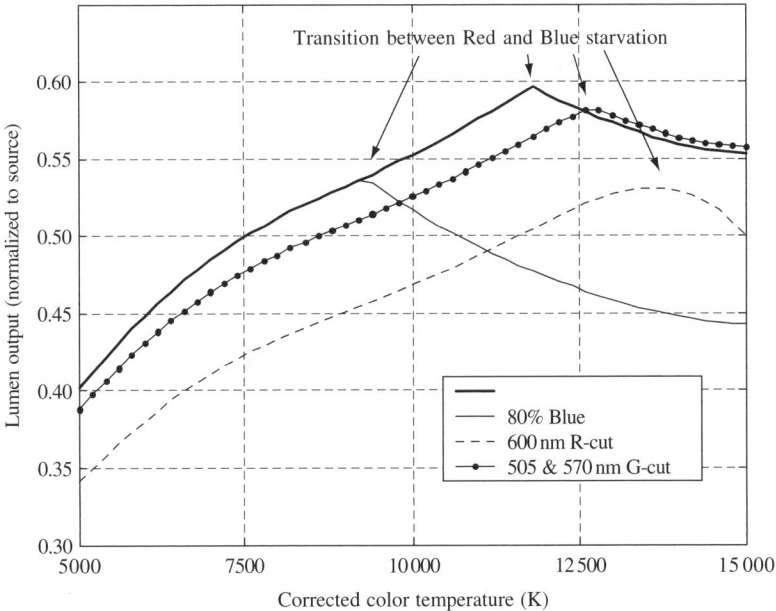

Figure 8.4 Lumen output of a three-panel color-corrected projection system as a function of white color temperature

mirrors, and front screens often significantly attenuate blue. The obvious kinks in the curves represent transitions between red-limited and blue-limited situations. The modified red-cut trace indicates the trade-off between output lumens and red color saturation. Typically a 5 nm shift in 50% cut value toward longer wavelength results in ∼15% decrease in corrected system throughput. In most common cases, the need to attenuate the green-rich source and to remove the 580 nm yellow spike can result in ∼50% loss due to color correction alone.

8.3 Color Management in Projection Systems

Color management in this context covers the separation of illuminating white light into desired color bands and their subsequent recombination. Separation is primarily carried out spatially and/or temporally. Spatial separation occurs when light is physically separated into one or more colored beams. Temporal separation occurs when continuous white illumination is separated sequentially in color, whereby any one pixel can modulate a specific color when synchronized with the illumination. A third separation option involves angular separation where colored beams are separated in angle, allowing physical separation just prior to the panel. Spatial recombination is simply the inverse of spatial separation, whereas temporal recombination occurs through the finite time over which retinal signals are integrated. Angular separation leads to separate RGB illuminated pixels, which are directly imaged onto the screen via the projection lens. Their recombination is through spatial averaging in the eye.

202 COLOR MANAGEMENT

8.3.1 Spatial Color Separation and Recombination

There are two approaches to spatial color separation and recombination. One splits and recombines colored beams outside of the modulating system and therefore does not affect contrast. A second approach is incorporated within the modulating system, which leads to efficient compact systems but does require careful polarization engineering.

8.3.1.1 Independent of the Modulating System [Itoh Y., 1997]

(a) Color Separation
The standard dichroic plate color splitting architecture is shown in Figure 8.5.

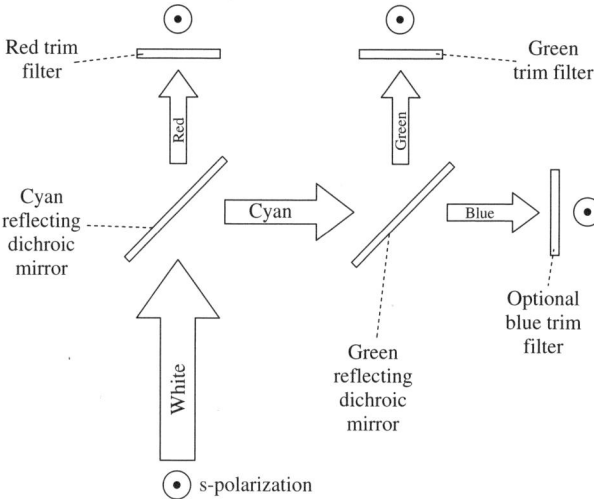

Figure 8.5 Typical RGB color splitting scheme using dichroic plate beam splitters

Whilst appearing straightforward in design, there are some subtle issues that should be noted:

- **Loose component specification.** Being part of the illumination system, the cosmetic and transmitted wavefront tolerances are loose, since relatively crude imaging of the light patch is required.

- **Preferably s-polarized.** In general, dichroics reflect s-polarized light more efficiently due to the increased Fresnel boundary reflections when incident off axis (see Chapter 4). Choosing s-polarization as the polarization preference minimizes cost through reduction in coating layers. Throughput is also maximized since the additional folding mirrors reflect s-polarization most efficiently.

- **Order of splitting.** For any given system, the weakest primary color should be split off first, since this reduces the number of elements it has to encounter prior to being modulated. In UHP systems this is red. The second split siphons off the green, as this is compatible with the X-cube recombination system, to be considered in the next section. The dichroic is typically a green/magenta beam splitter to reduce its number of coating layers.

- **Size and off-telecentricity.** The light travels through the system in air with relatively large extreme angles. To reduce the system size without introducing extra lenses, the light is typically off telecentric, and the angular performance of uniformly coated dichroic plates affects the color across the panel. To avoid this, two techniques are employed. First, coatings graded in layer thickness are used, such that filter cut-off wavelengths vary across a plate. This reduces spatial color variation, but does not in general remove it entirely. Therefore a second clean-up of the transmitted primary spectra is carried out by additional trim filters.

- **Trim filters.** These are positioned before the modulating system at normal incidence. They act to mask the spatially varying filter transitions produced by the beam splitters as shown in Figure 8.6. To be effective, the shortest pass band imposed by the beam splitters must exceed that of a trim filter, which forces the projected spectra to have spectral gaps between bands. For UHP illumination this does not significantly affect throughput since the photopic rich yellow light must be removed in any case in order to achieve an acceptable color gamut.

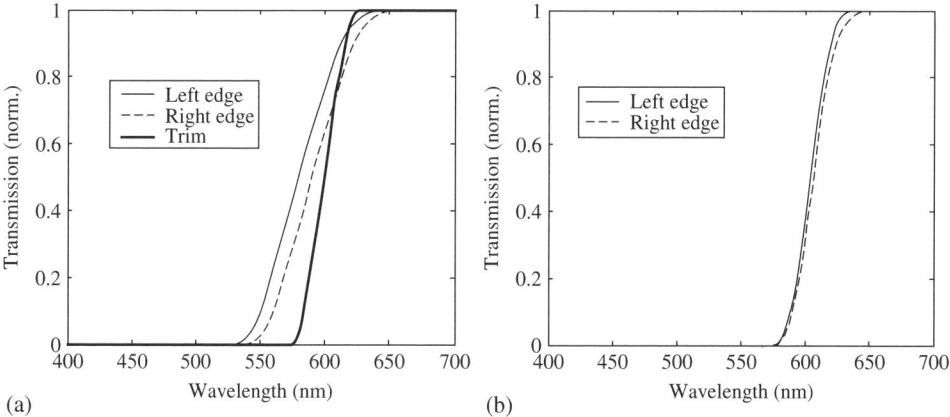

Figure 8.6 Effect of trim filter on primary red channel filter. Filter transitions as seen at the panel for (a) graded dichroic splitter and trim only, and (b) both in combination

(b) Color Recombination

The standard color image combining component is a dichroic-based X-cube shown in Figure 8.7. It consists of four glass prisms bonded to form a cube with crossing, dichroic-coated reflecting surfaces. By embedding the coatings within a cube of glass, the astigmatic aberration caused by imaging through tilted plates is avoided, while at the same time the imaging bfl (back focal length) is reduced. A short bfl results from having high-index material between the projection lens and panel, and by achieving three-color combination in a single cube volume. Physical registration of the quadrants of this element can be demanding, with poor registration resulting in image defects along a central line. In addition, waveguiding effects within the adhesive layer at the crossing seam can cause non-uniformity in

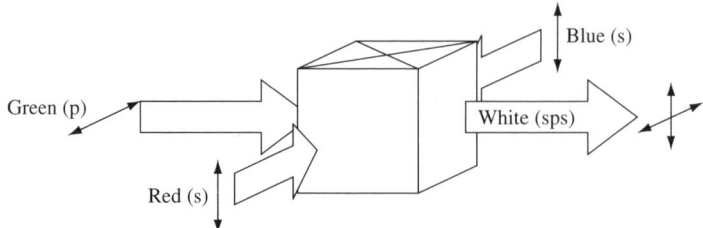

Figure 8.7 Schematic of an **sps** dichroic X-cube color combining element

the projected dark state. These defects are much more obvious in high-resolution displays. Fundamental dichroic properties (see Chapter 4) make efficient reflection of **p**-polarized light difficult and force the transmitted channel to be **p**-polarized. Similarly the reflected channels favor **s**-polarization. This makes channel-specific polarization manipulation a requirement with typical **sps** X-cubes, with the polarization of the transmitted green light being altered before, and sometimes after, the X-cube (see Chapters 9 and 10). The latter polarization transformation is often necessary to reduce color non-uniformities in the image, resulting from angular and polarization-dependent reflection off RPTV cabinet fold mirrors. In most cases an achromatic QWP is used to achieve the output conditioning; however, a *GM* (or *MG*) RSF (Retarder Stack Filter) is preferable if there is a significant polarization preference in both the fold mirrors and screen.

8.3.1.2 Integral to the Modulating System [Sharp G. D., 2002]

Two approaches can be considered. The first uses PBSs and RSFs to separate and recombine color channels while at the same time separating input and output beams as introduced in Chapter 6. Another less successful approach is to use dichroic color splitters in double pass directly adjacent to the panels. A PBS then separates the input and output beams prior to and following these splitting elements.

(a) RSF/PBS
This approach uses the beam splitting properties of a PBS to divide and recombine colored light by encoding complementary colored beams with orthogonal polarization states. Since this method is fundamentally based on polarization manipulation and its control, it naturally maintains the polarization integrity necessary for high contrast. The high angular performance of RSF components make compact and/or efficient, low-$f_{/\#}$ systems possible. The basic splitting and combining functions are shown in Figure 6.7, which can be combined to form the two-panel building block shown in Figure 8.8.

This specific example separates and combines blue and green light using two blue/yellow RSFs (*BY*1 and *BY*2). Unpolarized cyan input light first passes through an **s**-transmitting sheet polarizer. Blue light is then selectively transformed to **p**-polarization by the input *BY*1 RSF. The PBS then separates the orthogonally polarized colors, directing them toward reflective panels that spatially modulate their polarization states. The OFF-state corresponds to no change in polarization, where the unaltered portions of each beam retrace the path to the source. ON-state light from each channel, which is converted to the orthogonal polarization

COLOR MANAGEMENT IN PROJECTION SYSTEMS

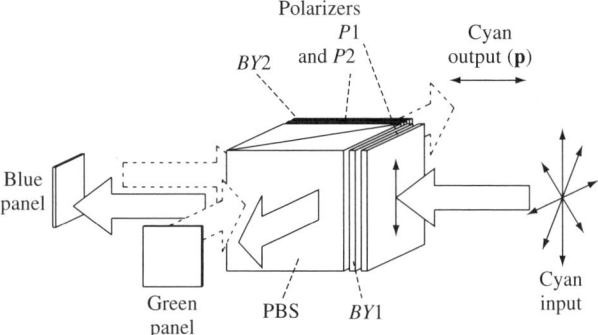

Figure 8.8 Blue/green shared PBS subsystem schematic

by the panel, exits through the fourth PBS port. In this way recombination of imagery is achieved. The need for an RSF and exit polarizer at the output is not essential in this isolated case, but frequently is in practice. For instance, the separation/recombination of a third color usually requires a uniform output polarization from the above subsystem.

As described, colored light is separated and combined with no cross-contamination between channels. In practice, imperfect components lead to possible color contamination and, more critically, low contrast. To determine the practical performance, it is necessary to analyze the system mathematically. Rigorous mathematical analysis of polarization transformation through optical systems is complex (see Chapter 3) and often reveals little of how elements affect performance. For the RSF/PBS systems a more intuitive method is to track optical intensities of orthogonal polarizations. This approach can also be applied to systems with components, such as PBSs, that are specified by their transmitted and reflected intensities, but do not have easily accessible phase information. Despite this limitation, this method can deliver quantitative results in systems where the following apply:

1. the phase between polarization components is near random and varies significantly as a function of propagation angle and wavelength; and

2. in situations where polarization transformations are either small or nearly complete, the latter relates to the performance of birefringent elements such as RSFs.

The method uses two-element column vectors to describe the polarization state of a propagating beam where the components are equal to the intensity of orthogonal **s**- and **p**-polarizations. For the specific case of Figure 8.8, the unpolarized cyan input beam is written:

$$I = \begin{pmatrix} 0.5 \\ 0.5 \end{pmatrix} \tag{8.1}$$

under the assumption that the input light is of unity intensity.

The input neutral **s**-polarizer $P1$ is then written as the matrix:

$$P1 = \begin{pmatrix} P1_s & 0 \\ 0 & P1_p \end{pmatrix} \approx \begin{pmatrix} 1 & 0 \\ 0 & \delta \end{pmatrix} \tag{8.2}$$

where δ is the finite, but small, leakage. $P2$ is written similarly.

COLOR MANAGEMENT

The non-ideal $BY1$ polarization filter is described by:

$$BY1 = \begin{pmatrix} (BY1)_p & (BY1)_x \\ (BY1)_x & (BY1)_x \end{pmatrix} \quad (8.3)$$

where $(BY1)_p$ represents the transmitted parallel polarizer yellow spectrum and its complementary crossed polarizer spectrum, $(BY1)_x$. A similar expression is used to describe the output $BY2$ filter.

The PBS performance is then described by the transmission and reflection matrices

$$\mathbf{T} = \begin{pmatrix} T_s & 0 \\ 0 & T_p \end{pmatrix} \quad \mathbf{R} = \begin{pmatrix} R_s & 0 \\ 0 & R_p \end{pmatrix} \quad (8.4)$$

Using this scheme, the OFF- and ON-states of the system can be derived.

System OFF-state Using the component performance descriptions above, the OFF-state of the subsystem in Figure 8.8 is given by the product:

$$OFF = P2 \cdot BY2 \cdot \mathbf{T} \cdot \begin{pmatrix} 1 & 0 \\ 0 & 1 \end{pmatrix} \cdot \mathbf{R} \cdot BY1 \cdot P1 \cdot I + P2 \cdot BY2 \cdot \mathbf{R} \cdot \begin{pmatrix} 1 & 0 \\ 0 & 1 \end{pmatrix} \cdot \mathbf{T} \cdot BY1 \cdot P1 \cdot I \quad (8.5)$$

where the central identity matrix represents the transformation of the LC in the OFF-state. The first term describes the predominantly green light that is initially reflected by the PBS, whereas the second represents blue light that is initially transmitted. The central triple products of the two terms are mathematically identical, which implies equivalent leakage from both channels, and allows simplification of the off-state leakage expression:

$$OFF = 2 P2 \cdot BY2 \cdot \mathbf{T} \cdot \mathbf{R} \cdot BY1 \cdot P1 \cdot I \quad (8.6)$$

Evaluating this yields:

$$\begin{pmatrix} \delta \cdot (BY2)_p \cdot R_s \cdot (BY1)_p + \delta \cdot (BY2)_x \cdot T_p \cdot R_s \cdot (BY1)_x + \delta^2 \cdot (BY2)_p \cdot T_s \cdot R_s \cdot (BY1)_x + \delta^2 \cdot (BY2)_x \cdot T_p \cdot R_p \cdot (BY1)_p \\ (BY2)_x \cdot T_s \cdot R_s \cdot (BY1)_p + (BY2)_p \cdot T_p \cdot R_p \cdot (BY1)_x + \delta \cdot (BY2)_x \cdot T_s \cdot R_s \cdot (BY1)_x + \delta \cdot (BY2)_p \cdot T_p \cdot R_p \cdot (BY1)_p \end{pmatrix} \quad (8.7)$$

This expression is clearly too complex to be of use. However, inserting the typical values $T_s \sim 0.001$, $T_p \sim 0.95$, $R_s \sim 1$, $R_p \sim 0.05$, $\delta \sim 0.001$, and neglecting any terms an order of magnitude lower than the largest terms, we have the much simpler expression:

$$\begin{pmatrix} 0 \\ (BY2)_x \cdot T_s \cdot (BY1)_p + (BY2)_p \cdot T_p \cdot R_p \cdot (BY1)_x \end{pmatrix} \quad (8.8)$$

Both of the remaining p-polarized terms can be large in the region where the filter spectra overlap. Figure 8.9 shows typical BY filter spectra, indicating the two relevant overlap regions. Adjusting the filter 50% transition points can reduce one overlap region at the expense of the other. Whereas the first term can be made negligible by selecting a PBS with low T_s performance, the second term remains significant. A low-T_s PBS should therefore be used, with the $(BY1)_x$ and $(BY2)_p$ spectra overlapping by no more than $\sim 10\%$ to achieve high contrast.

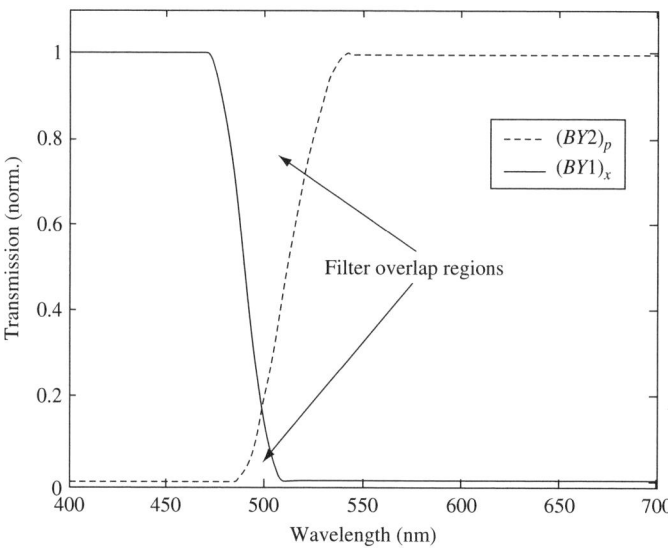

Figure 8.9 Spectral curves of the two BY RSFs

ON-state In the case where both the panels are in the ON-state, we have the following expression for the output spectrum:

$$ON = P2 \cdot BY2 \cdot T \cdot \begin{pmatrix} 0 & 1 \\ 1 & 0 \end{pmatrix} \cdot R \cdot BY1 \cdot P1 \cdot I + P2 \cdot BY2 \cdot R \cdot \begin{pmatrix} 0 & 1 \\ 1 & 0 \end{pmatrix} \cdot T \cdot BY1 \cdot P1 \cdot I \quad (8.9)$$

where the two terms represent the ON-state spectrum of each panel.

Expanding and simplifying these expressions by assuming the same typical values for component parameters yields:

$$\frac{1}{2}\begin{pmatrix} 0 \\ (BY2)_p \cdot T_p \cdot R_s \cdot (BY1)_p \end{pmatrix} \quad \text{and} \quad \frac{1}{2}\begin{pmatrix} 0 \\ (BY2)_x \cdot T_p \cdot R_s \cdot (BY1)_x \end{pmatrix} \quad (8.10)$$

for the separate channel output spectra.

From these expressions, it can be seen that in the absence of other system filter functions, the color of each channel is determined primarily by one of the filters, especially in the cases where system contrast requirements have forced minimal filter overlap. In this specific blue/green case, $(BY1)_x$ and $(BY2)_p$ determine the blue and green filter spectra respectively.

From the above expressions, the performance of the shared PBS reflective subsystem is determined by the external filters in the following way:

1. The filter that shares the PBS prism with a panel modulating color band Y affects the contrast of color Y, and therefore must have low leakage in that band. At the same time it affects the color and transmission of color band X of the panel opposite, and must have the correct 50% cut, and low ripple.

2. The filter that shares the PBS prism with a panel modulating color band X affects its contrast and must have low leakage. Its 50% cut and ripple affect the color band Y, which is modulated by the opposite panel.

208 COLOR MANAGEMENT

With this color management approach, color band contrast does not demand high PBS contrast; T_s/T_p up to 2% can often be tolerated, although to avoid overlap leakage, $T_s < 0.1\%$ is necessary in the filter transition region. Conversely, both high contrast and transmission are delivered by high-T_p PBSs.

Actual contrast and color coordinates delivered by this type of color management system will be left to the more representative three-panel systems of Chapter 10. This analysis will be used together with predicted panel contrast performance and other system effects.

(b) Dichroic-based Color Division and Recombination

These systems use a PBS to separate input and output white light, while relying on polarization-preserving dichroic splitters to separate and recombine color adjacent to the panels. Systems based on an X-cube [Takanashi I., 1993] and the Philips prism [Melcher R., 1998] have been considered (see Figure 8.10). Note that plate dichroic beam splitters cannot be used, as they introduce astigmatic aberrations. In general, these systems deliver low contrast, as dichroic splitters do not in general preserve polarization (see Chapter 4). Attempts have been made to compensate for polarization mixing but with limited success [Greenberg M., 2000].

A more successful approach is to use a two-color, dichroic color separator and combiner as shown in Figure 8.11. Here the **s**- and **p**-polarization states of a cube dichroic have the same geometrical angular dependence as a PBS if their coated surfaces are parallel (see Chapter 3).

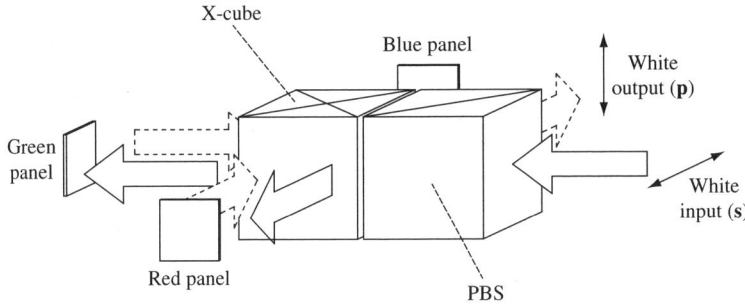

Figure 8.10 Example of a dichroic-based color division and recombination system

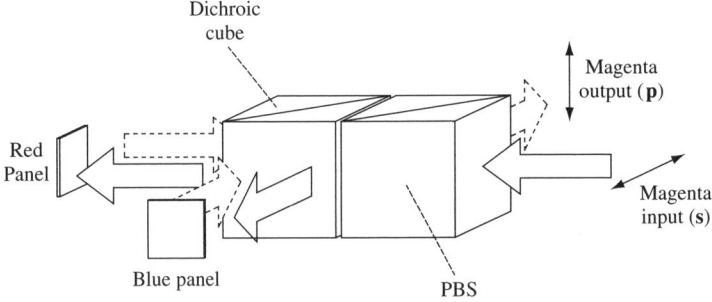

Figure 8.11 Schematic of a two-color, dichroic-based color division and recombination subsystem

If pure **p**-polarization as defined by a cube PBS is fed into the dichroic splitter, then in principle no polarization mixing can occur at the dichroic interface. The returning light will also remain **p**-polarized if the panels and QWPs transform the polarization correctly on reflection, leading to high contrast. In practice, embedded dichroic cube beam splitters tend to exhibit PBS behavior (see Chapter 4) over a large wavelength band, squeezing the bands over which the cube can be used in double pass. For this reason, a double pass dichroic cube splitter cannot be considered for adjacent color bands, since even red/blue discrimination produces unavoidable losses. These trade-offs will be discussed again in Chapter 10 in the context of a complete three-panel system.

8.3.2 Temporal Color Separation

The most common and successful temporal color separation scheme is the color wheel used in most DLP systems. They can be incorporated directly into temporal LC projection systems, but are not easily compatible with polarization conversion systems. An alternative approach, which can be directly inserted into PCS illumination systems, is to use a polarization-based LC color shutter. A third, scrolling bar, rotating prism option has also been commercially developed [Shimizu J., 1999]. In principle this third option avoids temporal light loss, but other related losses tend to negate much of the theoretical throughput benefits. Each will be considered in turn.

8.3.2.1 The Color Wheel

The color wheel, as the name suggests, is a motorized rotating wheel consisting of colored dichroic filter segments [Sampsell J., 1993a]. It is generally incorporated into a projection illumination system just prior to a light-pipe integrating rod as shown in Figure 8.12. A motor rotates the wheel at high angular speeds such that the transmitted light changes color as the different segments intersect the beam. The transition time between color fields is determined by the time taken for the join or "spoke" between two color filter sectors to pass through the focused light patch. Color mixing takes place during this transition period forcing the panel

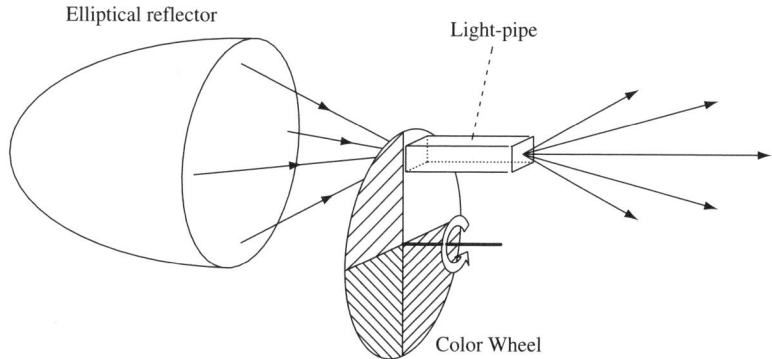

Figure 8.12 Schematic of a color-wheel-based temporal color modulation illumination system

in most cases to be OFF. To minimize the transition time and increase system efficiency, the beam size must be small compared to its radial position on the wheel.

The transition time τ, in seconds, is determined by the rotation speed of the wheel, ρ rpm, the dimension, d mm, of the input aperture of the light-pipe (effectively the diameter of the focused source), and the radial position, r mm, of the illumination on the wheel by:

$$\tau = \frac{30d}{\pi r \rho} \qquad (8.11)$$

Typical operating parameters are $\rho = 7200$ rpm, $d \sim 5$ mm, and $r \sim 30$ mm, which yield a transition time $\tau \sim 0.2$ ms. For a wheel with three equally sized RGB segments, this represents 7% loss due to transition blanking of the panel. To reduce τ and increase efficiency, wheel size could be increased, but with limitations imposed by overall system size and cost. Minimizing spot size increases aberrations from the elliptical reflector with diminishing returns. Also, smaller spot sizes increase energy densities and act to reduce the slopes of the primary dichroic filters through increased incidence angles. Moreover, the increased angles could exceed the critical angle of a total internally reflecting light-pipe.

Another important aspect of the color wheel, and indeed of temporal color separating systems, is the ability to overlap primary color bands. This allows, for example, the low-wavelength green cut-off to be below the blue cut-off if desired by alternative color standards. Temporal color separation also permits the introduction of extra color fields in addition to the required saturated primaries. One commonly used method to boost system output is the introduction of a white frame [Sampsell J., 1993b]. In this context, a white frame would be the projection of an unfiltered light spectrum for part of the frame period. In an ideal, infinitely fast, single panel system, illuminated by RGBW colored spectra with fraction α of the total full color period, the lumen output is:

$$(1-\alpha)\left(\frac{R_L}{3} + \frac{G_L}{3} + \frac{B_L}{3}\right) + \alpha W_L \qquad (8.12)$$

where R_L, G_L, B_L, and W_L are the luminous outputs for continuous red, green, blue, and white projected spectra respectively. The effect of introducing a white frame is to reduce the maximum brightness of the saturated primary colors relative to white. In TV systems where good color fidelity is desired, a white frame is typically not acceptable. In front projectors, white frames with $\alpha \sim 0.15$ are common since data projection is very forgiving of color rendition. Since the white frame is the unfiltered lamp spectrum and includes ~ 580 nm light, $W_L \approx 1.3(R_L + G_L + B_L)$. This implies that front projectors with a 15% white frame would yield $\sim 30\%$ more lumens (assuming the typical R:G:B corrected lumen mix of 20:75:5).

8.3.2.2 LC-based Temporal Color Modulators

LC color modulators are an alternative method of separating color temporally. Being LC based, they use polarized light and complement LC projection systems. LC color modulators use polarization filters that sandwich LC polarization switches. The polarization filters can be cholesteric LC based [Bachels T., 2001] or use the stacked birefringent approach described in Chapter 6. The simplest LC color modulator uses an achromatic LC polarization switch in series with a single primary color polarization filter [Sharp G., 1996]. To switch between three primary colors, a second stage is required consisting of a further LC cell and filter. This simple color switch approach suffers from the chromaticity of the LC cells and mixed color transition periods. A better suited approach to full-color LC projection systems uses an additive, internal polarizer free, color-band-independent ColorSwitch® (CS) [Sharp G., 2000]. Figure 8.13 shows a single CS red stage consisting of a zero twist, 0° aligned LC cell sandwiched between two RSFs that are inverses of each other (having the symmetry of Case IV, Section 6.5.3).

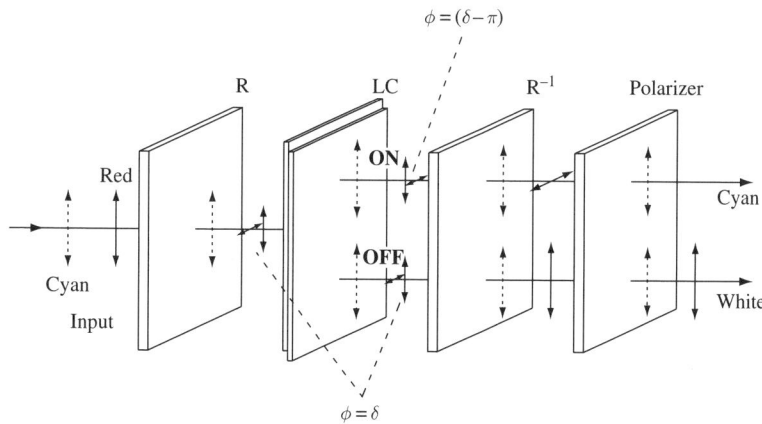

Figure 8.13 A red CS stage. Red (solid line) and cyan (dashed line) polarization are traced through the stage

The CS stage of Figure 8.13 operates on linearly polarized input white light. The first RSF creates a 45° oriented elliptical SOP for the color band to be switched, while leaving the polarization of the remaining spectrum unchanged. In the OFF-state, the LC retains all polarization states such that the second, inverse RSF returns all light to the original polarization. The overall structure is effectively isotropic for all wavelengths. In the ON-state, the LC retards one polarization component of the mixed band, such that the output RSF creates the orthogonal polarization state. The LC therefore transforms one band only, and through their independence, R, G, and B stages can be cascaded to form a three-color CS, without introducing internal polarizers.

Mathematically we can describe the operation of a CS stage for the transformed and stationary band using Jones' calculus described in Chapter 3.

An arbitrary phase Jones' matrix that describes the input polarization filter can be written as:

$$\mathbf{F}_t = \frac{1}{\sqrt{2}} \begin{pmatrix} e^{i\alpha} & e^{i\beta} \\ -e^{-i\beta} & e^{-i\alpha} \end{pmatrix} \quad (8.13)$$

for the transformed band, and:

$$\mathbf{F}_i = \frac{1}{\sqrt{2}} \begin{pmatrix} e^{i\gamma} & 0 \\ 0 & e^{-i\gamma} \end{pmatrix} \quad (8.14)$$

for the isotropic band, where α, β, and γ are arbitrary real numbers.

212 COLOR MANAGEMENT

The corresponding second inverse RSFs would then be:

$$\mathbf{F}_t^{-1} = \frac{1}{\sqrt{2}} \begin{pmatrix} e^{-i\alpha} & -e^{i\beta} \\ e^{-i\beta} & e^{i\alpha} \end{pmatrix} \quad \text{and} \quad \mathbf{F}_i^{-1} = \frac{1}{\sqrt{2}} \begin{pmatrix} e^{-i\gamma} & 0 \\ 0 & e^{i\gamma} \end{pmatrix} \tag{8.15}$$

A zero-twist LC modulator can be described by the following matrices representing the ON- and OFF-states:

$$\mathbf{ON} = \begin{pmatrix} e^{-i\pi \frac{\Gamma}{\lambda}} & 0 \\ 0 & e^{i\pi \frac{\Gamma}{\lambda}} \end{pmatrix} \quad \text{and} \quad \mathbf{OFF} = \begin{pmatrix} 1 & 0 \\ 0 & 1 \end{pmatrix} \tag{8.16}$$

where the retardance Γ is set at $\sim \lambda_0/2$ for the transformed band center wavelength, λ_0.

The truth table for a single CS stage is as given in Table 8.2, where **0** indicates that the input linear state is preserved, and **1** that the polarization is transformed to the orthogonal state. The nature of polarization ensures that stacked stages are independent, since the switch operates equivalently on the two orthogonal linear input states.

An RGB CS is shown schematically in Figure 8.14. To improve switching speed and off-axis FOV performance, paired parallel-aligned LC cells could be considered for the

Table 8.2 CS truth table

	LC on (driven)	LC off (ov)
Transformed band	$(0\ 1) \cdot \mathbf{F}_t^{-1} \cdot \mathbf{OFF} \cdot \mathbf{F}_t \cdot \begin{pmatrix} 1 \\ 0 \end{pmatrix} = \mathbf{0}$	$(0\ 1) \cdot \mathbf{F}_t^{-1} \cdot \mathbf{ON} \cdot \mathbf{F}_t \cdot \begin{pmatrix} 1 \\ 0 \end{pmatrix} = \mathbf{1}$
Isotropic band	$(0\ 1) \cdot \mathbf{F}_i^{-1} \cdot \mathbf{OFF} \cdot \mathbf{F}_i \cdot \begin{pmatrix} 1 \\ 0 \end{pmatrix} = \mathbf{0}$	$(0\ 1) \cdot \mathbf{F}_i^{-1} \cdot \mathbf{ON} \cdot \mathbf{F}_i \cdot \begin{pmatrix} 1 \\ 0 \end{pmatrix} = \mathbf{0}$

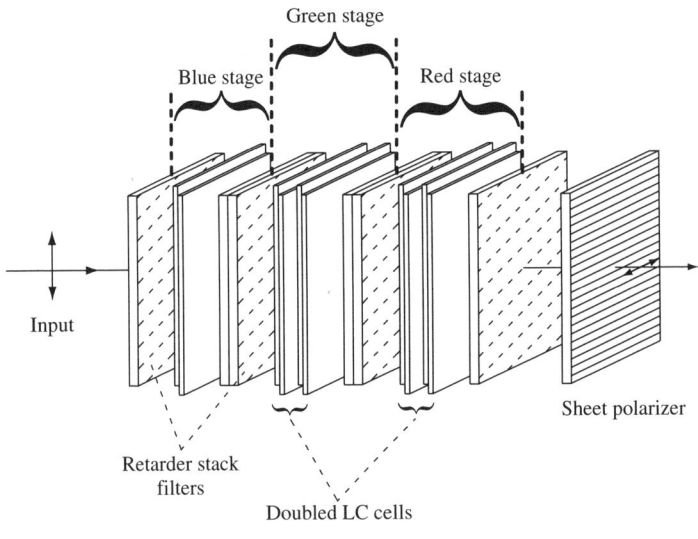

Figure 8.14 RGB CS schematic

green and red stages. The limiting LC switching speed is proportional to d^2, where d is the cell thickness (see Section 5.5). Since a near-constant total thickness is required to swing the necessary half-wave retardance, two cells of half thickness produce 4× faster switching than a single cell. Doubling the cell also improves FOV, since the central splayed regions of the two π cells rotate skew ray polarization in a manner that is opposite to the effect of the outer splayed regions. With double cells in the green and red stages, switching speeds of <0.3 ms are obtainable and saturated colors are achieved with $>f_{/2.0}$ illumination.

In contrast with the color wheel, the RGB three-stage CS can deliver all colors contained within the gamut defined by the individual stages with the appropriate set of voltages. This allows signal-dependent control of color. For example, switching between video and data content could allow for saturated fast RGB sequencing in the former case, and slower, brighter, less saturated RGBW in the latter. It also allows dimming, which would act in place of the auto-iris described in Section 2.5.6. Transmission of the CS is, however, lower than the color wheel, being further reduced in a system by the requirement of an extra input polarizer. The throughput ramifications of sequential color will be discussed in more detail in Chapter 11.

8.3.2.3 Hybrid Temporal and Spatial Scrolling Color Separation Systems

Two scrolling color modulation schemes have been developed for projection. The first is shown in Figure 8.15, whereby refraction through a rotating glass prism produces separated RGB stripes to scroll smoothly over the panel. Each focused stripe scans the panel until the corner of the cube intersects the light path. At this point there is a transition and a new stripe is formed at the top of the panel. Using a single cube in this manner is difficult in practice as the colored stripes tend to overlap as a result of the chromatic aberrations caused by the cube. A more successful implementation uses a split color path approach with a rotating prism for each color. This concept was developed by Philips and is used in the 55″ diagonal 55PL977S RPTV system [Shimizu J., 2001].

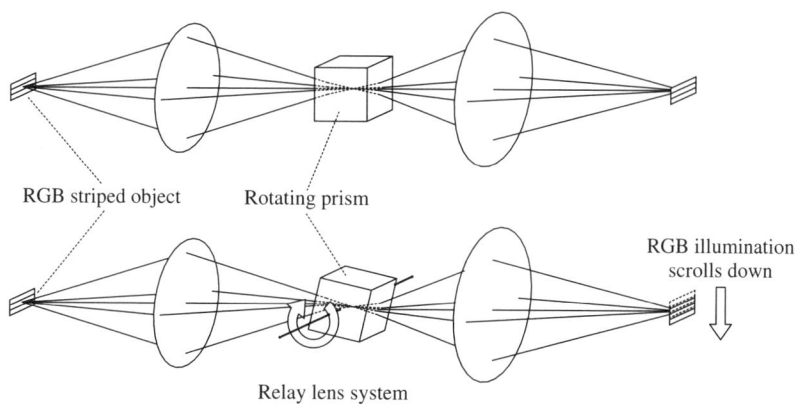

Figure 8.15 Schematic of a rotating prism, scrolling color illumination system

214 COLOR MANAGEMENT

The scrolling prism illumination system has the following attributes that make it very attractive for single panel systems:

- **No light loss.** All light from the source illuminates the panel at any instant and therefore there is in principle no light loss. In practice, separating RGB color bands increases the optical extent of the illumination by a factor of three. In an étendue-limited system this results in light loss. Bigger panels can reduce the loss, but with cost and system size implications.
- **Panel address compatibility.** The simplest addressing scheme for active matrix LC panels is line by line. The moving, non-illuminated boundary between color stripes allows LC lines to switch in sequence.
- **Reduction in color break-up.** Having all colors illuminating the panel simultaneously can reduce the color break-up artifact common to time sequential systems.

A second scrolling color system approach uses a spiral color wheel. Shown in Figure 8.16, the spiral color wheel is placed at the exit of a light-pipe such that the exit aperture imaged on the panel has stripes that scroll. Unlike the scrolling prism approach, there is no inherent light savings, although light recycling techniques [Dewald D., 2001] can be implemented with this approach as discussed in Chapter 11.

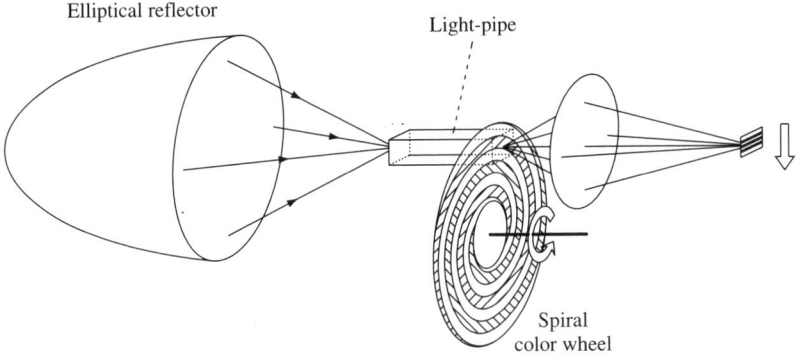

Figure 8.16 Spiral color wheel scrolling concept

8.3.2.4 Angle-encoded Color Separation

Angle-encoded color separation is not currently used in commercial systems due to practical limitations. It is only introduced briefly here. From a conceptual standpoint, the method encodes overlapping color beams with different propagation angles that are then separated at the panel by micro-optical elements into local pixel-sized RGB illuminated regions. Early systems used tilted dichroic mirrors in the path of a near-collimated beam as shown

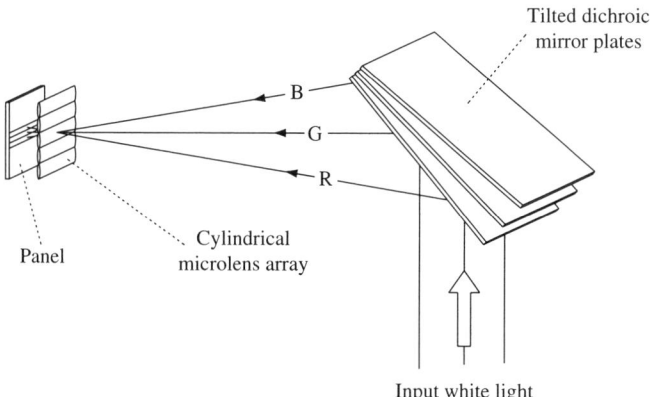

Figure 8.17 Schematic showing angle encoding of overlapping colored beams. Spatial color separation is carried out close to the modulating panel with microlens elements

in Figure 8.17 [Hamada H., 1994]. The superimposed angle encoded reflected beams are separated physically by cylindrical microlens elements adjacent to the panel. Implementing this in reflective architectures is problematic due to proximity of the microlens array to the panel, which acts to reflect and depolarize incident. In transmission, however, this approach can yield high transmission, again under the assumption that losses through étendue limitation are small.

Finally, an angular color separation system was developed by JVC [Suzuki T., 1999] for reflective panels using holographic micro-optical elements (see Figure 8.18). Holographic grating elements are situated on top of a single reflective panel to both separate and focus RGB illuminating beams onto separate pixels. The system was commercialized with relatively reasonable performance; however, color saturation, efficiency, and contrast do not achieve the levels currently demanded.

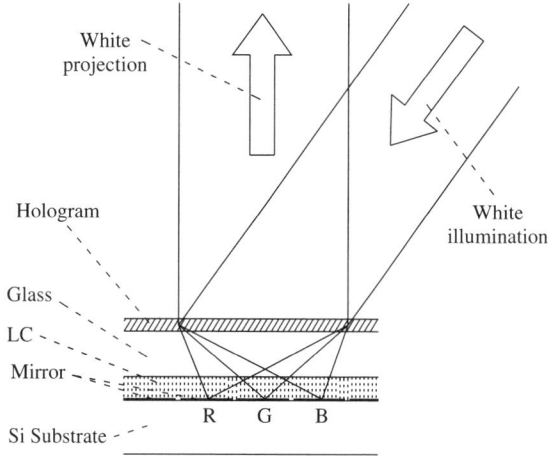

Figure 8.18 Holographic angular color separation

References

[Bachels T., 2001] T. Bachels, K. Schmitt, J, Fünfschilling, M. Stalder, H. Seiberle, and M. Schadt, Advanced electronic color switch for time-sequential projection, SID'01 Digest, pp.1080–1082, 2001.

[Dewald D., 2001] D. S. Dewald, S. M. Penn, and M. Davis, Sequential color recapture and dynamic filtering: a method of scrolling color, SID'01 Digest, pp.1076–1089, 2001.

[Greenberg M., 2000] M. R. Greenberg and B. J. Bryars, Skew ray compensated color separation prism for projection display applications, SID'00 Digest, pp.88–91, 2000.

[Hamada H., 1994] H. Hamada *et al.*, A new bright single panel LC-projection system without a mosaic color filter, IDRC'94, pp.422–423, 1994.

[Hunt R., 2004] R. W. G. Hunt, Displays: past, present, and future IDW'04 pp.3–6, 2004.

[Itoh Y., 1997] Y. Itoh, Y.-I. Nakamura, K. Yoneno, H. Kamakura, and N. Okamoto, Ultra-high-efficiency LC projector using a polarized light illuminating system, SID'97 Digest, pp.993–996, 1997.

[Melcher R., 1998] R. L. Melcher, P. M. Alt, D. B. Dove, T. M. Cipolla, E. G. Colgan, F. E. Doany, K. Enami, K. C. Ho, I. Lovas, C. Narayan, R. S. Olyha, Jr., C. G. Powell, A. E. Rosenbluth, J. L. Stanford, E. S. Schlig, R. N. Singh, T. Tomooka, M. Uda, and K. H. Yang, Design and fabrication of a prototype projection data monitor with high information content, *IBM J. Res. Dev.*, 42, pp.321–338, 1998.

[Sampsell J., 1993a] J. B. Sampsell, An overview of the digital micromirror device (DMD) and it applications to projection displays, SID'93 Digest pp.1012–1015, 1993.

[Sampsell J., 1993b] J. B. Sampsell, White light enhanced color field sequential projection, US Patent 5,233,385, 1993.

[Sharp G., 1996] G. D. Sharp and K. M. Johnson, High-brightness saturated-color shutter technology, SID'96 Digest, pp.411–443, 1996.

[Sharp G., 2000] G. D. Sharp, J. R. Birge, J. Chen, and M. G. Robinson, High throughput color switch for sequential color projection, SID'00 Digest, pp.96–99, 2000.

[Sharp G. D., 2002] G. D. Sharp, M. G. Robinson, J. Chen, and J. Birge, LCOS projection management using retarder stack technology, *Displays*, 23, p.139, 2002.

[Shimizu J., 1999] J. A. Shimizu, Single panel reflective LCD optics, Proceedings of IDW'99, pp.989–992, 1999.

[Shimizu J., 2001] J. A. Shimizu, Scrolling color LCOS for HDTV rear projection, SID'01 Digest pp.1072–1074, 2001.

[Süsstrunk S., 1999] S. Süsstrunk, R. Buckle and S. Swen, Standard RGB color spaces, Proceedings of the 7th Color Imaging Conference: Color Science, Systems, and Applications, pp.127–134, 1999.

[Suzuki T., 1999] T. Suzuki and S. Nakagaki, Color filter and color image display apparatus employing the filter, US Patent 5,999,282, 1999.

[Takanashi I., 1993] I. Takanashi, S. Nakagaki, I. Negishi, T. Suzuki, F. Tatsumi, R. Takahashi, and K. Maeno, Display apparatus, US Patent 5,239,322, 1993.

9

Transmissive Three-panel Projection System

9.1 Introduction

The majority of current commercially available LCD front and rear projectors incorporate high-temperature polysilicon (HTPS) transmissive panels in a standard three-panel, X-cube configuration. In this chapter, the generic architecture will be described only briefly, as it represents a combination of subsystems already covered. Emphasis will instead be placed on the analysis of system throughput and contrast. Accurate throughput numbers are notoriously difficult to predict, but an estimation can help highlight the significant contributing factors and also allows for comparisons with other systems. Low contrast on the other hand is a major concern in current transmissive systems. Methods to improve system contrast will therefore be emphasized, specifically the use of retardation film-based compensation technology. A rigorous contrast analysis of systems employing current transmissive panels with several proposed compensation schemes is presented.

9.2 Brief System Description

The now standard transmissive LC projection system (see Figure 9.1) was first introduced commercially by Seiko-Epson in 1997 [Itoh Y., 1997] and is essentially that used in the currently available, highly successful Sony Grand WEGA RPTV.

The illumination, color management, modulation, and imaging subsystems that make up this generic three-panel system have all been treated separately in previous chapters and require little if any repeated explanation here. Briefly, it consists of a UHP source,

---

Polarization Engineering for LCD Projection M. G. Robinson, J. Chen and G. D. Sharp
© 2005 John Wiley & Sons, Ltd

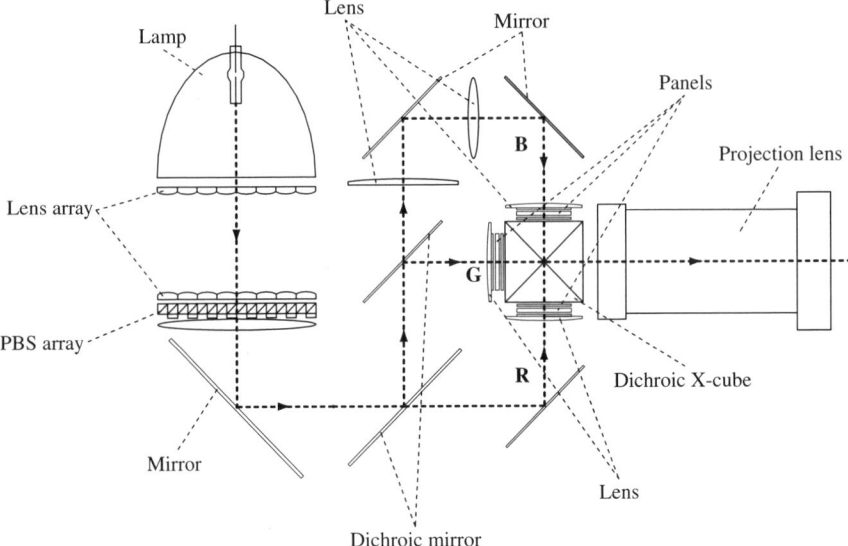

Figure 9.1 System schematic of the standard three-panel transmissive system

a parabolic reflector, and paired lens arrays for homogenization of the illumination as described in Section 2.5.5. It includes a polarization conversion system consisting of a linear array of polarizing beam splitter elements with 45° oriented half-wave retardation film at the output of alternate elements situated directly after the second lens array. A telecentric relay lens system sandwiches graded dichroic beam splitting plates that separate the illumination into three primary colored beams as described in Section 8.3.1.1(a). The separate colored illumination is incident on three transmissive TN panels that modulate light by manipulating polarization between sheet polarizers as described in Section 5.7.2. The panels are situated around an X-cube that combines the transmitted light (see Section 8.3.1.1(b)), which is collected and projected by a telecentric, long bfl (back focal length) projection lens, similar to that described in Section 2.5.3, to form an image on a screen.

The specific components and system design are an amalgamation of the ideas covered in the preceding chapters (most significantly Section 8.3.1.1) with the exception of two subtle aspects: the relayed blue illumination channel, and the polarization conversion within the red and blue channels. Both these issues are dictated by the X-cube:

- **Blue channel illumination relay.** From the physical layout of the panels surrounding the X-cube, the optical path lengths within the split illumination system cannot be made equal. To overcome this, a simple three-lens relay system is introduced into the lumen-rich blue channel.

- **Red and blue polarization conversion.** The requirement that the transmitted red and blue light reflecting by the X-cube are **s**-polarized dictates that the **p**-polarized output from the red and blue panels must be transformed. Typically, this is achieved with channel-specific half-wave retardation films oriented at 45°.

9.3 System Throughput

It is useful to consider the overall system throughput as a product of individual subsystem terms. This is especially convenient and more accurate when systems with similar components are directly compared (see later chapters).

The system throughput (i.e., its lumen output, Φ_{out}) can be expressed as the product:

$$\Phi_{out} = \Phi \eta_{ill} \eta_{cm} \eta_{cc} \eta_m \eta_{im} \tag{9.1}$$

where:

- Φ is the lumen output of the bare lamp
- η_{ill} is the efficiency of the illumination, which includes coupling to the reflector, étendue clipping, PCS losses, polarization efficiency terms, and Fresnel reflection losses from components, including relay lenses
- η_{cm} is the throughput of the color management system
- η_{cc} is a color correction factor related to removing unwanted yellow light and balancing the remainder to yield an acceptable white point
- η_m is the modulation system efficiency
- η_{im} is the imaging system throughput that can include the projection lens and folding mirrors, but not the screen.

9.3.1 Lamp Flux Output, Φ

Commercial ultra-high-pressure (UHP) mercury lamps deliver slightly greater than 60 lumens/watt [Moench H., 2003]. Lamps vary in power between ~100 and 250 W, which relates to ~6000 and 15 000 lm output. In general, lower wattage lamps are favored by RPTV manufacturers for their increased reliability: >10 000 hours is quoted for the Philips 100 W UHP lamp, whereas only 3000 hours is typical of 150 W lamps. For the sake of this calculation we will assume a 7200 lm output source (i.e., 120 W). The lamp spectrum is essentially that shown in Figure 2.17 with slight variations dependent on mercury pressure and lamp geometry. Since the measurement of lumens is dominated by the ~550 nm green spike, subtle variations in spectrum do not often show up in direct lumen output.

9.3.2 Illumination Efficiency, η_{ill}

This term includes light collection efficiency, Fresnel reflection losses, absorption of polarizing elements, and overfill of the panel. It is assumed that the contributing elements have near achromatic performance:

- **Collection efficiency.** The geometrical collection efficiency is determined by the quantity of light coupled into the finite étendue of the projection system, as introduced in

Chapter 2. Coupling efficiency can be directly measured using the technique described elsewhere [Stupp E. H., 1999] yielding efficiencies shown in Figures 2.20 and 2.21. Actual efficiencies can vary considerably from lamp to lamp according to the finite manufacturing tolerances of arc alignment within the reflector and electrode gap variation. The étendue is determined primarily by panel size and projection lens capture angles, with some additional reduction due to conversion systems (see Chapter 2 for details). Here, a typical transmissive PCS RPTV system consisting of a 0.7″ diagonal, 16:9 panel imaged with an $f_{/2.8}$ projection lens will be assumed (imaging system étendue $\sim 14\,\text{mm}^2$), as this is also typical of panel sizes and projection speeds of later comparative reflective systems. The system collection efficiency, assuming a 1 mm arc gap UHP lamp, is $\sim 50\%$ taken from Figure 2.21.

- **UV/IR filter.** A typical UV IR filter has a transmission of 97%.

- **Folding mirror.** Highly reflecting silver-based mirrors can now achieve efficiencies >99%, although more realistic cost-effective aluminum mirrors have efficiencies $\sim 95\%$. In a typical red-starved system, the transmission of the red channel (see Chapter 8), which contains two illumination fold mirrors, is approximately 90%.

- **Illumination lenses.** Fresnel reflection loss yields a collection lens transmission of $\sim 98\%$. Note that the lenses adjacent to the panels are not included here as they are typically coated with a dichroic trim coating and considered part of the color management and modulation system.

- **Panel overfill.** Overfilling the panel is necessary to avoid edge effects resulting from illumination non-uniformity. Typically RPTV panels have a 5% overfill in both dimensions resulting in a net throughput of 90%.

Taking the product of the individual throughputs, we get an overall Illumination, efficiency, η_{ill}, of 39%.

9.3.3 Color Management System Efficiency, η_{cm}

This term incorporates the passive losses of all components of the color management system, at the peak transmission of the dichroic filters. The filter transition bandwidths and slopes are considered part of the color correction term to be considered later, making this term independent of a particular system color point preference. This also allows direct comparison to reflective systems that use PBSs. To determine throughput, measured spectral transmissions of the dichroic components and polarizers of the S-E 5500c transmissive system yield practical numbers. They include the $f_{/2.8}$ average performance of the cyan and green reflecting dichroics (Figure 9.2), the $f_{/2.8}$ averaged transmission through the green and red channel, the trim dichroic coatings (Figure 9.3), the $\sim 90\%$ transmission of a typical output polarizer for the red and green channels (Chapter 4), and the $f_{/2.8}$ averaged transmission and reflection of the X-cube for the correct **sps** projected polarizations (Figure 9.4).

Figures 9.2, 9.3, and 9.4 indicate that the system-limiting red efficiencies of the dichroic splitter, trim filter with clean-up polarizer, and X-cube are 98%, 86%, and 94%, respectively. Removing the $\sim 10\%$ loss due to the clean-up polarizer (to be included in the modulating system) yields an overall color management system efficiency of $\sim 88\%$.

SYSTEM THROUGHPUT 221

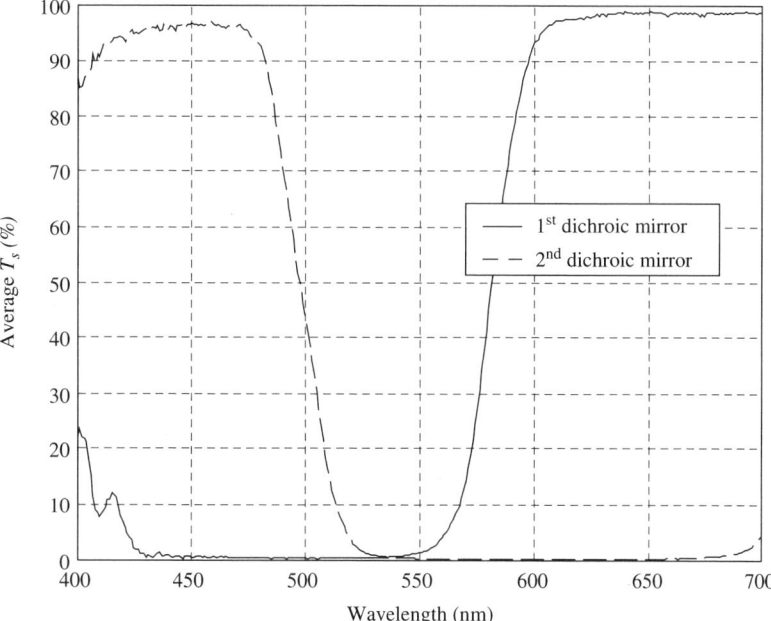

Figure 9.2 Average transmission of s-polarized light through both input dichroic splitting plates

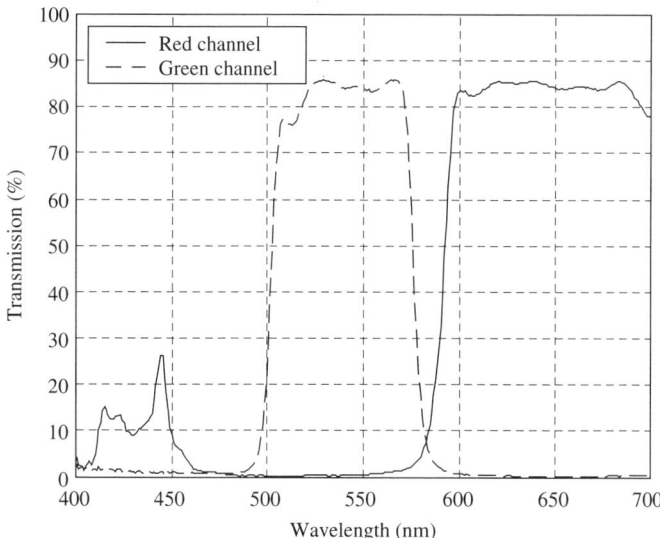

Figure 9.3 Transmission through red and green channel trim field lenses. Polarizers and half-wave plates are attached

222 TRANSMISSIVE THREE-PANEL PROJECTION SYSTEM

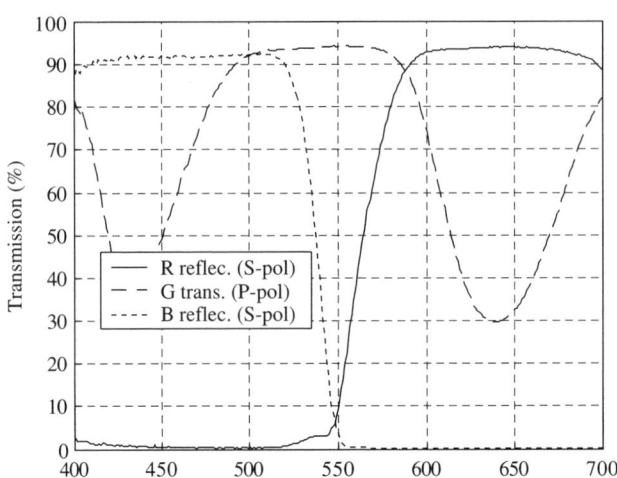

Figure 9.4 The averaged mixed polarization output of the S-E 5500c X-cube. Red efficiency ~94%

9.3.4 Color Correction Efficiency, η_{cc}

This term covers the optical losses incurred in creating acceptable primary colors and a balanced white:

- **Primary colors.** For any given system, the choice of primary colors can vary as discussed in Chapter 8. For the example system introduced here (S-E 5500c) the colors are derived from the net filtering of each primary channel on the source spectrum. Taking the measured component spectral traces of Section 9.3.3, the normalized filter traces shown in Figure 9.5 can be derived, from which the maximum lumen efficiency is determined as follows:

$$\eta_{pc} = \frac{\int [(R(\lambda) + G(\lambda) + B(\lambda)) \cdot S(\lambda) \cdot y(\lambda)] d\lambda}{\int [S(\lambda) \cdot y(\lambda)] d\lambda} \quad (9.2)$$

where $S(\lambda)$ is the standard UHP source spectrum. Calculation in this case yields a primary color correction factor of 79%.

- **White correction.** Once the primary filtering functions have been determined, the individual channels must be modulated to form an acceptable white as described in Section 2.2.4. For the typical red-starved UHP system considered here, there is considerable loss dependent upon the desired color temperature. Relevant white correction efficiency factors η_{wc} are listed in Table 9.1 for various white color temperatures. They are derived assuming that the desired white point must lie on the black-body curve (or Planckian locus), which is a good assumption for RPTV systems designed to show video content. Taking an average RPTV white color temperature of 8000 K, the net color correction factor η_{cc} is ~55%. This factor will apply even if subsequent components adjust the balance of the primaries from that of the bare lamp spectrum, just so long as the system remains red starved.

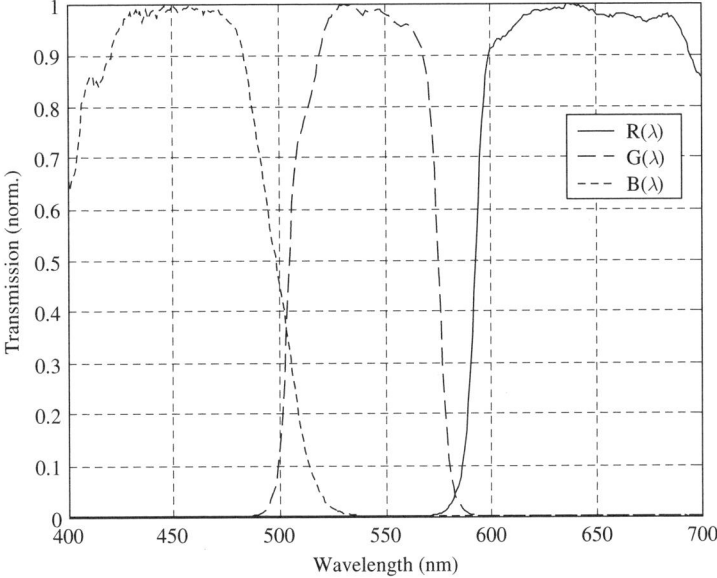

Figure 9.5 Primary filtering functions of the S-E 5500c example system

Table 9.1 White correction efficiencies for various corrected white color temperatures

Corrected color temperature (K)	White correction efficiency η_{wc}, (%)
6 500	63
8 000	70
9 500	76
12 000	77

9.3.5 Modulation System Efficiency, η_m

The modulation system efficiency includes the panel efficiency and the transmission of both polarizer and analyzer. The throughput of the HTPS panel structure (see Figure 1.14) is determined by passive Fresnel reflections, conducting electrode (ITO) absorptions, LC mode efficiencies, and the aperture ratio.

- **Fresnel surface reflections.** Good AR coatings and index-matched ITO layers can reduce reflections to <2% overall.

- **ITO absorption.** A good compromise between conductivity and transparency reduces absorption to <2%.

- **LC mode efficiency.** For the standard 90° TN mode, LC mode efficiency can be close to 100% for a given wavelength. Chromaticity over the red spectrum does, however, result in an effective loss of ~1% in corrected white lumens.

- **Aperture ratio.** Manufacturers are continually improving their panel aperture ratios. The current panels have an aperture ratio of 53%, whereas the next generation will deliver 60% for 0.7″ diagonal 1280×720 (12 µm pixel pitch) panels [Epson, 2004]. The use of microlens array technology can improve the effective aperture ratio but often at the expense of contrast [Ou C., 2001]. Since contrast is critical in RPTV products microlens arrays are not typically used.

The overall panel efficiency is then 57%. Assuming ~90% red efficiency for both polarizer and analyzer, the overall modulating system is estimated at ~46% efficient.

9.3.6 Imaging System Efficiency, η_{im}

Imaging efficiency includes passive losses in the lens and Fresnel reflection losses in the fold mirrors. Here it does not include screen throughput as this is heavily dependent upon specific screen technology and gain:

- **Projection lens reflections.** Good wideband AR coatings (~0.75% reflection) on each of the ~10 elements of a typical projection lens leads to an overall throughput of ~85%.

- **Fold mirrors.** The reflectivity off the fold mirrors is ~92% for the typical large Al (aluminum) mirror and between 92 and 98% for the small mirror dependent upon its makeup. Assuming an all Al external fold mirror approach, an overall efficiency is ~85%.

Overall imaging system efficiency η_{im} is then ~72%.

9.3.7 Total System Lumen Output, Φ_{out}

The system lumen efficiency budget is summarized in Table 9.2.

The actual output of the S-E 5500c is rated at 650 lm, which is not inconsistent with the calculated 450 lm since fold mirrors would not be included in the front-projector specification and larger panels (~1″ diagonal) with microlenses are actually used.

At this point it is important to consider the cumulative errors of the above analysis. In the above ~15 efficiency terms, there could be an average ~2% error in each. This means that the final lumen number could have a 30% error, which is often witnessed in practice. Systems made from nominally identical elements often vary in output performance by several tens of percent. However, despite the approximate nature of this calculation, the above analysis is useful to appreciate the larger loss terms, and also to compare specific subsystem losses between otherwise identical systems, as will be demonstrated in later chapters.

Table 9.2 Estimated output of a 120 W three-panel transmissive HTPS projection system

Subsystem	Efficiency term	Efficiency
Illumination	Collection–étendue mismatch	50%
	UV/IR	97%
	Mirror	90%
	Lenses	98%
	Panel overfill	90%
Subtotal, η_{ill}		**39%**
Color management	Dichroic plate beam splitter	98%
	Polarizer + trim dichroic	96%
	X-cube	94%
Subtotal, η_{cm}		**88%**
Color correction	Primary colors	79%
	8000 K white	70%
Subtotal, η_{cc}		**55%**
Modulation system	Fresnel reflections	98%
	Panel ITO absorption	98%
	LC mode	99%
	Panel aperture ratio	60%
	Polarizer and analyzer	81%
Subtotal, η_m		**46%**
Imaging	Projection lens reflections	85%
	Fold mirrors	85%
	Screen	N/A
Subtotal, η_{im}		**72%**
System total		**6.3%**
Lamp lumen output		7200
Incident on screen		**450 lm**

9.4 Contrast

The contrast of an HTPS transmissive system is determined by the LC panel, which is sandwiched between two sheet polarizers, and is therefore isolated from the remainder of the system. Its performance at normal and off-normal incidence angles are both critical for good contrast. Normal incidence light can be controlled in polarization by suitable choice of LC mode with optional small compensating elements to negate any residual retardation (see Chapter 5). Off-normal performance is much more complex as demonstrated by the viewing angle issues still present in direct-view LCDs such as laptop screens.

In an $f_{/2.5}$, 90° HTPS projection system, contrasts up to 500:1 are achievable in principle when scattering is well controlled [Okamoto N., 2001]. However, current RPTV systems demand contrasts closer to 1000:1, making it necessary to increase panel contrasts through compensation techniques.

The viewing angle of direct-view LCDs was once an important performance issue, but significant progress was made in the late 1990s [Wu S., 2001; Bos P., 2002]. Compensation techniques can be categorized into three groups: internal, external, and hybrid. Internal approaches modify the structure of the LC by introducing, for example: multidomains (MDs) [Takatori K., 1992; Yang K., 1992; Chen J., 1995; Vithana Y., 1995; Nam S., 1997], in-plane switching (IPS) [Oh-e M., 1995; 1997; Lee S., 1997; Kim K., 1998a], fringe field switching (FFS) [Lee S., 1998], ITO patterning [Lien A., 1993; Koma N., 1995; Kim K., 1998b], and surface protrusion methods [Ohmuro K., 1997; Takeda S., 1998]. External approaches introduce additional birefringent compensating layers to one or both sides of a conventional LCD panel. For projection LCD panels, internal approaches are not considered suitable since resolution is high, forcing pixels ($<20\,\mu m$) to be physically small compared to the cell gap. Under these conditions, disclination lines in multidomain structures cause scattering, which reduce both system contrast and brightness. Furthermore, the main purpose of multidomain technology is to avoid gray-scale deformation in TN displays at high incidence angles. In a projection system telecentric imaging is used, so gray-scale deformation does not manifest itself as viewing angle deficiency. What is more critical in projection systems is the net system contrast, defined as an average over all illumination angles, which is much more effectively controlled by external compensating layers.

In general, external approaches involve laminating a phase compensator [Ohmuro K., 1997; Takeda S., 1998] or diffusive optics to the outside of an LCD panel. Since the FOV (Field Of View) of an LCD results from the residual birefringence of the black state, with magnitude dependent upon ray direction, a good external film compensation scheme should be one that acts to cancel residual birefringence, making the display isotropic for all incident rays. Ideally, the phase cancellation should occur for all angles over the operating temperature range and for all visible wavelengths.

Many film compensation schemes have been developed to enhance the FOV of LCD displays. Since the LC mode used in all current HTPS projection systems is essentially the conventional 90° TN, we will focus on specific compensation schemes related to this mode, bearing in mind the design criteria relevant to projection systems. These include material durability, manufacturability, and cost, together with the relaxation of large viewing angle compensation demanded by direct-view approaches.

Figure 9.6 shows the LC profile of a 90° TN mode in the (driven) OFF-state. It can be approximated by three regions, which include splayed regions near the boundaries and a homeotropic region in the middle of the cell. The homeotropic region implies that a simple negative c-plate compensation scheme could offer some compensation. However,

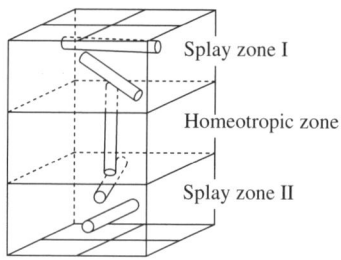

Figure 9.6 LC director profile of a TN LCD mode in its driven OFF-state

the splayed region between compensator and homeotropic LC limits the success of this approach (mathematically, the Jones' matrices do not commute). In order to access this central portion, some form of o-plate compensation is also required. At the very least, this is a homogeneous layer (e.g., negative uniaxial), which for projection incidence angles is an effective first-order approximation. The ideal compensation is one with the exact director profile of the LC layer, but mirrored and with negative anisotropy. LC polymers coated on a TAC substrate (providing c-plate compensation) offer the best solution for direct view. However, the benefits of FOV compensation can be quickly lost in projection, when typical in-plane issues, such as scatter and inhomogeneous optic axis distribution, are presented. These can occur as-fabricated, or are thermally induced under intense illumination.

In the following, computer modeling is used demonstrate the effectiveness of compensation schemes assuming that the illumination is uniform over a finite $f_{/2.5}$ cone. Material parameters of a typical Merck LC fluid ZLI4792 are used, with a typical dark OFF-state voltage of 6 V. The industry standard first minimum 90° TN configuration is assumed with 5° pretilt.

9.4.1 Negative c-plate Compensation

Negative c-plate compensation is an ideal compensation scheme for a perfect homeotropically aligned LC (see Section 7.3.1). Since the splayed region of the energized 90° TN layer represents a relatively small proportion of the overall LC in its OFF-state, it is worthwhile considering simple c-plate compensation first. From a material standpoint, c-plates can be made in a variety of ways, including spin coating or casting of polymers or by deposition of inorganic material layers (e.g., SiO_2 and TiO_2 layers) (see Section 4.2). Inorganic crystals such as sapphire can also be used, with excellent optical quality and UV stability. However, the small retardances required correspond to a sapphire thickness of tens of microns, which is not practical.

Figure 9.7 shows the simulated system contrast at 550 nm for a first minimum TN device versus the retardation (Δnd) of a sapphire c-plate, where the sheet polarizer and analyzer are assumed ideal. This shows that a negative c-plate gives almost no improvement over the uncompensated panel.

9.4.2 Splayed Negative Birefringent Film Compensation Scheme

To correct the splayed regions of the LC, a splayed negative compensator can be considered, such as the wide-view (WV) film manufactured by Fuji Film [Mori H., 1997; 1998; Sergan T., 1998; Vermeirsch K., 1999]. The film consists of a splayed discotic layer with a TAC substrate as shown in Figure 9.8. Discotic monomers are web coated on a TAC substrate, first with a thin layer of rubbed PI (polyimide). At the air–film interface, the discotic layer is naturally splayed at an angle ($\theta = 40°-68°$), which is determined by the surface tension and controlled by adjusting processing temperature, humidity, and/or rubbing pressure. Minimum elastic energy then results in the discotic molecule orientation evolving linearly within the bulk.

To ideally compensate the structure shown in Figure 9.6, the discotic layer exactly mirrors that of the LC. Furthermore, the TAC substrate usually has ∼60 nm negative retardation aligned normal to the substrate which provides additional compensation for the central

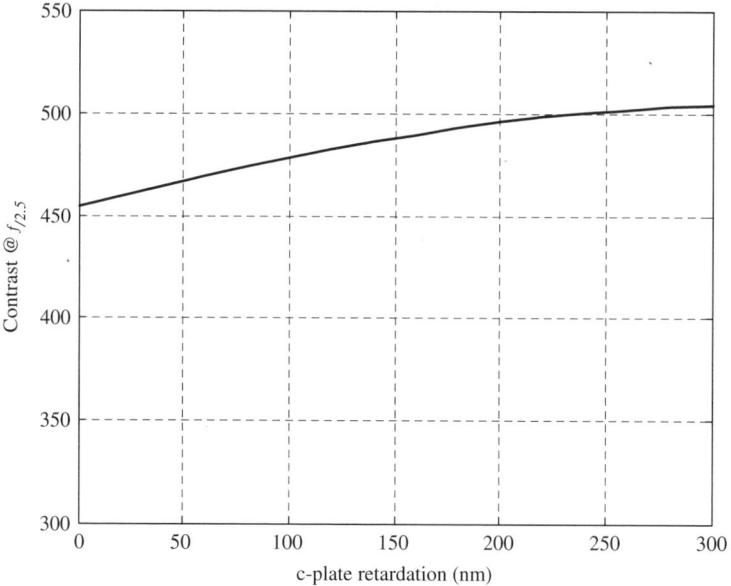

Figure 9.7 System contrast of a uniform $f_{/2.5}$ illuminated 90° TN HTPS projection system compensated by a single negative c-plate

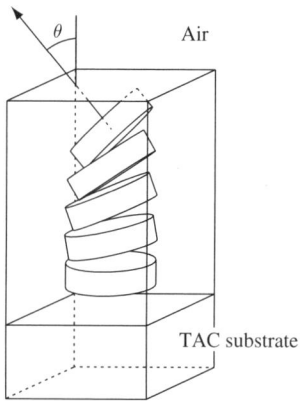

Figure 9.8 The structure of Fuji WV discotic compensation film

homeotropic region. By placing discotic films on both sides of a transmissive TN display, the direct-view cone is expanded close to ±60°. Figure 9.9 shows schematically the optimum orientation of a double WV film compensation scheme. Note that the film is bonded to the panel using a pressure-sensitive adhesive.

The parameters of a discotic film needed for high system contrast can be optimized for any given wavelength, provided that the LC cell information is known. In three-panel HTPS systems, dispersion is not as critical as for a full-color transmissive panel, since only monochrome compensation is required. Figure 9.10 is the system contrast versus the air

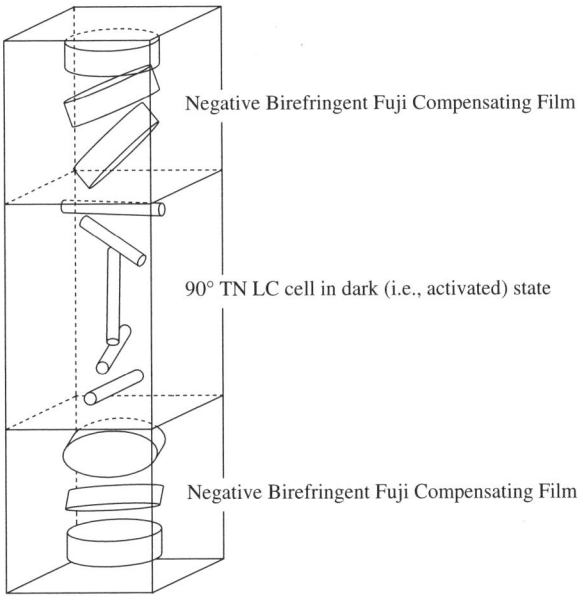

Figure 9.9 WV film compensation scheme for a TN display

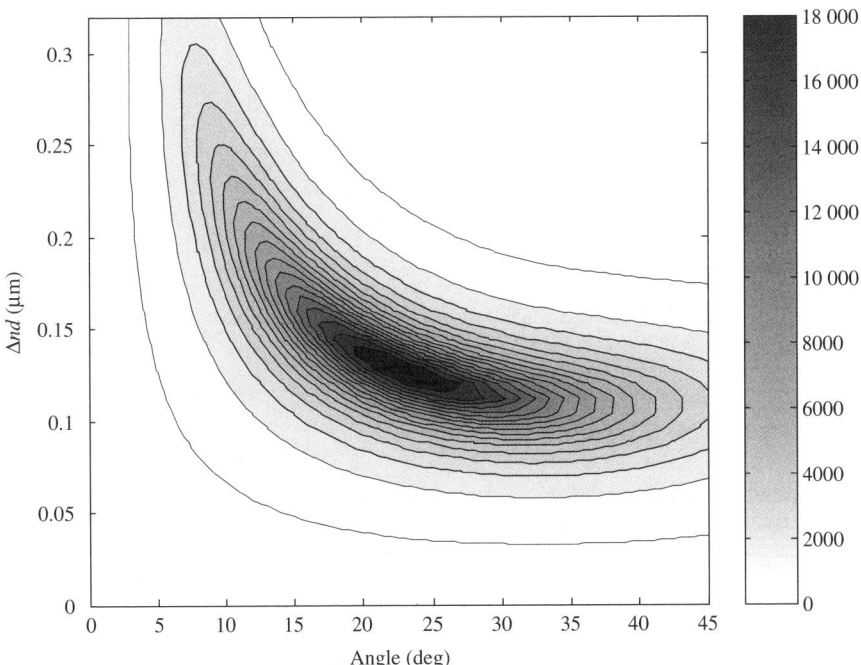

Figure 9.10 System contrast at $f_{/2.5}$ as a function of the air interface discotic angle and Δnd

interface tilt of the discotic molecule and retardance (Δnd) at 550 nm. A TAC layer of -60 nm retardance is assumed part of each compensating layer.

From Figure 9.10, the optimum values for this discotic compensation approach are $\Delta nd = 0.13 \mu m$ at $24°$, which improves the system contrast from 450:1 to 18 000:1 at $f_{/2.5}$.

9.4.3 Negative o-plate Compensation

An alternative to LC polymer for projection uses homogeneous negative o-plates. There are two potential configurations: one placing a negative o-plate on each side of the TN display, as above, and another whereby crossed negative o-plates are placed on a single side (Figure 9.11). Figure 9.12 shows system contrast versus retardance and tilt angle for the first configuration, with a very similar plot obtained for the second. These compensation schemes are nearly as effective as discotic films for projection systems using $\Delta nd \sim 0.11 \mu m$ and an optic axis tilt of $\sim 12°$.

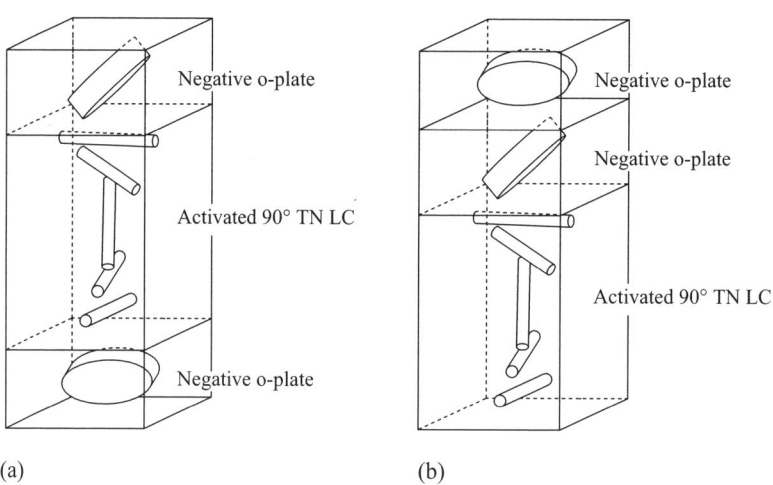

Figure 9.11 Two compensation schemes using negative o-plates

9.4.4 Positive o-plate Compensation Scheme

Due to the relative abundance of durable positive uniaxial materials, compensation schemes using them have been also considered. At first glance, it is not clear that a positive anisotropic material can compensate a positive anisotropic LC layer. However, from Chapter 7, we established that crossed identical a-plates behave similar to a single negative c-plate in specific viewing planes. A first attempt would therefore be to mimic the c-plate compensation scheme of Section 9.4.1 above. However, the modest performance of the c-plate scheme, and the clear enhancement through using o-plates, lends themselves to the crossed positive o-plate scenario shown in Figure 9.13. Figure 9.14 shows simulated system contrast for the green channel at $f_{/2.5}$ versus Δnd and tilt angle.

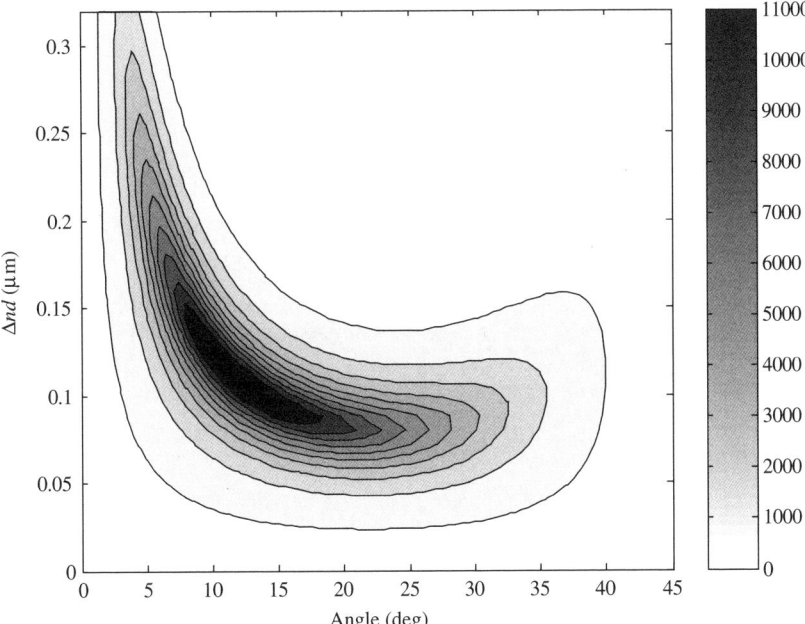

Figure 9.12 System contrast of TN display compensated by two negative o-plates as shown in Figure 9.11(a)

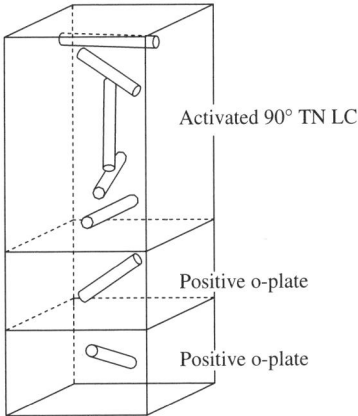

Figure 9.13 Compensation scheme using positive o-plates

This compensation scheme can be used to significantly increase the contrast of a transmissive TN LC panel. In this case the optimized Δnd and tilt angle are ~90 nm and ~28° respectively.

While negative birefringent approaches can be explained by simple negation of the positive residual structure, the positive approach is less intuitive. The mechanism can be visualized

232 TRANSMISSIVE THREE-PANEL PROJECTION SYSTEM

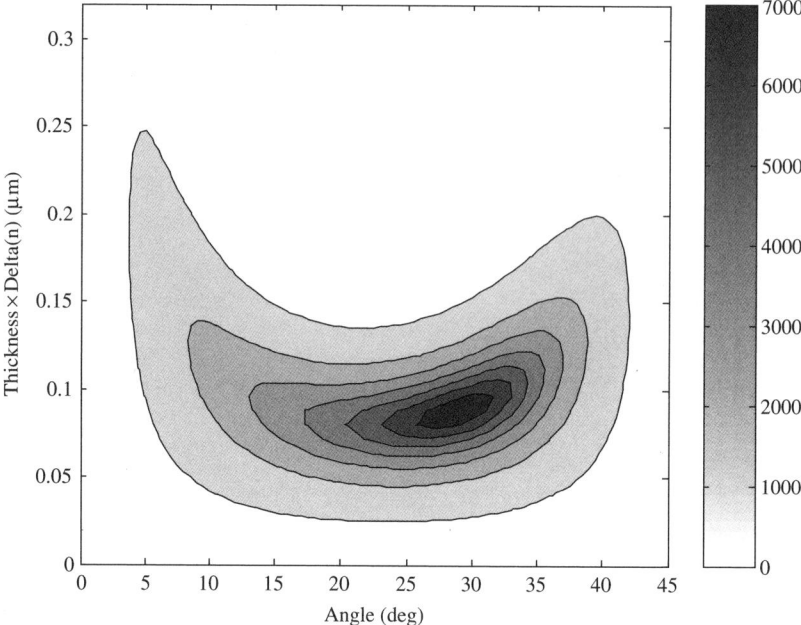

Figure 9.14 System contrast of TN display compensated by two positive quartz o-plates as shown in Figure 9.13

by considering the polarization transformation of rays propagating in the two orthogonal planes containing the polarizer and analyzer axes. The three distinct layers of the driven LC structure (Figure 9.6) and the two tilted compensator layers then appear to a ray as three retarders with axes oriented at $+\varphi°$, $0°$ and $-\varphi°$ with respect to the **p**-axis, and perform successive polarization transformations (i), (ii), and (iii) on the Poincaré sphere as shown in Figure 9.15. The closed circuit of the transformation triangle for the specific ray at 5°

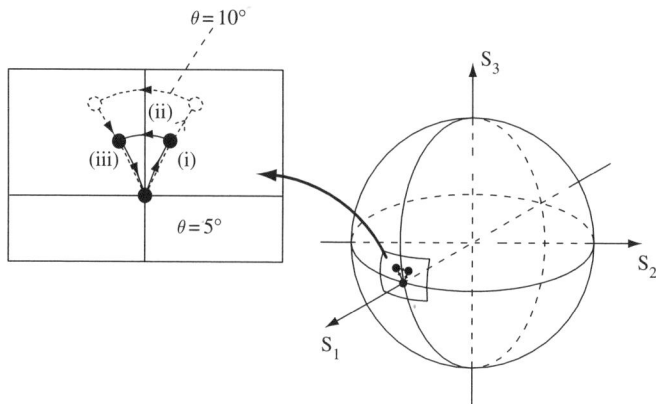

Figure 9.15 Three-transformation visualization of the positive o-plate compensation scheme

in Figure 9.15 results in zero net polarization change and high contrast. Furthermore, the transformations scale similarly as a function of angle as demonstrated by the dashed triangle representing a 10° incident ray.

For durability reasons, quartz is a good candidate for this application, but again the thickness requirements are not practical. An alternative option is to use positive uniaxial LC polymer materials.

9.4.5 Liquid Crystal Polymer (LCP) Compensation Scheme

LCPs are similar to discotic films, whereby an LC monomer is coated onto an LC anchoring layer such as a rubbed PI layer or a polarized UV-exposed LPP (Linearly Polarized Polymer), forming a splayed aligned molecular structure (Figure 9.16). A stable LCP film is then formed through cross-linking by further UV exposure. The pretilt angle at the anchoring layer and air interface layer can be controlled by suitable choice of LCP material, alignment layer, and process conditions. The pretilt angle θ_1 at the air interface is usually determined by the LCP material itself and its process conditions and is typically close to 0°. The pretilt angle θ_2 at the opposing surface is controlled by the alignment layer. PI usually offers 0°–5° pretilt angle. A much wider range of pretilt angles (0° to 90°) can be achieved if LPP alignment is adopted. In this latter case, UV exposure strongly affects the pretilt angle.

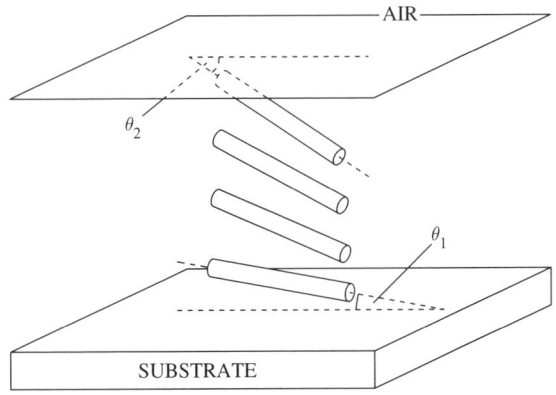

Figure 9.16 Configuration of an LCP film

There are four possible configurations for splayed LCP compensation layers shown in Figure 9.17, with simulation indicating that configuration III offers the best performance. Figure 9.18 plots the system contrast of a 90° TN display compensated by the bilayer structure shown in Figure 9.17(III). The optimum pretilt angle (~ 2 × average splay angle) and retardance Δnd are equal to 50° and 125 nm respectively. The system contrast can be further optimized if non-identical splayed compensation films are used as in the case of the uniformly tilted positive o-plate solution. But once again, the more attractive symmetrical solution should yield sufficient performance.

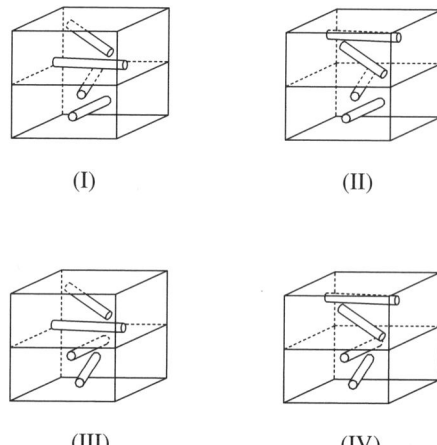

Figure 9.17 Four possible configurations for the splayed LCP compensation bilayer

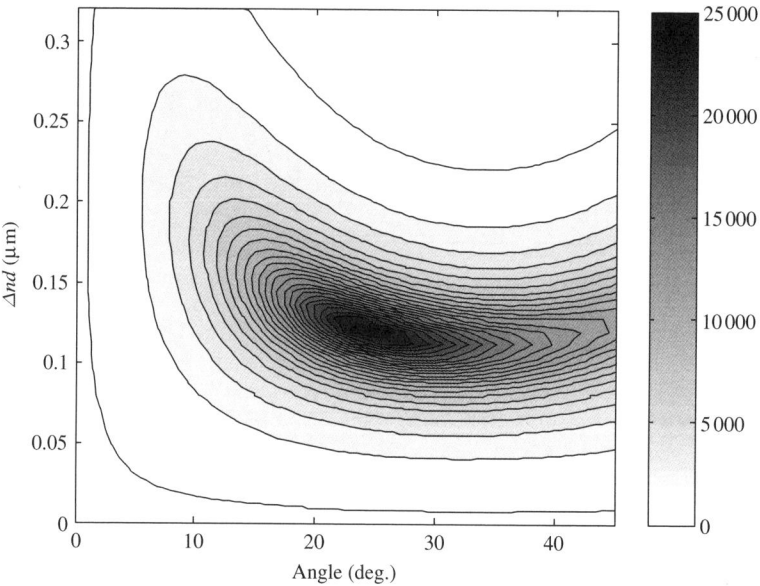

Figure 9.18 System contrast versus retardance and average splay angle of a 90° TN LCD compensated by the bilayer LCP compensation scheme shown in Figure 9.17(III)

The dominance of the transmissive three-panel system is threatened by the higher contrast, resolution, and brightness performance delivered by DLP systems and the reflective LCOS projection systems of the next chapters. The above compensation schemes offer a performance enhancement that allows HTPS-based systems to remain competitive with emerging technologies.

References

[Bos P., 2002] P. J. Bos, LCD field of view issues, SID'02 Seminar, 2002.

[Chen J., 1995] J. Chen, P. J. Bos, D. R. Bryant, D. L. Johnson, S. H. Jamal, and J. R. Kelly, Four-domain TN-LCD fabricated by reverse rubbing or double evaporation, SID'95 Digest, pp.865–868, 1995.

[Epson, 2004] http://www.Epson-ed.info/Exhibition/FPDi2003/prdct_e/HTPS.html

[Itoh Y., 1997] Y. Itoh, Y.-I. Nakamura, K. Yoneno, H. Kamakura, and N. Okamoto, Ultra-high-efficiency LC projector using a polarized light illuminating system, SID'97 Digest, pp.993–996, 1997.

[Kim K., 1998a] K. H. Kim, K. H. Lee, S. B. Park, J. K. Song, S. N. Kim, and J. H. Souk, Domain divided vertical alignment mode with optimized fringe field effect, 18th International Display Research Conference, Asia Display'98, pp.383–386, 1998.

[Kim K., 1998b] K. H. Kim, S. B. Park, J. U. Shim, J. H. Souk, and J. Chen, New LCD modes for wide-viewing-angle applications, SID'98 Digest, p.1085, 1998.

[Koma N., 1995] N. Koma, Y. Baba, and K. Matsuoka, No-rub multi-domain TFT-LCD using surrounding-electrode method, SID'95 Digest, p.869, 1995.

[Lee S., 1997] S. H. Lee, H. Y. Kim, I. C. Park, B. G. Rho, J. S. Park, H. S. Park, and C. H. Lee, Rubbing-free, vertically aligned nematic liquid crystal display controlled by in-plane field, *Appl. Phys. Lett.*, 71, p.2851, 1997.

[Lee S., 1998] S. H. Lee, S. L. Lee, and H. Y. Kim, Electro-optic characteristics and switching principle of a nematic liquid crystal cell controlled by fringe-field switching, *Appl. Phys. Lett.*, 73, pp.2881–2883, 1998.

[Lien A., 1993] A. Lien and R. A. John, Multi-domain homeotropic liquid crystal display for active matrix application, Euro Display'93, p.21, 1993.

[Moench H., 2003] H. Moench, H. Giese, U. Hechtfischer, G. Heusler, A. Koerber, F.-C. Noertemann, P. Pekarski, J. Pollmann-Retsch, A. Ritz, and U. Weichmann, UHP lamps with increased efficiency, SID'03 Digest, pp.758–761, 2003.

[Mori H., 1997] H. Mori, Y. Itoh, Y. Nishiura, T. Nakamura, and Y. Shinagawa, Novel optical compensation film for AMLCD's using a discotic compound, Proceedings of the SID Symposium, pp.941–944, 1997.

[Mori H., 1998] H. Mori and P. J. Bos, Designing concepts of the discotic negative birefringence compensation films, Proceedings of the SID Symposium, pp.830–833, 1998.

[Nam S., 1997] M. S. Nam, J. W. Wu, Y. J. Choi, K. H. Yoon, J. H. Jung, J. Y. Kim, K. J. Kim, J. H. Kim, and S. B. Kwon, Wide-viewing-angle TFT-LCD with photo-aligned four-domain TN mode, SID'97 Digest, pp.933–936, 1997.

[Oh-e M., 1995] M. Oh-e, M. Ohta, S. Arantani, and K. Kondo, Principles and characteristics of electro-optical behavior with in-plane switching mode, Asia Display'95, p.577, 1995.

[Oh-e M., 1997] M. Oh-e, M. Yoneya, and K. Kondo, Switching of negative and positive dielectro-anisotropic liquid crystals by in-plane electric fields, *J. Appl. Phys.*, 82, pp.528–35, 1997.

[Ohmuro K., 1997] K. Ohmuro, S. Kataoka, T. Sasaki, and Y. Koike, Development of super-high-image-quality vertical-alignment-mode LCD, SID'97 Digest, p.845, 1997.

[Okamoto N., 2001] Norihisa Okamoto, Developments in p-Si TFT LCD projectors, SID'01 Digest, pp.1176–1179, 2001.

[Ou C., 2001] C. R. Ou, W. J. Young, T. J. Chen, W. T. Lee, H. Y. Lin, and C. C. Lin, Microlens: a projection system point of view, Asia Display/IDW'01, p.1331, 2001.

[Sergan T., 1998] T. A. Sergan and J. R. Kelly, Improvement of the optical characteristics of a TN LCD using negative in-plane and switched discotic films, Proceedings of the SID Symposium, pp.694–697, 1998.

[Stupp E. H., 1999] E. H. Stupp and M. S. Brennesholtz, *Projection Displays*, John Wiley & Sons, Ltd, Chichester, 1999.

[Takatori K., 1992] K. Takatori, K. Sumiyoshi, Y. Hirai, and S. Kaneko, A complementary TN LCD with wide-viewing-angle grayscale, Japan Display'92, Proceedings of the 12th international display research conference, pp.591–594, 1992.

[Takeda S., 1998] S. Takeda, S. Kataoka, T. Sasaki, H. Chida, H. Tsuda, K. Ohmuro, Y. Koike, T. Sasabayashi, and K. Okamoto, A super-high image quality multi-domain vertical alignment LCD by new rubbing-less technology, SID'98 Digest, p.1077, 1998.

[Vermeirsch K., 1999] K. Vermeirsch, K. D'havé, and B. Verweire, Characterization of the biaxiality of the Fuji-wide view film, SID'99 Digest, pp.677–680, 1999.

[Vithana Y., 1995] Y. Vithana, Y. K. Fung, S. H. Jamal, R. Herke, P. J. Bos, and D. L. Johnson, A well-controlled tilted-homeotropic alignment method and a vertically aligned four-domain LCD fabricated by this technique, SID'95 Digest, p.873, 1995.

[Wu S., 2001] S. T. Wu and D. K. Yang, *Reflective Liquid Crystal Displays*, p.305, John Wiley & Sons, Ltd, Chichester, 2001.

[Yang K., 1992] K. H. Yang, Two-domain 80°-twisted nematic liquid crystal display for greyscale applications, *Jpn. J. Appl. Phys.*, Part 2, 31, pp.L1603–1605, 1992.

10

Three-panel Reflective Systems

10.1 Introduction

The previous chapter covered in detail what has become the standard three-panel transmissive architecture. Its commercial success has driven down the cost of individual components and insured their availability. For this reason, the first three-panel LCOS system developed was a derivative of the transmissive system, with each modulating subsystem (i.e., polarizer/panel/analyzer) being replaced by a PBS and LCOS panel. This so-called "3×PBS/X-cube" architecture has been further developed from its original format by employing alternatives to the MacNeille PBS cubes. A further development of the three-panel transmissive system is the off-axis LCOS system. Here the modulator consists of a separate sheet polarizer and analyzer, with off-normal panel illumination used to create separate paths. More recently, architectures employing RSFs have been successfully introduced. This chapter will describe these approaches, concentrating on actual systems currently being developed for the RPTV market. The initial description of the color management using standard illumination optics will be cursory as it has been dealt with in detail elsewhere. Emphasis will be on the subtle polarization effects of the modulating systems and the restrictions they place on the system design and its final performance. Other approaches that have yet to deliver satisfactory performance at the prototype level have been left out intentionally. For further information on these systems, the reader is directed towards other publications [Stupp E. H., 1999].

10.2 3 × PBS/X-cube System

The generic system, shown schematically in Figure 10.1, is based on three cube MacNeille PBSs surrounding an X-cube. It was first developed by IBM and Nikon in the mid 1990s [Melcher R., 1998] (see Chapter 1) and later commercialized by JVC [Stirling R., 2000; Imaoka H., 2001].

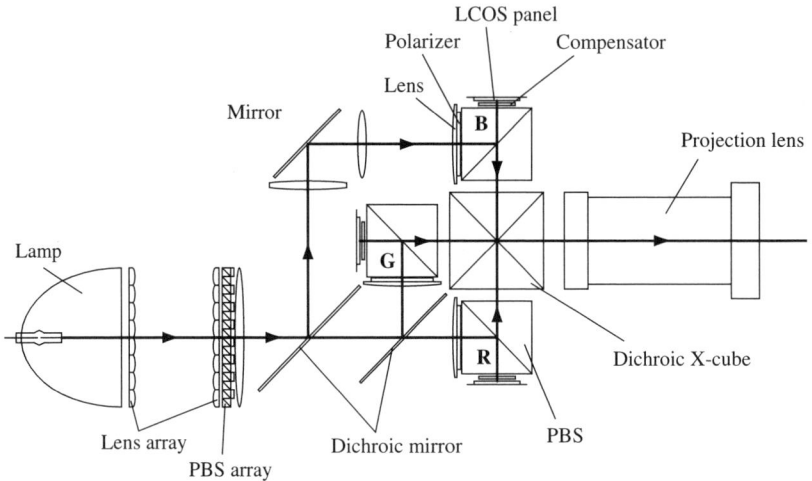

Figure 10.1 Schematic of the 3 × PBS/X-cube reflective three-panel architecture

10.2.1 Description of Basic Operation

Predominantly **s**-polarized white light enters the system typically from a homogenized, polarization-converted UHP source. Blue light is first separated with a graded dichroic mirror and directed via mirrors and path matching relay optics to a PBS. A second dichroic mirror then taps the green light from the red and directs the separate illuminating beams to separate PBSs. Coated lenses, together with pre-polarizers, are situated adjacent to the input of the PBSs to restore telecentricity prior to reflection from the PBS and to ensure clean input polarization. For transmission and durability reasons, the polarizers are typically wire grid components, although for cost purposes film-type absorption polarizers can sometimes be used in the red and green channels. The reflected light from each PBS is then incident on an LCOS panel via any compensating elements, one being a QWP to correct for the skew ray polarization mixing of the MacNeille prism. Modulated light from each channel is analyzed in transmission by each PBS before entering the combining X-cube. At the exit of the red and blue PBSs are wavelength-specific HWPs oriented at 45° which transform the exiting **p**-polarized light to **s**- for efficient reflection by the X-cube. A projection lens images the superimposed R, G, and B modulated beams onto a screen, via any folding mirrors in the case of an RPTV system.

10.2.2 Comparison to Transmissive System

The metrics are as follows:

- **Increased engine size.** Having PBSs as part of the modulating system increases the overall system size. Engine size is not a major issue in RPTV systems, but can be a deciding factor for front projectors.

- **Increased back focal length (bfl).** The additional projection path through the PBS, and the larger X-cube necessary to accommodate it, increase the bfl requirement on the projection lens.

- **Working $f_{/\#}$.** Accommodating the larger bfl with a cost-effective projection lens typically limits systems to higher $f_{/\#}$, increasing the optical mismatch between source and system and decreasing optical throughput. RPTV systems are, however, not so demanding of optical throughput.

- **ANSI contrast.** The introduction of reflective LC panels tends to decrease ANSI contrast through back reflection of projected light from surfaces within the projection path as described in Chapter 2.

- **Uniformity.** In both systems, dark-state uniformity is affected primarily by thermally induced stress birefringence in components making up the modulating system. The large glass PBSs can cause non-uniformity, but the more recent introduction of highly elastic glasses effectively solves this issue. In the transmissive system, absorption by the polarizers and the panel black matrix can cause thermal gradients, which can introduce birefringence in substrate materials.

- **Cost.** As with all systems, final costs are heavily dependent upon the quantity cost of key components. Currently, the expected additional cost of three PBSs is expected to be greater than the cost savings associated with smaller LCOS panels.

- **Resolution.** The conflicting requirements of smaller, lower cost panels with high aperture ratio for high transmission make transmissive panels unattractive at high resolutions. Though LCOS does have resolution limits, 0.7″ diagonal, 1080i panels are commonplace and higher resolutions have been demonstrated.

- **Speed.** In three-panel systems, subframe (<16 ms) switching speed is not essential, as frame-to-frame images are highly correlated. However, fast-moving scenes displayed with slow panels tend to give the sensation of streaking and associated blurring. Reflective panels are typically $\sim 4\times$ faster than their transmissive counterparts due to the double pass through the material, which halves the cell gap for any given LC mode and material.

- **Brightness.** For systems with similar panel size and lamp arc length, the loss in brightness of the reflective system through increase in illumination $f_{/\#}$ is offset by the greater losses of the transmissive modulating system (see later for details). Increasing the lamp power can improve throughput of reflective systems, but is limited in transmissive systems due to reliability concerns.

- **Contrast.** Improved planarized reflection layers with normally black (VA) LC modes have pushed the sequential on-screen contrast of reflective systems in excess of 1000:1. Scattering from electrode structures in transmissive panels limits contrast to typically $<1000:1$.

240 THREE-PANEL REFLECTIVE SYSTEMS

Of the above metrics, brightness and contrast are considered key and require more detailed analysis.

10.2.3 Brightness

The system throughput can be directly compared to the transmissive system of Chapter 9 by considering the transmission of the weakest color (red) through the modulation system. This assumes both systems use similar color management, panel sizes, lamps, and system $f_{/\#}$. Similar to the throughput analysis of the previous chapter, the modulating system throughput can be written:

$$\eta_m = \eta_{pol}\eta_{PBS}\eta_p \qquad (10.1)$$

where:

- η_{pol} is the transmission of the input clean-up polarizer, which in cases where a wire grid polarizer is used is ~90%

- η_{PBS} is the cone averaged transmission $\bar{T}_p$ of the PBS, which for a highly transmitting MacNeille cube is ~95%

- η_p is the panel efficiency that includes reflection and LC mode losses. Currently panels are typically ~70% efficient.

Combining these yields an overall modulation system efficiency of 60%, which, when compared to the 46% efficiency for the three-panel transmissive system of the previous chapter, represents a 30% increase in system throughput.

10.2.4 Contrast

The contrast of the system is determined by the MacNeille PBS/LCOS modulating system contrast, which can be written as the expanded sum of contributing leakage terms given by equation (7.23) of Section 7.8:

$$\frac{1}{C_{sys}} \approx \frac{1}{C_{coating-only}} + \frac{1}{C_{pre-cond}} + \frac{1}{C_{AR}} + \frac{1}{C_{compensation}} + \frac{1}{C_s} \qquad (10.2)$$

Each term can be estimated for the MacNeille-based, 3×PBS/X-cube architecture:

- $C_{coating-only}$ describes the contrast of a system in which all polarization is geometrically compensated and the panel is perfect. Assuming a high contrast pre-polarizer and no clean-up analyzer, i.e., $C_{P1} > 1000$ and $C_{P2} = 1$ in equation (7.26), the leakage due to PBS coatings is dominated by the average transmission of s-polarized light, or:

$$\frac{1}{C_{coating-only}} \approx \langle T_s \rangle \approx \frac{1}{10\ 000} \qquad (10.3)$$

- $C_{pre-cond}$ is a consequence of the geometrical mismatch between clean-up o-type polarizers and MacNeille PBSs, which can be calculated analytically using equation (7.27). Perfect clean-up sheet polarizers provide input light with near-perfect linear polarization, but since the polarization axes of cube MacNeille PBSs are rotated for skew rays, there is a mismatch leading to an average of ~0.25% PBS **p**-polarized light entering the system (see Section 7.3.3.2). For a perfect ($T_s = 0\%$, $T_p = 100\%$) PBS coating this would not matter, since only **s**-polarized light would be reflected. However, for PBS coatings that have a finite reflection of **p**-polarized light, $R_p \sim (1 - T_p)$, this light will be directed toward the panel and escape from the system in the off-state.

The reflection of **p**-polarized light as a function of incidence angle, $R_p(\theta)$, can be measured, simulated, or obtained from the vendor for any given cube (see Chapter 4), with values for any intermediate angles determined by suitable interpolation. Taking the data from Figure 4.17, and averaging over a green wavelength band (530 nm < λ < 560 nm), an FOV mapping of R_p can be obtained by geometrical calculations as shown in Figure 10.2(a).

When multiplied by the **p**-component of Figure 7.23(a), an intensity map of the **p**-polarized light on the panel can be obtained, as shown in Figure 10.2(b). This light escapes the system with minimal loss and represents a contrast limit $C_{pre-cond} \sim 35\,000{:}1$. In the 3 × PBS/X-cube system, the demands on the PBS to have both low T_s and high T_p often lead to a compromised R_p performance, one which is lower than the particular low R_p cube of Figure 4.17 and can lead to:

$$\frac{1}{C_{pre-cond}} \approx \frac{1}{15\,000} \tag{10.4}$$

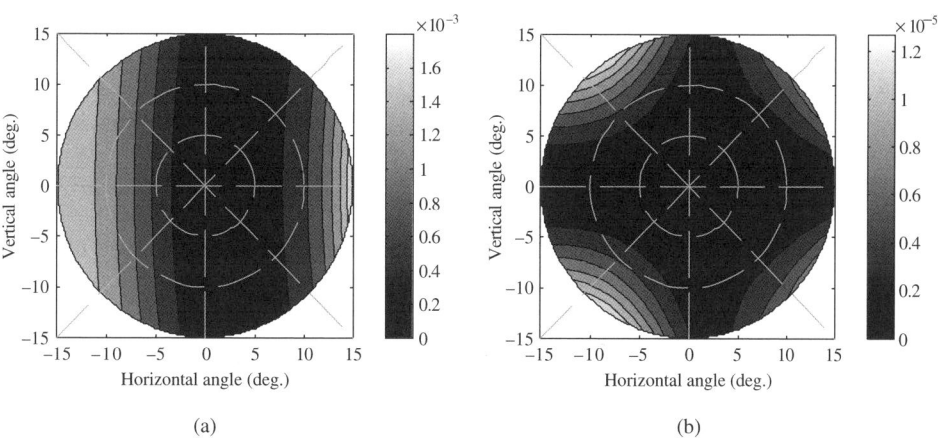

Figure 10.2 (a) FOV plot of R_p, (b) FOV of $R_p \times$ **p**-component

- C_{AR} is given by equation (7.21). In the case where a separate index-matched low-birefringence QWP compensator is used, the PBS AR coating and the adjacent AR coating of the QWP dominate and sum to give R_f. Assuming good <0.3% reflecting broadband AR coatings, $R_f \sim 0.006$. Since the second term of equation (7.21) is negligible in comparison to the first:

$$\frac{1}{C_{AR}} \approx \frac{1}{8000} \tag{10.5a}$$

- $C_{compensation}$ is leakage after compensation of the LC mode, in combination with the PBS reflection geometry as discussed in detail in Section 7.4. In a test system with a 0° oriented QWP and mirror, the combined leakage at $f_{/2.8}$ is ~40 000:1 (Section 7.3.3.1). For a <10° pretilt VA mode panel at $f_{/2.8}$ compensated by a uniaxial QWP, the contrast from Figure 7.31 is estimated as:

$$\frac{1}{C_{compensation}} \approx \frac{1}{9000} \quad (10.5)$$

- C_s is essentially the contrast of the bare panel, assuming all polarization effects are corrected. Though difficult to measure at high levels, good panels can deliver contrasts exceeding 3000:1. This has been achieved using a wire grid PBS with input and output clean-up polarizers and effective panel compensation. The limiting factor at this level is depolarization due to scatter from imperfect surfaces, or:

$$\frac{1}{C_s} \approx \frac{1}{3000} \quad (10.6)$$

Combining these terms yields an estimated system contrast of 1400:1 for the 3×PBS/X-cube architecture with good panels. Practical systems can achieve >2000:1 with, for example, higher $f_{/\#}$ systems, those employing non-cylindrically symmetric angular illumination distributions, improved PBS and AR coatings, and better raw panels. This calculation, as with throughput, highlights the individual terms covered in Chapter 7 as they relate to a real system, and can be useful for relative system comparisons.

10.2.5 Systems Upgrades

A recent version of the MacNeille 3×PBS/X-cube architecture employs tilted PBS cubes. This improves coating performance by matching the coating and substrate material properties better, as discussed in Chapter 4. The slightly reduced incidence angle can also act to decrease geometrical effects, making the system contrast higher for any given compensation scheme and panel. Furthermore, the reference system architecture that uses the latest core uses a dichroic plate X-mirror to equalize path lengths and avoid relaying the blue illumination. The architecture [Ishino H., 2004] is shown in Figure 10.3.

10.2.6 Alternative PBS Solutions

More recently there have been alternative modulating systems that offer improved performance over the basic reflective 3×PBS/X-cube system. The first uses wire grid plate (WGP) PBSs and a second uses multilayer birefringent cube (MBC) PBSs.

10.2.6.1 Wire-grid PBS System [Arnold S., 2001; Pentico C., 2003]

A wire grid 3×PBS/X-cube reference architecture is shown in Figure 10.4.

In plate PBS systems, imagery must be reflected from the surface to avoid astigmatism, which forces a more complex system. The design shown in Figure 10.4 is necessary to

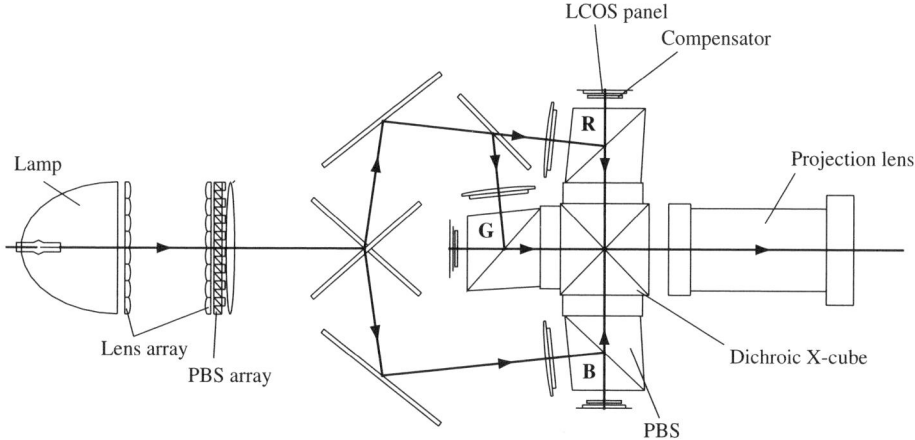

Figure 10.3 Modified MacNeille 3 × PBS/X-cube architecture

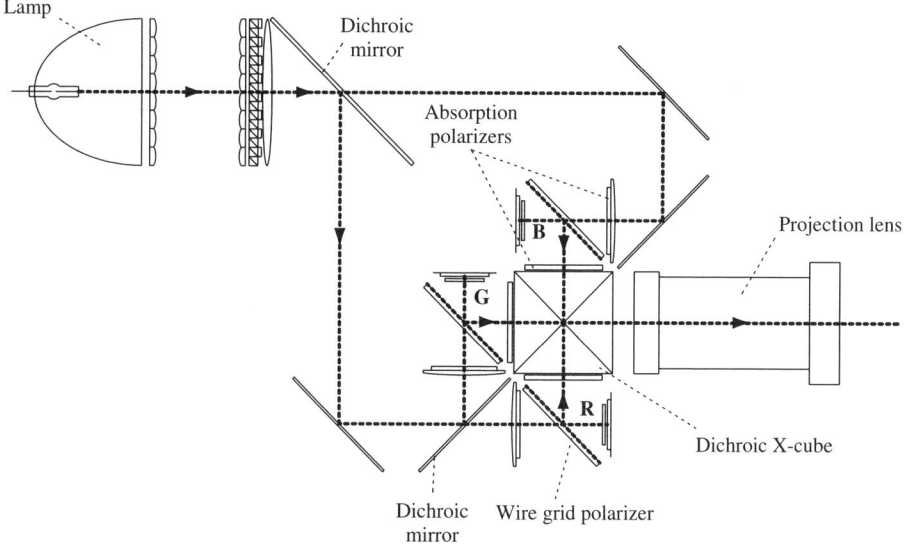

Figure 10.4 WGP PBS three-panel LCOS system

avoid mechanical and optical interference, and exacerbated by the telecentric requirement for PBS self-compensation and high contrast. One advantage of this system is that the matched illumination paths avoid optical relay of the blue channel. The increased path length in air acts to increase system size and bfl, but operation at $f_{/2.5}$ is possible with commercial projection lens designs. The major benefit over the MacNeille version is in system contrast, as the wire grid version does not require geometrical polarization correction. One drawback is transmission since pre- and post-polarizers are necessary to achieve the high contrast, where the clean-up on the output side is best an absorption sheet polarizer to avoid reduction

244   THREE-PANEL REFLECTIVE SYSTEMS

in ANSI contrast. Its performance can be compared to the conventional $3 \times$ PBS/X-cube system similar to that of the previous sections:

- **Increased engine size.** The optical path in air tends to increase the overall system size, although not having to accommodate glass cubes can alleviate mechanical interference.
- **Increased bfl.** The optical path in air produces an increased bfl.
- **Lower working $f_{/\#}$.** The high angle performance of the wire grid polarizers makes operating at $f_{/2.5}$ attractive, despite demanding more complex projection lens designs. The lower chromatic and spherical aberration correction does help in this regard.
- **Comparable ANSI contrast.** The number of reflecting surfaces in the projection path of the two systems is comparable.
- **Improved uniformity.** Removing bulk material prevents stress-induced birefringence effects and in general improves uniformity.
- **Resolution.** Here the quality of the imaging system can limit the resolution at the screen. While the wire grid system can in principle be very good, mechanical registration of the plates, and maintenance at high temperatures, can cause misalignment of the pixels, which becomes more challenging with higher resolution panels. Also, the localized heat generated by the metal absorption can cause the plate to deform due to thermal expansion. Deformation of the plate by only a few waves per inch can severely damage the image quality at the screen, causing smearing.
- **Reduced brightness.** By again considering the modulating element in isolation, a comparative throughput can be estimated. Its efficiency can be written as:

$$\eta_m = \eta_{in\_pol} \cdot \eta_{PBS\_t} \cdot \eta_p \cdot \eta_{PBS\_t} \cdot \eta_{out\_pol} \quad (10.7)$$

where:

- η_{in_pol} is the transmission of the input clean-up polarizer; in cases where a wire grid polarizer is used, this is $\sim 90\%$
- η_{PBS_t} is the **s**-transmission of the wire grid PBS and is typically $\sim 90\%$ (Figure 4.23)
- η_p is the usual $\sim 70\%$ panel efficiency factor
- η_{PBS_r} is the $\sim 94\%$ **s**-reflection off the wire grid PBS (Figure 4.23)
- η_{out_pol} is the $\sim 90\%$ transmission of the output absorption clean-up sheet polarizer.

The overall modulation system efficiency is therefore 48%, which is comparable to the three-panel transmissive system efficiency (46%), but is $\sim 30\%$ lower than the MacNeille equivalent. Removing the input clean-up polarizer would regain 10% of the efficiency loss, but could compromise the very good contrast performance. Removal of the output absorption polarizer is not an option due to modest ($\sim 3\%$) R_p.

- **Increased sequential contrast.** The contrast performance of the wire grid PBS modulating system is very good. The individual contrast terms $C_{coating-only}$, $C_{pre-cond}$, C_{AR} all exceed >100 000:1, and $C_{compensation}$ > 50 000:1 assuming dual clean-up polarizer with a- and c-plate panel compensation. Therefore, as stated above, the C_s term dominates. Current VA panels can deliver up to 3000:1 contrast in such systems, beyond which other system effects from insufficient light shielding etc. will dominate.

10.2.6.2 MBC PBS System [Bruzzone C., 2003]

A bonded 3×PBS/X-cube was recently introduced, shown in Figure 10.5. It can be considered a direct replacement of the MacNeille version in the system shown in Figure 10.1. As for the WGP PBS case above, the advantage of this technology over the MacNeille version stems from the lack of geometrical polarization rotation. It is also capable of high **p**-transmission and therefore low R_p, possibly sufficiently low to avoid an input clean-up polarizer. On the downside, it is based on a polymer film embedded in a glass cube, resulting in possible stress birefringence issues. Note that phase distortion in the reflected channel due to poor flatness of the film forces the image to be transmitted by the PBS. This latter point is unfortunate as the inability to reflect the image precludes its use in a shared PBS color management system. Future encapsulating techniques could potentially overcome this limitation. The organic nature of the film is also a reliability concern, although recent reports suggest that this may be less of an issue in the future [Bruzzone C., 2004].

Figure 10.5 Picture of the Vikuiti optical core (Courtesy C. Bruzzone)

The modulation system throughput of this architecture can be estimated at 68% if no input clean-up polarizer is used and ∼62% in the typical configuration, making it one of the highest transmitting systems reported. Contrast can be high, with ∼8000:1 measured using QWPs and mirrors (see Figure 4.21).

The overall system contrast of an actual system can be expected to be a combination of the ∼8000:1 system contrast with a ∼3000:1 panel contrast since all other terms of equation (10.2) are negligible. Simple calculations based on reciprocal contrast numbers yield ∼2200:1.

246 THREE-PANEL REFLECTIVE SYSTEMS

10.2.6.3 Off-axis System [Bone M., 1999]

The off-axis system was one of the first LCOS systems to be commercialized. Shown schematically in Figure 10.6, this approach separates the input and output beams incident on a panel in angle. The great advantage of this approach is the ability to avoid double pass through pre- and post polarizing elements such as PBSs, compensators, etc. On the downside, the off-axis projected light requires expensive off-axis projection optics [Bone M., 2000], and registration of the panel images at the screen can be a problem as the system runs off telecentric to avoid optical interference between illumination and projected beams. With off-telecentric systems, image magnification is related to the distance of the panel to the lens, a condition absent from the constant magnification telecentric case. Furthermore, the lateral color distortion is more difficult to control in off-axis lens designs.

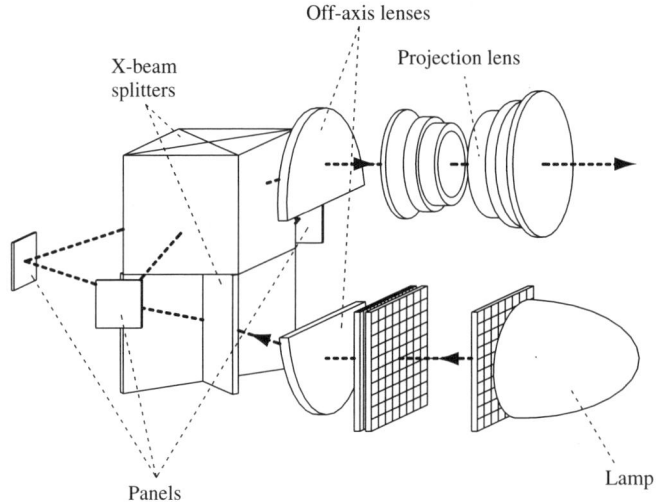

Figure 10.6 Off-axis system schematic

Its contrast and throughput can be estimated:

- **Contrast.** The contrast of the bare off-axis system is determined by the modulating system for each of the color channels introduced in Section 7.7.2. With perfect panels, and with an HWP polarizer compensation, the system should deliver ~8000:1 [Bone M., 2001]. Introducing panels reduces the contrast to a level limited by the compensation of the panel. Using a negative c-plate with VA panels, the contrast can be expected to reach >1000:1. Early commercial systems based on normally black TN modes yielded ~300:1 contrast.

- **Throughput.** The transmission of the modulating system is that of two sheet polarizers, double pass through the field lens, and panel loss. With typical values, $\eta_m \sim 56\%$, which is ~10% lower than the 3×PBS/X-cube architecture due essentially to the extra sheet polarizer. In the commercially released Everest off-axis front-projection system, 1100 lumens is the specified output from a 150 W UHP lamp.

10.3 Polarization Color Filter Systems

The simplest approach to RSF based color management uses a single PBS to split and combine light between two panels as described in Section 8.3.1.2. For a three-panel system at least one additional beam splitter is required to separate and combine light from a third reflective panel. One of the first architectures to exploit this approach uses a single PBS [Heine C., 2001; Van Doorselaer G., 2002]. Shown in Figure 10.7, it uses an input green/magenta (*GM*) RSF (Retarder Stack Filter) and single PBS to separate green light from magenta, and then a dichroic cube to split and combine the red and blue spectra. An output *GM* filter recombines the polarization of the projected light prior to a clean-up polarizer, as for the subsystem described earlier.

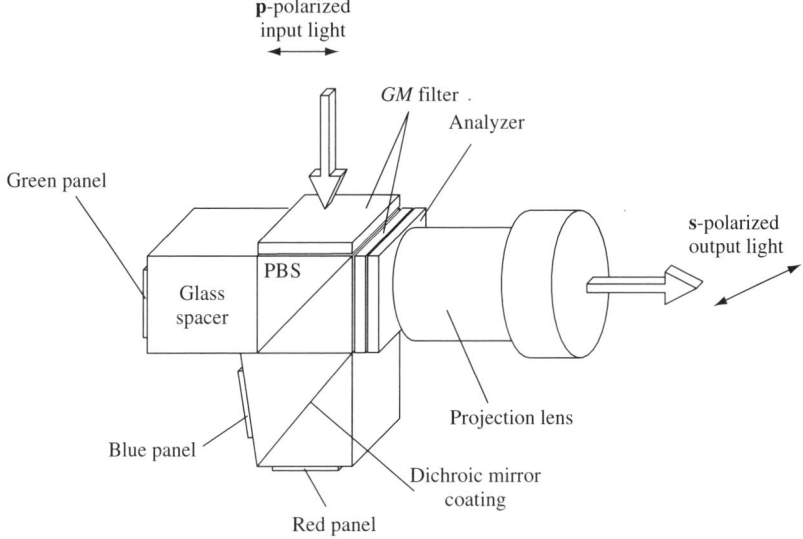

Figure 10.7 Color corner architecture

Originally, the reflecting planes of the PBS and the dichroic were parallel for geometrical polarization matching reasons. Mechanical interference, and the need to use high-index, low-stress birefringence glass, however, made a higher incidence dichroic cube more attractive, with minimal performance degradation due to the non-ideal reflecting plane geometry. Unfortunately, polarization mixing by the dichroic cube causes significant, non-uniform magenta leakage. Furthermore, the dichroic cube's highly separated **s**- and **p**-bands significantly reduce the throughput of the limiting magenta light. These two factors make it difficult to achieve the required contrast and brightness targets of current projection systems.

An improved system is an architecture which uses four RSFs and four PBSs [Robinson M., 2000]. Shown in Figure 10.8, this system can achieve high contrast and acceptable throughput by using the shared PBS directly adjacent to the red and blue panels.

The input **p**-polarized white light passes through a green/magenta (*GM*) RSF, which transforms magenta into **s**-polarized light leaving green unaffected. A first PBS then separates the colors, and a second splits and combines blue and green with a pair of input/output

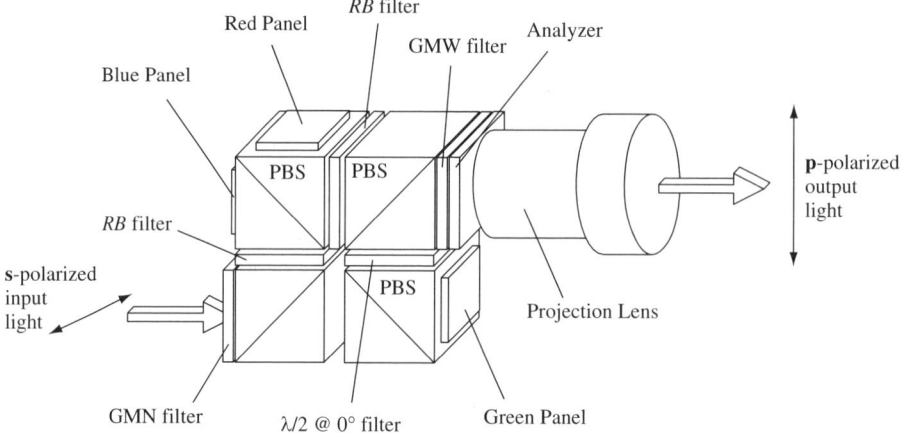

Figure 10.8 Color Quad™ (CQ) three-panel architecture

red/blue *RB* filters. The green light is incident on a panel and deflected toward a combining output PBS. A final GM filter combines the projected output into a single polarization for "clean-up".

This system is capable of achieving high performance and is very tolerant of component specifications. In particular, the coating performances of the PBSs can be modest, particularly for **s**-transmission, allowing cost-effective low-index glass to be used. The sequential contrast can be very high, although there is a bias toward green, which can make the OFF-state magenta in hue. This architecture has been successfully commercialized by JVC in the SX21 front projector and specifies 1500 ANSI lumens with a 250 W lamp at a sequential contrast >800:1. In prototype systems, >2000:1 is readily achievable.

10.3.1 The CQ3 Three-PBS Architecture [Robinson M., 2004]

Recent improvements over the four-PBS system have been made by first removing the output polarizer to increase transmission, image clarity, and durability, and second by replacing the input PBS with a dichroic mirror plate. The subtleties of this latest design make it very attractive from a performance/cost standpoint. Here, this particular modification is presented in detail as a specific example of this generic color management approach.

The three-PBS architecture is shown schematically in Figure 10.9. Predominantly **s**-polarized light enters the architecture from the left as drawn and is separated into blue transmitted and yellow reflected beams. The blue light polarization is cleaned up with a linear polarizing element before reflecting onto a blue modulating panel. When transformed by the panel into **p**-polarized light, it is analyzed in transmission through the adjacent PBS before further rotation by a *BY* RSF into **s**-polarized light. Finally, blue imagery is reflected from the output PBS and is imaged onto a screen. The green and red illumination is cleaned up prior to manipulation by an input *RC* filter. The shared PBS then operates as described in the previous chapter to separate and recombine the R and G channels, such that the modulated light is transmitted through the output PBS to join the modulated blue light.

POLARIZATION COLOR FILTER SYSTEMS

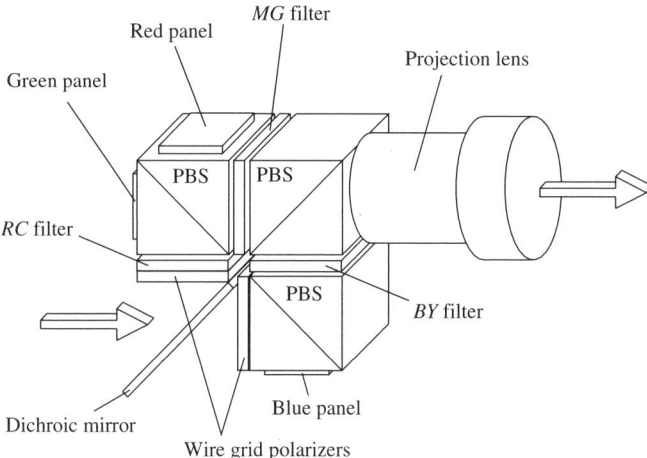

Figure 10.9 CQ3 three-PBS architecture

The cost advantages stem primarily from a reduction in the PBS and polarization filter count from the four-PBS version. Its performance is also improved in many aspects:

- **Increased throughput.** Transmission is increased primarily from avoiding the need for an output absorption sheet polarizer. Although the average transmission of a neutral sheet polarizer is ~90%, in a color balanced system its effect on the brightness of the projected image is determined by transmission of the weakest color. In RPTV systems this is often blue, making the throughput improvement consistent with the 15–20% attenuation that sheet polarizers deliver at wavelengths <500 nm. Further improvements have been made over blue-starved systems by reducing the number of PBSs that blue light must pass through.

- **Better image quality.** Removing the free-standing sheet polarizer at the entrance to the projection lens avoids any phase deformation and any resulting soft focus within the image. Also, imaging the photopically dominant green panel in transmission avoids phase distortion through reflection at the PBS coating surface.

- **Improved dark-state color.** The contrast of the isolated blue channel is greater that that of the red and green color channels sharing the single PBS. This compensates for lower contrast blue panels and creates a more naturally colored gray OFF-state. The four-PBS version tends to exhibit a blue or magenta dark state.

- **Higher ANSI contrast.** Reducing the projection path component count and associated reflections improves ANSI contrast, without preventing the polarization isolation techniques of an exit QWP (see Chapter 7).

- **Better uniformity.** Thermally induced stress birefringence is controlled by higher index, lead glass PBSs. Also, the free-standing input *RC* filter does not exhibit stress through differential expansion associated with encapsulating glass. Furthermore, the air space surrounding the input dichroic splitter allows the temperature of the higher illuminated input components to be cooled effectively with modest airflow. The net result is a much more uniform colored projected image, particularly at the more sensitive lower gray levels.

250 THREE-PANEL REFLECTIVE SYSTEMS

The overall contrast of the CQ3 is determined by component performance and can achieve a greater performance than the four-PBS system, but without the tolerance to PBS design as will be seen in the following system analysis.

10.3.2 System Analysis

(a) ON-states

The three-PBS architecture can be analyzed using the intensity-based vector approach outlined in Chapter 8. Using this formulation, we can track the light through each of the different channels for the projected ON-states by representing propagation through each component by a matrix multiplication. The projected red, green, and blue spectra are:

$$R_{ON} = \mathbf{T}^3 \cdot MG \cdot \mathbf{R}^2 \cdot \begin{pmatrix} 0 & 1 \\ 1 & 0 \end{pmatrix} \cdot \mathbf{T}^2 \cdot RC \cdot \mathbf{P} \cdot \mathbf{D}_r \cdot \begin{pmatrix} 1 \\ \delta_{PCS} \end{pmatrix} \quad (10.8)$$

$$G_{ON} = \mathbf{T}^3 \cdot MG \cdot \mathbf{T}^2 \cdot \begin{pmatrix} 0 & 1 \\ 1 & 0 \end{pmatrix} \cdot \mathbf{R}^2 \cdot RC \cdot \mathbf{P} \cdot \mathbf{D}_r \cdot \begin{pmatrix} 1 \\ \delta_{PCS} \end{pmatrix} \quad (10.9)$$

$$B_{ON} = \mathbf{R}^3 \cdot BY \cdot \mathbf{T}^1 \cdot \begin{pmatrix} 0 & 1 \\ 1 & 0 \end{pmatrix} \cdot \mathbf{R}^1 \cdot \mathbf{P} \cdot \mathbf{D}_t \cdot \begin{pmatrix} 1 \\ \delta_{PCS} \end{pmatrix} \quad (10.10)$$

where:

$\mathbf{T}^j = \begin{pmatrix} T_s^j & 0 \\ 0 & T_p^j \end{pmatrix}$ is the transmission power matrix of the jth PBS

$\mathbf{R}^j = \begin{pmatrix} R_s^j & 0 \\ 0 & R_p^j \end{pmatrix}$ is the reflection power matrix of the jth PBS

$MG = \begin{pmatrix} (MG)_p & (MG)_x \\ (MG)_x & (MG)_p \end{pmatrix}$ is the magenta/green filter power transfer matrix with a similar expressions for the RC and BY filters

$\mathbf{P} = \begin{pmatrix} P_s & 0 \\ 0 & P_t \end{pmatrix}$ is the power transmission of the clean-up input polarizers

$\mathbf{D}_r = \begin{pmatrix} D_{r,s} & 0 \\ 0 & D_{r,p} \end{pmatrix}, \mathbf{D}_t = \begin{pmatrix} D_{t,s} & 0 \\ 0 & D_{t,p} \end{pmatrix}$ are the reflection and transmission matrices of the dichroic mirror, and δ_{PCS} is the power fraction of **p**-polarized light entering from the imperfect PCS system.

The internal polarization transformation matrix represents the ideal polarization manipulation at the panel.

Expanding these expressions and keeping the dominant factors, we get the following expressions for the projected red, green, and blue filter spectra:

$$R_{ON} = D_{r,s} T_p^2 T_p^3 P_s (MG)_x (RC)_x \quad \text{p-polarized} \quad (10.11)$$

$$G_{ON} = D_{r,s} T_p^2 T_p^3 P_s (MG)_p (RC)_p \quad \text{p-polarized} \quad (10.12)$$

$$B_{ON} = D_{t,s} T_p^1 P_s (BY)_x \quad \text{s-polarized} \quad (10.13)$$

POLARIZATION COLOR FILTER SYSTEMS   251

As these expressions represent spectra, they are best visualized graphically. Figures 10.10, 10.11, and 10.12 show typical individual component spectra and the subsequent projected spectra for red, green, and blue respectively. It is assumed that the wire grid clean-up polarizer and the PBS transmission are achromatic and their transmission values are at 100% and 95% for this exercise.

As shown in Figures 10.10, 10.11, and 10.12, the filter spectra for each of the primary colors can depend heavily, and often exclusively, on a single filter to determine the color.

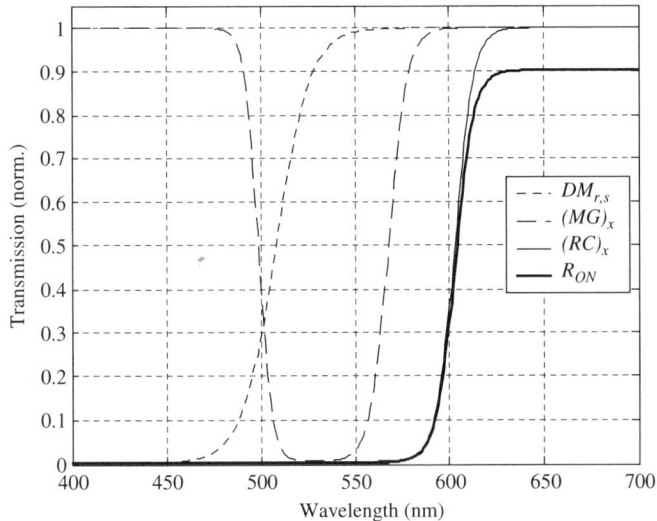

Figure 10.10   Red filter spectrum and relevant component spectra

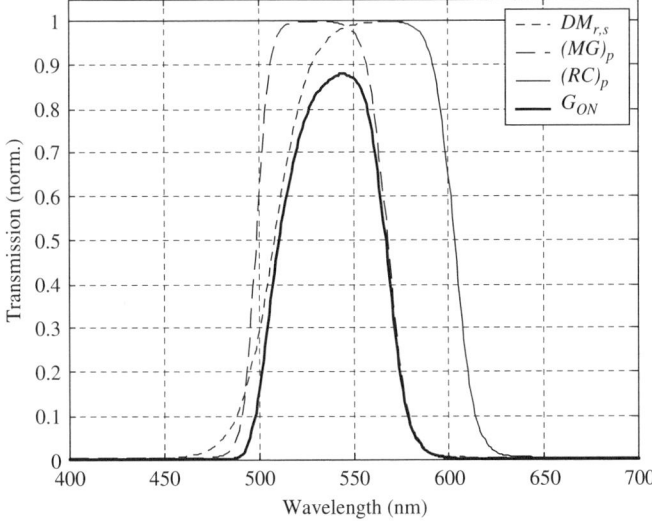

Figure 10.11   Green filter spectrum and relevant component spectra

252 THREE-PANEL REFLECTIVE SYSTEMS

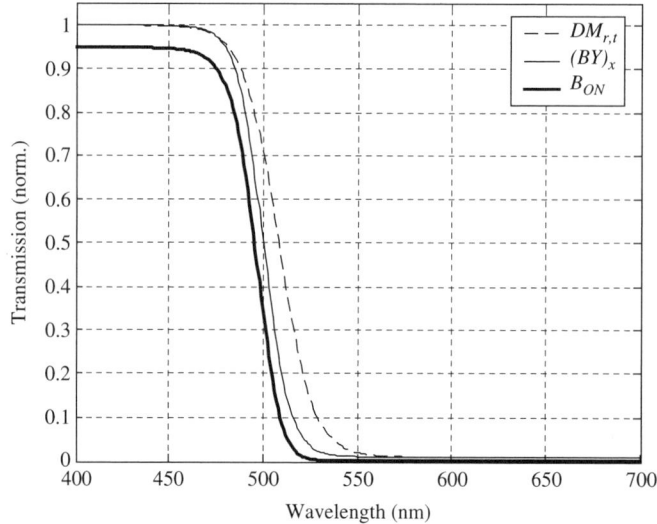

Figure 10.12 Blue filter spectrum and relevant component spectra

This is particularly true of red, which is simply the source filtered by the *RC* filter between crossed polarizers. For green, the critical long yellow cut-off wavelength is determined by the *MG* filter, and the shorter, more forgiving, cyan 50% point by the dichroic mirror. Blue is affected slightly by both the dichroic mirror and the internal *BY* filter.

(b) OFF-states

The OFF-states, and consequently the sequential contrast, of the three-PBS system can be analyzed similar to the ON-states. Using the same notation as above we can derive the following expressions for the three-PBS OFF-states:

$$R_{OFF} = \mathbf{T}^3 \cdot MG \cdot \mathbf{R}^2 \cdot \mathbf{T}^2 \cdot RC \cdot \mathbf{P} \cdot \mathbf{D}_r \cdot \left(\frac{1}{\delta_{PCS}}\right) \tag{10.14}$$

$$G_{OFF} = \mathbf{T}^3 \cdot MG \cdot \mathbf{T}^2 \cdot \mathbf{R}^2 \cdot RC \cdot \mathbf{P} \cdot \mathbf{D}_r \cdot \left(\frac{1}{\delta_{PCS}}\right) \tag{10.15}$$

$$B_{OFF} = \mathbf{R}^3 \cdot BY \cdot \mathbf{T}^1 \cdot \mathbf{R}^1 \cdot \mathbf{P} \cdot \mathbf{D}_t \cdot \left(\frac{1}{\delta_{PCS}}\right) \tag{10.16}$$

Since $\mathbf{R}^2 \cdot \mathbf{T}^2 \equiv \mathbf{T}^2 \cdot \mathbf{R}^2$ the leakages from the red and green channels are mathematically equivalent and a combined red/green leakage, GR_{OFF}, can be considered with each panel contributing half. The primary leakage terms are:

$$GR_{OFF} = L_1 + L_2 \quad \text{p-polarized} \tag{10.17}$$

where:

$$L_1 = 2D_{r,s} P_s T_p^2 T_p^3 T_s^2 (RC)_p (MG)_x \tag{10.18}$$

and:

$$L_2 = 2D_{r,s}P_sT_p^2T_p^3R_p^2(RC)_x(MG)_p \qquad (10.19)$$

The leakage of blue light is lower by an order of magnitude from that of the green/red channels, and is often below that of scattering effects within the optical components. However, care should be taken to avoid the *BY* transition being at a slightly longer wavelength than that of the dichroic mirror. Conversely, in the case of the green and red channels, the expression for the leakage accurately predicts OFF-state spectra. This yellow leakage is dominated by the performance of the *MG* and *RC* filters together with the *GR* PBS, which is best visualized as in Figure 10.13. This shows the leakage spectra due to the first and second terms (scaled by 1000) of the above expression, respectively, together with the relevant component spectra.

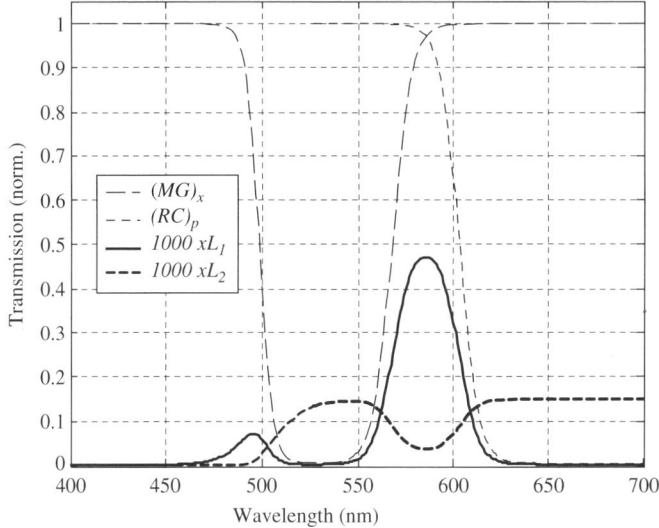

Figure 10.13 Leakage due to separated *RC* and *MG* filters corresponding to the two dominant yellow leakage terms. It assumes an average $T_p^2 \sim 3\%$, with *MG* and *RC* filter leakages $\sim 0.5\%$

From Figure 10.13 it can be seen that the two terms correspond to the separation and overlap of the *MG* and *RC* filter spectra, weighted by either the **s**-transmission or **p**-reflection of the adjacent *GR* PBS cube. The **p**-reflection of the cube affects leakage throughout the green and red wavelengths, with a slight enhancement where the filters overlap. The **s**-transmission affects only the yellow light in the separation region around 575 nm. For this reason it is best to optimize the cube for high T_p and to compensate for possible large T_s by notch filtering.

(c) Yellow Notch Filter Option
A yellow notch (YN) filter would ideally have high transmission throughout the projected green and red wavelength bands and zero transmission in the narrow intermediate yellow wavelength region. As discussed in Chapter 6, this is not easily achievable with angularly

sensitive dichroic interference filters. Using a YN RSF requires in general an extra polarizer. However, in systems that use a PCS it is possible to insert a YN RSF anywhere between the exit of the PBS and the clean-up polarizer situated directly in front of the RC filter. The width of the notch is unaffected by the polarization integrity of the PCS, but the level of the blocking at the null is directly proportional to the amount of **p**-polarized light present. In practice, systems that have a PCS with >5:1 polarization ratios should use a YN RSF since it reduces the T_s specification on the GR cube by a factor of almost five. Incorporating a YN into the system affects the primary red and green colors since the short red and long green filter transitions would be determined by the product of $(MG)_p$ and $(RC)_x$ with the YN filter spectra. From a filter tolerance standpoint it is advantageous to have the 50% transition points matched between the three filters: the MG and RC are matched to the short and long 50% wavelength transitions of the YN, respectively. Chromatic analysis indicates that the critical yellow wavelength cut-off should match and be ~592 nm for RPTV systems adhering to standard gamuts.

(d) System Throughput

The three-PBS system, as for all shared PBS architectures, combines the color management and modulating systems. Its throughput is:

$$\eta_{cm} \cdot \eta_m = \eta_{dich\_r} \cdot \eta_{in\_pol} \cdot (\eta_{PBS\_t})^2 \cdot \eta_p \cdot (\eta_{RSF})^2 \qquad (10.20)$$

which equates to ~56% with the typical values 98%, 90%, 97%, 70%, and 98% for η_{dich_r}, η_{in_pol}, η_{PBS_t}, η_p, and η_{RSF} respectively. Compared to the ~41% of the transmissive system, it represents ~37% higher system transmission and yields the highest transmission value of all three-panel reflective systems.

(e) System Contrast

Equation (10.2) describing the contrast of the MacNeille-based 3×PBS/X-cube again can be used to describe the dark-state leakage of the three-PBS system, with equivalent values attributed to the last three terms. The first two terms describe the system leakage as described in part (b) above.

$C_{coating-only}$ is the system contrast as described above and in the case where a low-T_s PBS or YN filter is used can be written:

$$C_{coating-only} \approx \frac{1}{2\langle R_p \rangle \langle \delta_{RC} \rangle} \qquad (10.21)$$

where δ_{RC} is the leakage of the RC filter as seen between crossed polarizers in the photopically dominant green region of the spectrum. Typically good filters have $\delta_{RC} \approx 0.003$ and for typical PBSs $\langle R_p \rangle \approx 0.02$, giving:

$$\frac{1}{C_{coating-only}} \approx \frac{1}{8500} \qquad (10.22)$$

$C_{pre-cond}$ is finite if the polarization entering the non-compensated filter is not geometrically pre-conditioned, as is the case for the CQ3 system shown in Figure 10.9. It is determined analytically as for the 3×PBS/X-cube case, but results in twice the leakage since both panel ports contribute. Assuming low-R_p cubes we would get:

$$\frac{1}{C_{pre-cond}} \approx \frac{1}{20\,000} \qquad (10.23)$$

Combining all terms yields an estimated system contrast of 1400:1. Contrast improvements can be achieved with better cubes and better filters together with color-specific absorption filters in the shared panel ports. Introducing a green rejecting filter adjacent to the red panel can double the theoretical value for $C_{coating-only}$, boosting the theoretical system contrast to 1500:1. Using an o-plate to remove the pre-condition loss can further boost contrast to 1700:1.

10.4 Three-panel LCOS System Comparison

The architectures presented here are all either under development, or already commercialized. As yet there is no standard architecture, as they all offer certain advantages and relative weaknesses. The key metrics are still considered sequential contrast, throughput, and cost, with the last being very dependent on mass manufacturing cost reductions that are difficult to predict. Table 10.1 summarizes key system parameters.

Table 10.1 Theoretical system comparison

System property	3 × PBS/X-cube			Off-axis	CQ3
	MacNeille	WGP	MBC		
Volume (cubes)	9	14	9	12	4
Contrast with 3000:1 panels	1400	3000	2200	1000	1400
Transmission (normalized to greatest)	0.96	0.79	0.98	0.87	1

Further factors include uniformity, ANSI contrast, reliability, and physical stability required to maintain panel registration. It remains to be seen which, if any, system becomes a standard for three-panel LCOS systems.

References

[Arnold S., 2001] S. Arnold, E. Gardner, D. Hansen, and R. Perkins, An improved polarizing beam-splitter LCOS projection display based on wire-grid polarizers, SID' 01 Digest, pp.1282–1285, 2001.

[Bone M., 1999] M. Bone, M. Francis, P. Menard, M. Stefanov, and Y. Ji, Novel optical system design for reflective CMOS technology, Projection Displays V, SPIE 3634, pp.80–86, 1999.

[Bone M., 2000] M. F. Bone and D. G. Koch, De-centered lens group for use in an off-axis projector, US Patent 6,076,931, 2000.

[Bone M., 2001] M. Bone, Method to improve contrast in an off-axis projection engine, SID'01 Digest, pp.1180–1183, 2001.

[Bruzzone C., 2003] C. L. Bruzzone, J. Ma, and D. J . W. Aastuen, High-performance LCoS optical engine using Cartesian polarizer technology, SID'03 Digest, pp.126–129, 2003.

[Bruzzone C., 2004] C. L. Bruzzone, J. J. Schneider, and S. K. Eckhardt, Photostability of polymeric Cartesian polarizing beam splitters, SID'04 Digest, pp.60–63, 2004.

[Heine C., 2001] C. Heine and S. Linz-Dittrich, Color management system for 3-panel LCOS projectors, SID'01 Digest, pp.1189–1192, 2001.

[Imaoka H., 2001] H. Imaoka, S. Moriya, T. Suzuki, F. Koyama, R. Takahashi, Y. Ishizaka, and M. Takada, Projection-type display apparatus having polarized beam splitters and an illuminating device, US Patent 6,174,060, 2001.

[Ishino H., 2004] H. Ishino and T. Inoue, QUALIA-004 full high-definition home theater projector using silicon crystal reflective display (SXRD) technology, IDW'04, pp.1687–1688, 2004.

[Melcher R., 1998] R. L. Melcher, Mototsugu Ohhata, and Kunio Enami, High information content projection display based on reflective LC on silicon light valves, SID'98 Digest, paper 5.1, 1998.

[Pentico C., 2003] C. Pentico, M. Newell, and M. Greenberg, Ultra high contrast color management system for projection displays, SID'03 Digest, pp.130–133, 2003.

[Robinson M., 2000] M. G. Robinson, J. Korah, G. Sharp, and J. Birge, High contrast color splitting architecture using color polarization filters, SID'00 Digest, pp.92–95, 2000.

[Robinson M., 2004] M. G. Robinson, J. Chen, and G. D. Sharp, Optimization of three-panel liquid crystal on silicon video projection systems, IDW'04, pp.1683–1686, 2004.

[Stirling R., 2000] R. D. Sterling and W. P. Bleha, D-ILA™ technology for electronic cinema, SID'00 Digest, pp.310–313, 2000.

[Stupp E. H., 1999] E. H. Stupp and M. S. Brennesholtz, *Projection Displays*, John Wiley & Sons, Ltd, Chicherter, 1999.

[Van Doorselaer G., 2002] G. Van Doorselaer, D. Cuypers, H. De Smet, J. Van den Steen, A. Van Calster, K. Ten, and L.-Y. Tseng, A XGA VAN-LCoS projector, Eurodisplay, p.205, 2002.

11

Single and Dual Panel LC Projection Systems

11.1 Introduction

The advantage of reducing the number of panels in a projection system is primarily in cost reduction. Reducing the panel count has an obvious benefit to the bill of materials, but there is a further cost saving in assembly and net component count. The performance consequences can vary, including loss of throughput and color quality, color break-up artifacts, and contrast. This chapter first introduces single and dual panel systems based on sequential color, which includes those using scrolling illumination. A generic reflective single panel system will be used initially to introduce throughput and chromaticity issues common to sequential-color systems, followed by specific example systems. Dual panel systems are then introduced and their performance compared to the single and three-panel approaches. For the sake of completeness, previously commercialized single panel systems based on spatial color separation will be presented briefly at the end.

11.2 Generic Color Sequential Single Panel Reflective LC System

11.2.1 System Description

A generic approach to implementing single panel LC projection systems is to follow the approach taken by the successful DLP systems [Pettitt G., 2001]. Shown schematically in Figure 11.1, the single panel DLP system modulates light by deflecting it micromechanically

---

Polarization Engineering for LCD Projection M. G. Robinson, J. Chen and G. D. Sharp
© 2005 John Wiley & Sons, Ltd

258   SINGLE AND DUAL PANEL LC PROJECTION SYSTEMS

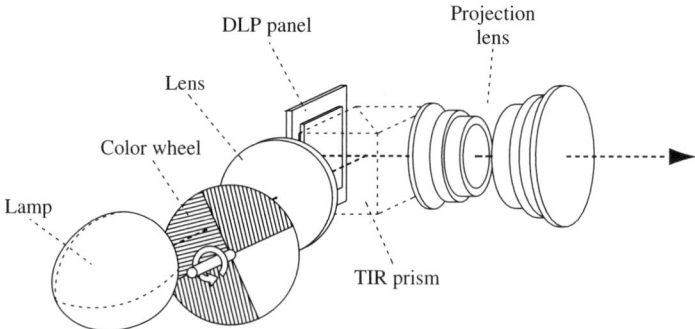

Figure 11.1 Schematic of a single panel DLP system

at each pixel into or out of the capture of the projection lens. The panel is illuminated off-axis using a TIR prism, which separates the angularly distinct input and output beams. Homogenization is carried out with a light-pipe, and the necessary temporal color modulation achieved with a rotating dichroic filter color wheel typically situated at the entrance of the pipe.

In its simplest form, a generic single panel LC system would replace the TIR prism in Figure 11.1 with a PBS, and the DLP with a fast responding LCOS panel [e.g., Jepsen M., 2001; Anderson J., 2002]. In doing this, however, the throughput is significantly reduced over the DLP system. Greater than 50% is lost by simply polarizing the source, and further temporal and efficiency losses are expected with the slower responding, less reflective LCOS technology. Throughput is therefore a major issue regarding the viability of sequential-color LCOS systems.

11.2.2 Single Panel LCOS System Throughput

Consistent with the preceding chapter, the throughput of a single panel system can be estimated, as for the transmissive system of Chapter 9:

1. **Illumination efficiency, η_{ill}.** Since the generic single panel system introduced here has no polarization conversion system (PCS) there is an efficiency loss consistent with Figure 2.20. Assuming a 0.7" panel size at $f_{2.8}$ ($\sim$14 mm^2/sr) this would reduce the source system matching throughput to $\sim$33% from $\sim$50%. Introducing a PCS would clearly be beneficial, but is not straightforward in color-wheel-based systems. The most compatible option is to use a polarization converting light-pipe of the type shown in Figure 2.16. However, low conversion efficiencies favor the more conventional PCS approach shown in Figure 11.2 despite the overhead of extra optical elements. The overall illumination efficiency would restore the 50% matching efficiency at the expense of system size cost and an extra $\sim$2% loss resulting from additional optical elements.

   A conventional PCS system is more compatible with the features of an LC color switch, as shown in Figure 11.3. In this case, however, an additional $\sim$10% is lost due to a pre-polarizer, and a further $\sim$10% from the internal losses of the ColorSwitch™ (CS).

   Further increases in illumination efficiency can be obtained by decreasing the illumination $f_{/\#}$ from 2.8 to 2.5 (14 to 17.5 mm^2/sr). From Figure 2.21, this implies a possible 5% increase in throughput for systems using a PCS.

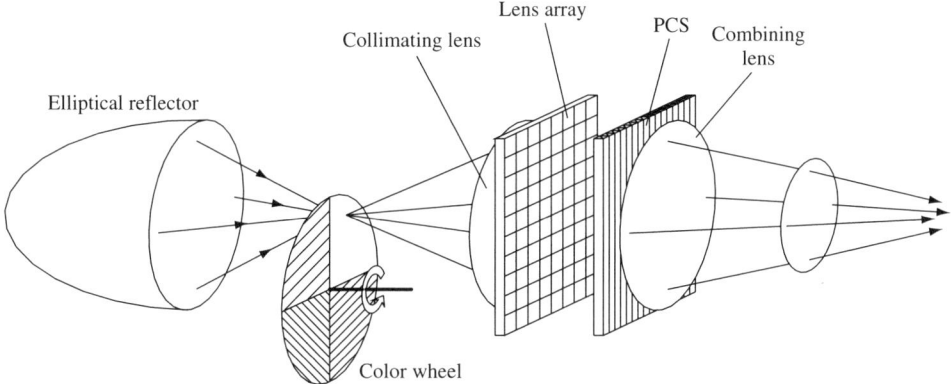

Figure 11.2 Single panel color wheel illumination with PCS

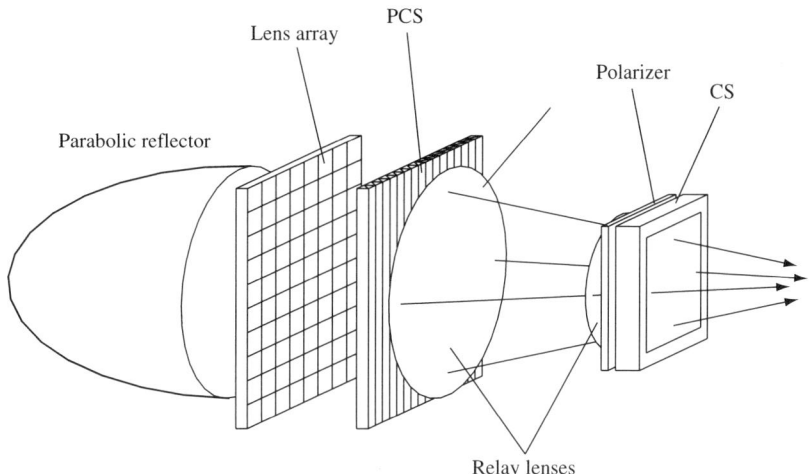

Figure 11.3 Single panel CS illumination with PCS

2. **Color management system efficiency, η_{cm}.** Introducing temporal RGB color separation reduces the available light by 67%, assuming equal duration for each of the color subframes. This factor can be improved, however, with temporal color balancing: that is, increasing the red-starved color field at the expense of the abundant green. To be consistent with the analysis of Chapter 9, temporal color balance will be treated in the color correction term to follow. A further loss of ~7% results from unusable transition times between color fields (see Section 8.3.2.1). On the bright side, there are fewer passive losses (~4%) due to fewer dichroic elements.

3. **Color correction efficiency, η_{cc}.** In the simplest single panel system, sequential-color fields have the same duration and color balance is achieved by attenuating the panel, particularly in the green frame, as shown in Figure 11.4(a). In this case the throughput is one-third of that obtained in a three-panel system with the same projected primary

260 SINGLE AND DUAL PANEL LC PROJECTION SYSTEMS

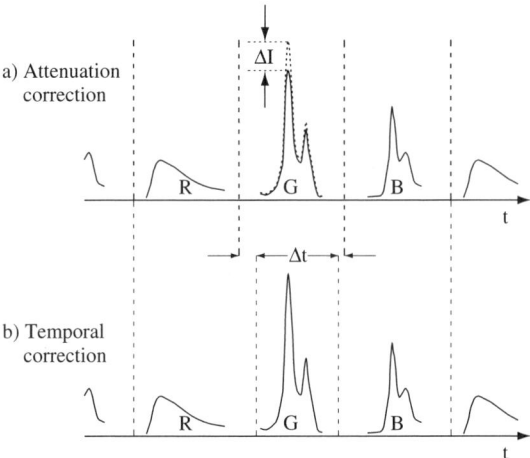

Figure 11.4 Attenuation and tempral white color balance correction

spectra. A more efficient scheme involves varying the period of the color fields, which also improves sequential contrast. This scheme is shown in Figure 11.4(b). Also, there are more optimum green defining filter cut-off values, as the maximum corrected color white output is obtained with the most desaturated green color point allowed. Adhering to the EBU gamut, the optimum filter values are ~480 and ~575 nm, compared to the original 535 and 565 nm, for the cyan and yellow cut respectively. The lumen output, normalized to the unfiltered source, for the two color balancing schemes with optimal color defining filters is shown in Figure 11.5 as a function of white output temperature. At 8000 K, a 36% increase in output from a one-panel system can be obtained by using temporal correction techniques with a desaturated green. Under these optimal conditions, 45% of the three-panel, 8000 K, color correction efficiency can be achieved. A further 30% throughput enhancement could be obtained by introducing a white frame, as discussed in Section 8.3.2.1. However, RPTV systems rely on the sum of the lumen content from each primary equaling that of white to ensure good reproduction of color from standard video signals. White frames are therefore not typical of these systems.

4. **Modulation system efficiency, η_m.** Since the generic system incorporates a MacNeille cube PBS this factor is ~60%, since it is equivalent to that of the 3×PBS/X-cube architecture of Section 10.2.1. In practical systems, the efficient VA mode of the typical 3×PBS/X-cube system is not sufficiently fast as yet to switch between color fields. Most single panel LC modes are therefore typically TN or MTN, which results in a further 5–10% LC mode loss. More significant is the replacement of a cube PBS by a more achromatic, higher contrast, and high angle tolerant (yet lower transmission) wire grid PBS. In this case, the ~60% efficiency of the MacNeille-based system is reduced to the ~48% factor of Section 10.2.6.1.

5. **Imaging system efficiency, η_{im}.** This is expected to be similar to the transmissive system at ~72%.

GENERIC SINGLE PANEL REFLECTIVE LC SYSTEM

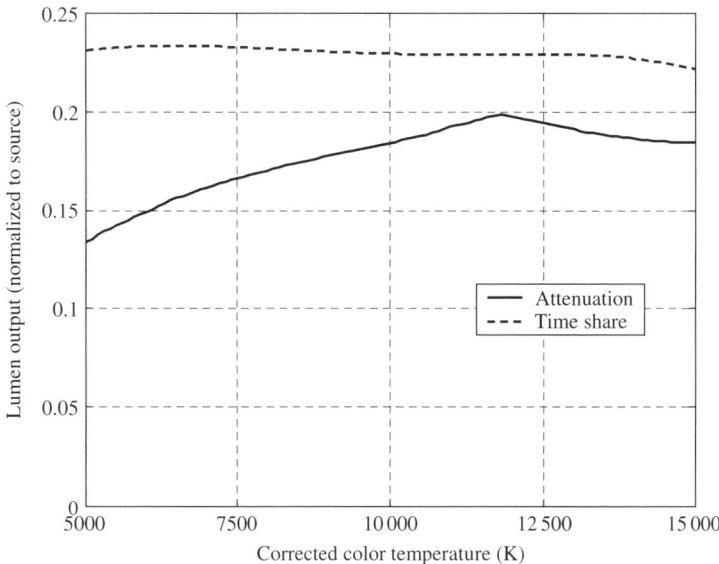

Figure 11.5 Single panel lumen output as a function of corrected white color temperature using channel attenuation and temporal white color balance schemes

Table 11.1 gives a summary of the system throughput estimations for both MacNeille and wire-grid-based systems using a color wheel and with temporal corrected $f_{/2.5}$ illumination.

11.2.3 System Contrast

In a similar manner to the $3 \times$ PBS/X-xube systems of Chapter 10, the contrast of the single panel system introduced here is determined by the modulating system assuming fast TN LCOS panels (e.g., 45° TN). Once again the contrast can be described by equation (10.2):

$$\frac{1}{C_{sys}} \approx \frac{1}{C_{coating-only}} + \frac{1}{C_{pre-cond}} + \frac{1}{C_{AR}} + \frac{1}{C_{compensation}} + \frac{1}{C_s}$$

Here, 0.25% reflectivity is assumed for all AR coatings

Case 1. Wire grid plate (WGP) PBS:

$C_{coating-only} > 100\,000:1$
$C_{pre-cond} > 100\,000:1$
$C_{AR} \sim 5000:1$ (maximum of the reflection curve in Figure 7.42)
$C_{compensation} \sim 3200:1$ (a-plate only—Figure 7.33); $\sim 10\,000:1$
 (c- and a-plates—Figure 7.34)
$C_{scattering} \sim 3000:1$

which yields $C_{sys} \sim 1150:1$ for a-plate compensation only, and $C_{sys} \sim 1500:1$ for a- and c-plate compensation.

Table 11.1 Single panel LCOS system throughput

Subsystem	Efficiency term	Efficiency (generic)	Efficiency (wire grid, color wheel, PCS, temporal color balance)
Illumination	Collection–étendue mismatch	33%	50%
	UV/IR	97%	97%
	Fresnel/absorption losses	98%	96%
	Panel overfill	90%	90%
Subtotal, η_{ill}		**28%**	**42%**
Color management	Dichroic transmission	98%	98%
	Temporal loss (ideal RGB)	33%	33%
Subtotal, η_{cm}		**32%**	**32%**
Color corrections	Primary colors	79%	71%
	8000 K white	70%	97%
Subtotal, η_{cc}		**55%**	**69%**
Modulator	Fresnel reflections	98%	98%
	Polarizer	90%	90%
	LCOS panel	70%	70%
	PBS	96%	80%
Subtotal, η_m		**59%**	**49%**
Imaging	Projection lens reflections	85%	85%
	Fold mirrors	85%	85%
	Screen	N/A	N/A
Subtotal, η_{im}		**72%**	**72%**
System total		**2.1%**	**3.3%**
Lamp lumen output		7200	7200
Incident on screen		**151 lm**	**238 lm**

Case 2. MacNeille PBS:

$C_{coating-only} > 10\,000{:}1$ (see Section 10.2.4)
$C_{pre-cond} > 15\,000{:}1$ (see Section 10.2.4)
$C_{AR} \sim 3300{:}1$ (Figure 7.42 maximum coupled with QWP reflection)
$C_{compensation} \sim 2200{:}1$ (Figure 7.36)
$C_{scattering} \sim 3000{:}1$

which yields $C_{sys} \sim 800{:}1$.

Note that both estimates assume achromatic panel compensation, which can be achieved with a wideband QWP in the MacNeille case, and a conventional low-retardance, in-plane, biaxial compensator in the WGP case.

11.2.4 Temporal System Issues

Peculiar to time sequential color systems are certain specific issues that include temporal color cross-talk, light recirculation, and color break-up.

GENERIC SINGLE PANEL REFLECTIVE LC SYSTEM

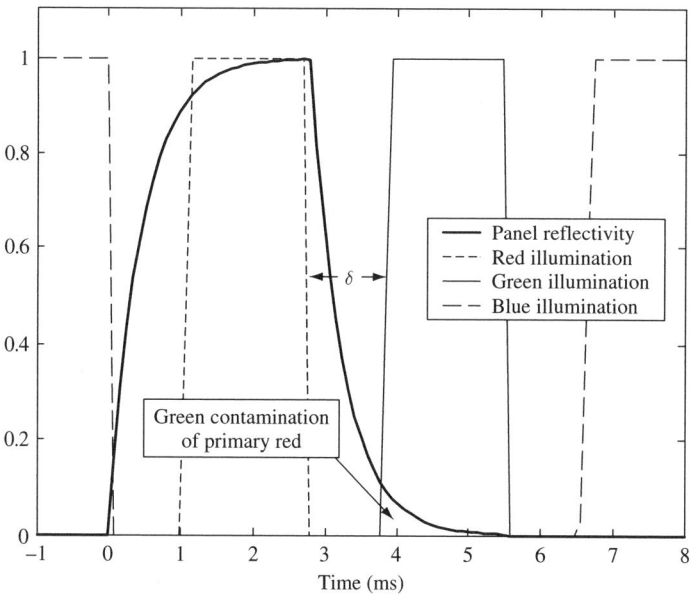

Figure 11.6 Temporal color mixing from finite LC switching time

11.2.4.1 Temporal Color Cross-talk

The assumption so far has been that color fields are modulated with a finite transition time (see Chapter 8) during which the panel is blanked instantaneously. In reality, the typical 0.2 ms blanking period can be comparable to the switching speed of the fastest panels. This implies that the panel is not fully off during the transition between fields, leading to primary color contamination. This is shown diagrammatically in Figure 11.6.

Assuming a typical exponential LC switching characteristic, the color cross-talk can be calculated as a function of final system efficiency assuming similar temporal switching characteristics for all color fields. The results are shown in Figure 11.7 for different LC 10–90% switching periods and assumes ~0.2 ms color wheel transition times. For fast (<0.5 ms) panels, efficiencies can be ~95% ($\delta \sim 0.4$ ms) for an acceptable 1% color cross-talk.

11.2.4.2 Light Recirculation

Systems that use a spiral color wheel can in principle make use of reflected light through recirculation [Hoffman J., 1995; Dewald D., 2001]. The basic concept is shown in Figure 11.8.

Light is focused onto an entrance aperture of a light-pipe and after traversing its length is emitted as a homogenized beam. It is subsequently polarized using a wire grid polarizer. At this point, white light is incident on a spiral color wheel, which reflects light of the rejected color back to the light-pipe. For example, light incident on a red transmitting region of the wheel is imaged onto the panel, but reflects the unwanted cyan. The unwanted complementary color passes back into the rod, together with the oppositely polarized white-light, and becomes homogenized on the return path. The light that is incident on the entrance

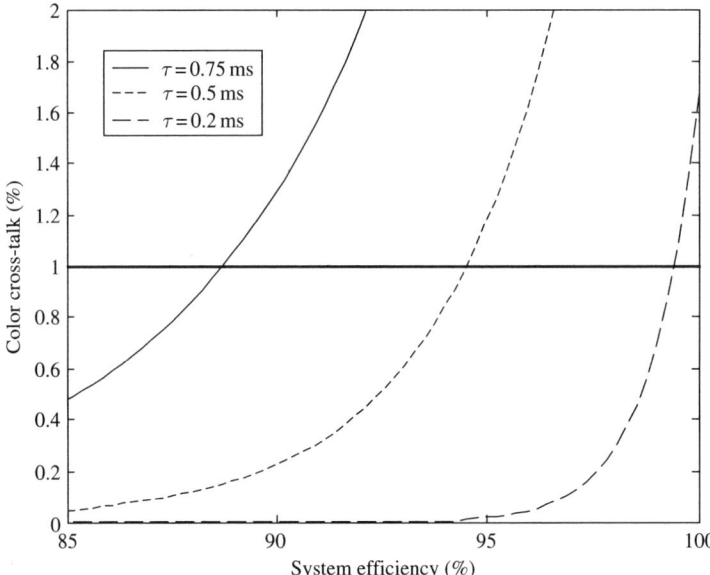

Figure 11.7 Color cross-talk versus system efficiency for different LC 10–90% switching times; 360 Hz color frame frequency is assumed

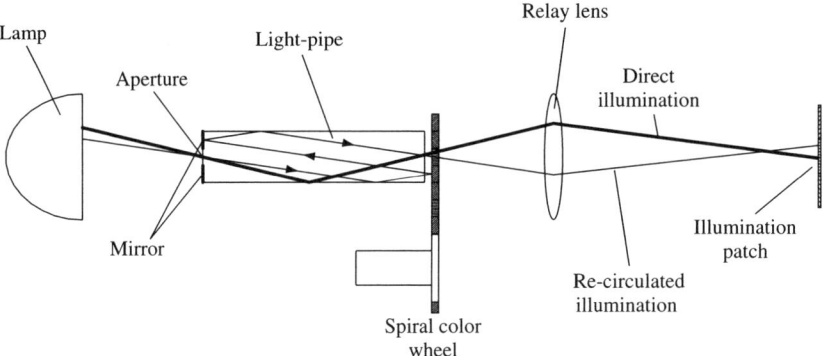

Figure 11.8 Light recirculation light-pipe scheme

aperture escapes and is lost, but the mirror surrounding the aperture reflects a proportion of this light back toward the wheel, where some fraction can contribute to the illumination. Introducing a QWP at the entrance or exit of the pipe allows greater polarization recovery as described in Section 2.5.5(b).

In practice, these schemes are difficult to implement as they can introduce non-uniformities. This is especially so in LCOS systems, where light is back reflected from modulated areas of the panel. In this case the illumination becomes data dependent and incorrect color and brightness representation occurs.

11.2.4.3 Color Break-up

Color break-up is an image artifact associated with all color sequential displays. It is most noticeable when a mixed color object is viewed in isolation on a low-luminance (black) background. When viewed directly, the object appears to be colored correctly and stationary. When the eye moves, however, the object appears to have indistinct but brightly colored edges. More disconcerting are the color flashes that are perceived when the eye is transitioning between positions on the screen, or to regions outside the displayed image. This artifact is a consequence of the image of an object shifting on the retina between the sequential-color frames (see Figure 11.9). Each color frame is therefore seen as a separate primary-colored object resulting in a color flashing sensation. Figure 11.10 is a diagram showing cross-sections of successive images of a white object on a moving retina as a function of time. Each line is a snapshot in time and the perceived color sensation is the integration of each of these images as depicted by the line at the bottom of the figure. In the case shown, the eye perceives a slight coloring of the edges of the object. In the case where either the frame rate of the color modulation is lower or the speed of the retina is greater, the color separation is complete and flashes of all primary colors are seen. The extent to which the flashes are observed is the difference between what is sensed at any one position on the retina and what is expected.

Several human factor studies related to this effect have been carried out, the most quantitative results being given by Post *et al*. [Post D., 1997; 1998]. This experimental-based research studied the specific case of color break-up of a moving object of brightness L (cd/m^2), contrast M (($L_{max} - L_{min}$)/($L_{max} + L_{min}$)), and velocity V (degrees/sec) displayed on a color sequential display of RGB field rate R (Hz). The observer is told to follow the white objects thus controlling eye movement rate. The general conclusion of the research

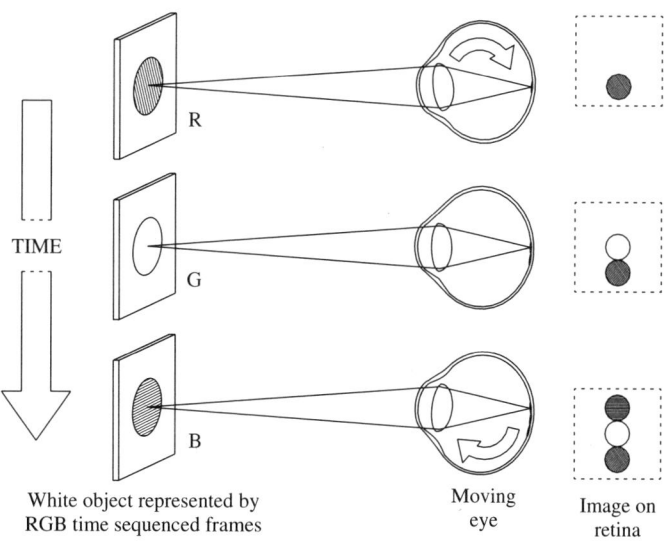

Figure 11.9 Movement of the eye separates a white object into separate colored objects when viewed on a color sequential display

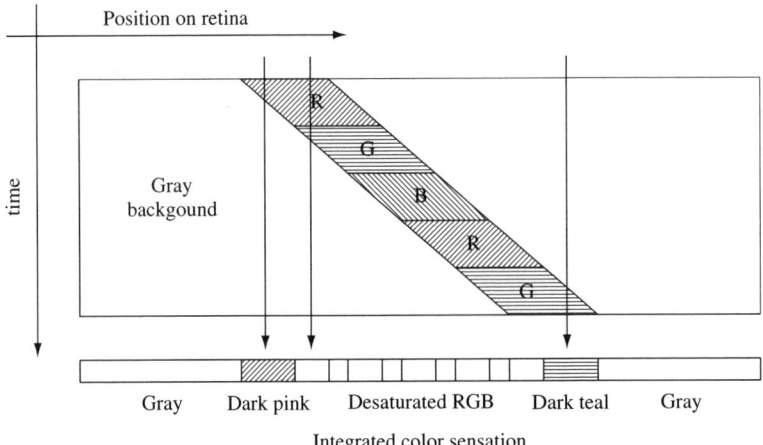

Figure 11.10 The sensation on the retina of separated colored images which integrate to give the sensation of colored edges

is summarized by equation (11.1), which describes maximum field rate R_{min} below which color break-up is observable:

$$R_{min} \approx 36 L^{0.08} M^{1.06} V^{0.69} \quad (11.1)$$

This expression indicates that break-up will be seen for high-brightness objects ($L \sim 600 \text{ cd/m}^2$) on a dark background with system contrasts $\sim$1000:1 ($M \sim 1$) and with typical $\sim$500 Hz field rates when the eye moves at $\sim$20°/sec. Although objects within a scene would not be expected to move with this speed, the eye often saccades at many times this speed, producing flashes outside the direct viewing of the eye. In video scenes where contrast is much lower and mixed colors are expected, little or no break-up is visible to most people.

While this gives a threshold at which color break-up is seen, it is also possible to estimate the obviousness of the color flashing sensation. Taking the scenario depicted in Figure 11.10, the difference between what is expected and what is perceived can be quantified. A worst case is the sensation of a photopically rich green color when viewing a white object on a dark background. Assuming the eye saccades at a rate fast enough to separate the individual colors, the separated green light would be integrated with that of the background to yield the spectral illumination of the retina $Sr(\lambda)$ given by:

$$Sr(\lambda) = \tau \cdot G(\lambda) + (T - \tau) \cdot BG(\lambda) \quad (11.2)$$

where τ is the period of the green illuminating subframe ($= 1000/R$ ms), $T (\gg \tau)$ is the integration time of the eye, $G(\lambda)$ is the primary green illumination spectrum, and $BG(\lambda)$ is the averaged background spectrum.

The perceived color of this spectrum can be calculated in the usual way to give a y coordinate (i.e., the one that is most different from the expected white) of:

$$y \approx \frac{\tau \cdot Y_G + T \cdot Y_{BG}}{T \cdot (X_{BG} + Y_{BG} + Z_{BG}) + \tau \cdot (X_G + Y_G + Z_G)} \quad (11.3)$$

where Y_G is the Y tristimulus value for the green spectrum, and $Y_G = \int G(\lambda) \cdot y(\lambda) \, d\lambda$ etc.

Taking 0.7 as a typical primary green y coordinate, $y = 0.3$ for a gray background, and an eye integration time T of 100 μs, equation (11.3) reduces to:

$$y \approx \frac{C\tau + 100}{300 + 1.4C\tau} \qquad (11.4)$$

where C is the ratio Y_G/Y_{BG}, and $X_{BG} \sim Y_{BG} \sim Z_{BG}$ since a typical uniform gray color is assumed.

This color coordinate differs from the expected white y coordinate (~ 0.3) by Δy, where:

$$|\Delta y| \approx \frac{60\gamma + 1}{30 + 140\gamma} \qquad (11.5)$$

where $\gamma = C/R$.

Setting a limit of $\Delta y \sim 0.1$ as being noticeably green, the frame rate R would have to be $20\times$ the contrast value. Color break-up is always noticeable under extreme conditions with feasible color field rates. However, it can be reduced to an acceptable level by increasing color field frequency and reducing contrast of both the display and its environment. Also for a given field rate, signal processing techniques can help reduce but not remove color break-up, particularly in the context of moving images [Kurita T., 2001; Van Dijk R., 2004]. Current DLP systems run at >500 Hz, which reduces the artifact to a level only objectionable to a small proportion of the population. LC systems struggle to operate at 500 Hz due to lower modulator switching speed, but should deliver ≥360 Hz performance.

11.3 Example Single Panel Color Sequential Systems

11.3.1 Scrolling Color System

This system was commercialized by Philips and is shown schematically in Figure 11.11 [Shimizu J., 1999; 2001]. It uses a split path system with synchronized rotating prisms in each of the RGB paths to produce scrolling horizontal bars on a single line-by-line addressed, 45° TN, 1.3″, 720p LCOS panel. Marketed as the Model #55PL977S, this 55″ RPTV system is specified at 400:1 contrast and projects ~200 lumens.

11.3.2 Field Sequential Single Panel System

Single panel sequential color projectors, shown schematically in Figure 11.12, have been commercialized [www.microdisplay.com]. Operating at a ~500 Hz field rate and with a reported contrast of >400:1, this system delivers between 150 and 200 ANSI lumens, suitable for modest screen sizes ~ 40″–50″. The cost savings of the single panel approach make it competitive with direct-view approaches such as TFT LCD and plasma. The fast field rate is possible with a 45° TN LCOS panel. Initial systems have 1280×720 resolution suitable for displaying 720p video. Future systems will harness the high-resolution capability of LCOS to deliver 1080i compatible resolution.

268 SINGLE AND DUAL PANEL LC PROJECTION SYSTEMS

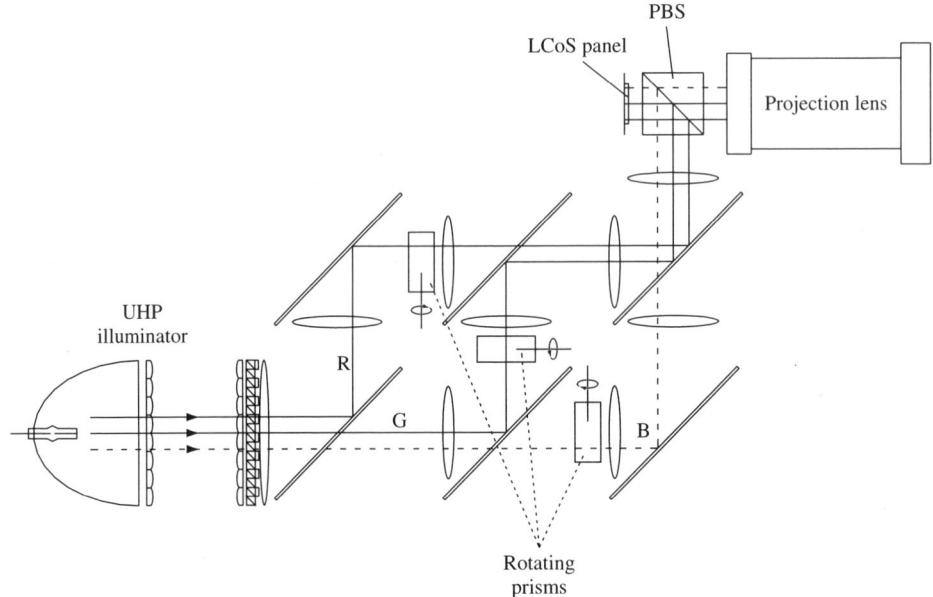

Figure 11.11 Schematic of the scrolling color, single panel system commercialized by Philips

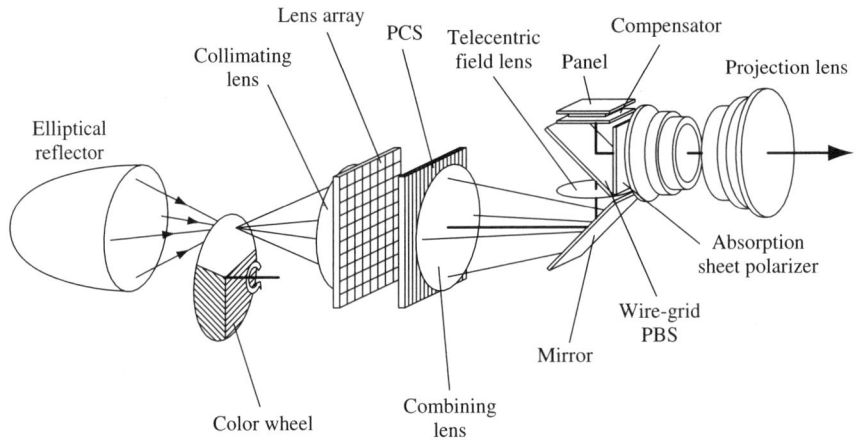

Figure 11.12 Field sequential single panel system

11.4 Two-panel Systems

Two-panel systems use a combination of spatial and temporal color separation. They offer an attractive approach with reduced size and cost relative to three-panel systems, and are capable of delivering high lumen outputs due to the imbalance of the native UHP spectrum. First introduced by Texas Instruments using two DLP panels [Poradish F., 1999], this approach places red continuously on one panel, while blue and green are time sequenced on a second panel. Early demonstrations showed a lumen output close to that of a three-panel system, with significantly reduced color break-up for a given color field rate.

11.4.1 White Color Balance

The lumen output of a two-panel system, relative to single and three-panel systems, is primarily determined by the color correction efficiency, η_{cc}. The relative lumen output of one-, two-, and three-panel systems as a function of corrected white temperature is shown in Figure 11.13. The primary color spectra in the three-panel case are filtered UHP spectra using the filter cut-off values of Table 8.1. In the one- and two-panel cases, time-sharing correction is assumed with a more optimal desaturated green defined by a 480–575 nm pass band filter. Two-panel color correction involves time sharing between blue and green fields.

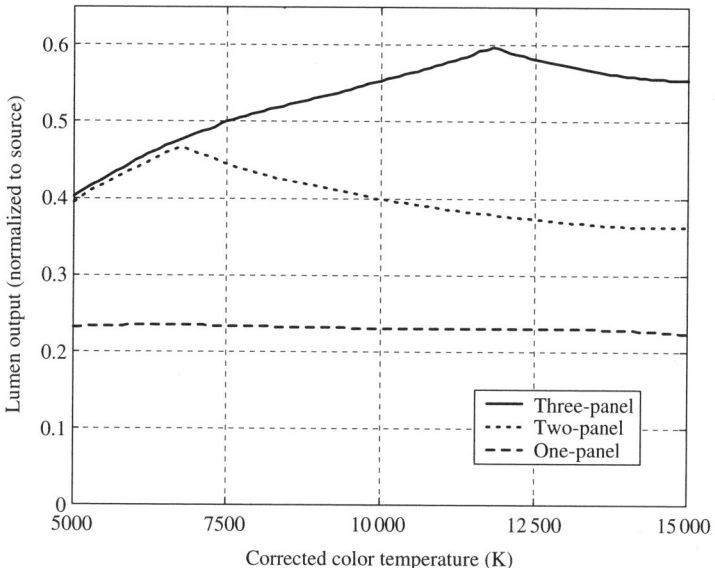

Figure 11.13 Maximum lumen output with color correction alone of one-, two-, and three-panel systems

From Figure 11.13 it is evident that a two-panel system is significantly brighter than a single panel system and can be equal to a three-panel system at low color-corrected temperatures. With a typical 8000 K correction, the two-panel is ~85% the brightness of a three-panel equivalent, and is almost twice that of a single panel system.

11.4.2 Color Break-up

The magnitude of color break-up in two-panel systems can be compared to that in a single panel system operating at similar field rates by considering the difference color sensation

270 SINGLE AND DUAL PANEL LC PROJECTION SYSTEMS

during fast eye saccades. Assuming again the worst-case scenario of a white object on a black background, equation (11.3) can be modified to describe the *y* coordinate of the bright yellow flash observed, since red is continuously on:

$$y \approx \frac{\tau \cdot Y_Y + T \cdot Y_{BG}}{T \cdot (X_{BG} + Y_{BG} + Z_{BG}) + \tau \cdot (X_Y + Y_Y + Z_Y)} \quad (11.6)$$

Taking the same approach of Section 11.2.4.3, the color difference between this yellow ($y \sim 0.5$) and the expected white ($y \sim 0.3$) is:

$$|\Delta y| \approx \frac{40\gamma + 1}{30 + 200\gamma} \quad (11.7)$$

Equation (11.5) implies that for the same level of detection ($\Delta y \sim 0.1$), the field rate R should be 10 × the contrast C, or a factor of two improvement over the single panel case. In other words, for a given contrast display and environment a two-panel system field rate can be half that of a single panel for the same level of break-up.

11.4.3 Two-panel Architectures

To implement a two-panel LC projection system it is necessary to use a temporal color management approach with a method of spatially separating the static red channel from the dynamic green/blue. Methods include the color wheel and LC color switch, and a 2×PBS/dichroic cube or shared PBS approach for the spatial separation and recombination kernel. The 2×PBS/dichroic cube kernel is the two-panel version of the 3×PBS/X-cube architecture of Chapter 10. These options can be interchanged, but here we will present a color wheel, 2×PBS/dichroic cube approach as a first example system, followed by a system using an LC color switch with a shared PBS.

11.4.3.1 Color Wheel, 2 × PBS/Dichroic cube, Two-panel System

A system using this approach has been demonstrated [Pentico C., 2003] using two wire grid PBS in the spatial color separating and combining kernel shown in Figure 11.14.

The complete system using a magenta/yellow color wheel with a conventional PCS is shown in Figure 11.15.

Briefly, the system operates by first temporally separating modulated magenta/yellow illumination into red with either blue or green using an input 45° dichroic beam splitter. The light is essentially **p**-polarized allowing the separated beams to be transmitted through WGP PBSs. The red light is continuously incident on a red modulating LCOS panel with s-polarized reflection analyzed and reflected by the WGP PBS toward the exit cube dichroic. To achieve acceptable contrast, the reflected red light is further analyzed with an absorbing analyzer at the entrance of the dichroic cube (DC). The coating of the DC acts to reflect yellow light, which is often unused in projection systems. In the modulated channel, blue or green light follows a similar optical route via a second WGP PBS, absorbing analyzer, and output yellow trim DC.

The inherent contrast of this kernel is high (>13 000:1 [Pentico C., 2003]) due to the performance of the WGP PBS modulating system, although the lower contrast of the fast-responding panel reduces final performance. Estimated contrast would be similar to the

TWO-PANEL SYSTEMS 271

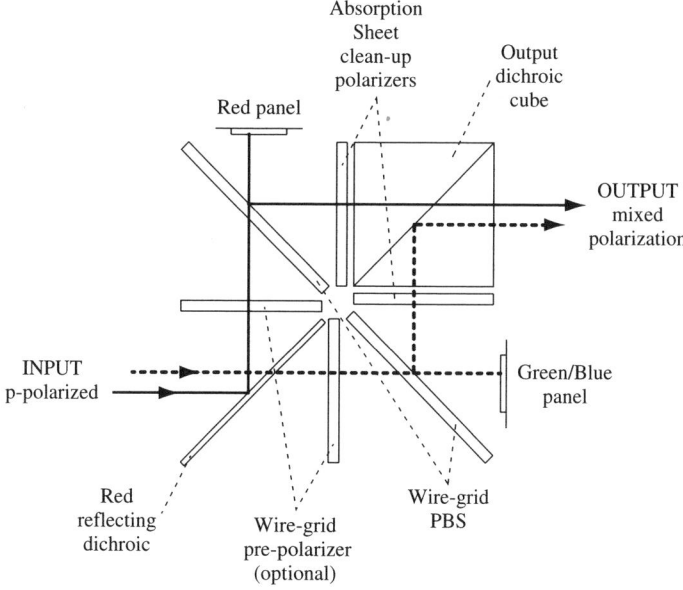

Figure 11.14 Wire grid, 2 × PBS/dichroic cube, two-panel kernel

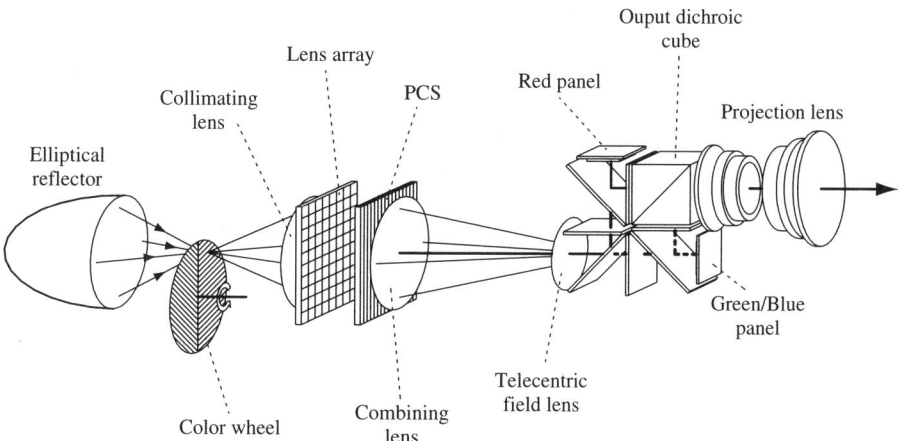

Figure 11.15 Schematic of the color wheel and wire grid kernel, two-panel LC projection system

value of ~1500:1 for the WGP PBS single panel system (Section 11.2.3). Practical systems, however, have yet to achieve >1000:1 due to non-optimum color balance and larger than 0.25% AR reflectivity. The transmission of the kernel is ~69%, excluding panel losses, which makes it effectively twice that of the equivalent single panel system of Table 11.1 from the ~100% increased color correction term. This gives a projected output of ~470 lumens, which is comparable to the HTPS three-panel system. In practice, this estimated output has been achieved with a 150 W source and not the 120 W UHP used in the theoretical estimation.

11.4.3.2 Shared PBS, Two-panel System

Using the shared PBS color management approach outlined in Chapter 8, a single PBS cube can be used to separate and recombine two color spectra [Chen J., 2001]. A two-panel system using this kernel and an LC color switch is shown schematically in Figure 11.16.

In the shared PBS two-panel system, a two-stage LC color switch modulates white light from a polarization-converted UHP lamp into time sequential yellow and magenta **s**-polarized beams. The resulting beam is incident on a red/cyan (*RC*)RSF, which transforms the red light to **p**-polarization. Unwanted yellow light remains **s**-polarized with either the accompanying green or blue light. The oppositely polarized beams can then be compensated with an optional o-plate to correct for geometric depolarization effects prior to separation by the PBS. Red light is incident on a panel, compensated by a red QWP, whereas the green or blue light is reflected toward the sequentially modulated panel. Image beams from each path are reunited to form an output beam mixed in both color and polarization. This is compensated and combined in polarization with an output RSF, which also acts to transform the **p**-polarized unwanted yellow light to **s**. The combined output is then analyzed by an absorption sheet polarizer, which is necessary to achieve high red contrast by avoiding unwanted **p**-reflections. Absorbing the unwanted light, rather than reflecting it back toward the panels, helps with ANSI contrast. A third task for this exit polarizer is to absorb the unwanted yellow light, thus ensuring saturated primary colors.

The intrinsic contrast of the modulating kernel is determined by the performance of the polarization filters together with PBS reflection of **p**-polarized light, and transmission of **s**-polarized light, as in the three-PBS case of Chapter 10. An estimation would place the system contrast at 800:1 since it corresponds directly to the compensated single panel PBS architecture of Section 11.2.3. In practice ∼500:1 has been achieved to date, limited by similar issues to those of practical WGP PBS two-panel systems. The output of this system can be comparable to the wire grid system and delivers ∼450 lumens with a 150 W UHP source.

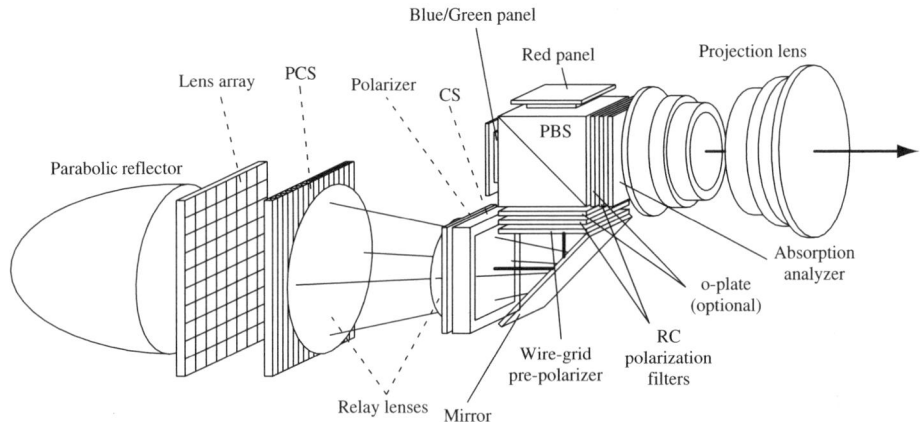

Figure 11.16 Shared PBS, LC color switch (CS) system schematic

11.5 Commercialized Single Panel Projection Systems Based on Spatial Color Separation

Three systems of note that use spatial color separation have been commercialized. These are as follows.

11.5.1 Angular Color Beam Separation with Panel-based Microlens Arrays

Developed by Sharp [Hamada H., 1994; 1996], this system, shown in Figure 11.17, offers high brightness from a single transmissive panel [Ling L., 2001]. It uses the color management scheme described in Section 8.3.2.4. While an innovative and high-performance system when released in the early 1990s, it was found to be sensitive to arc flicker, and additionally required 3× the number of pixels for any given resolution. It was subsequently withdrawn from the market.

11.5.2 Holographic Micro-optic Color Separation

In the late 1990s a 61″ (AV-61S901 1280×720) RPTV system was introduced by JVC that used the holographic color separation described in Section 8.3.2.4. The 1.22″ diagonal panel had an intrinsic resolution of 3840×720 pixels [Sterling R., 2000]. Manufacturing

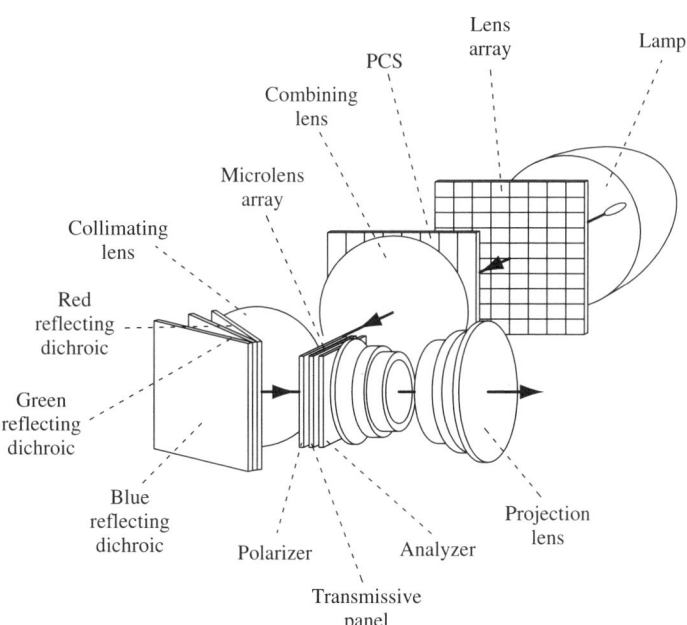

Figure 11.17 Schematic of the angular color beam separation system employing a panel-based microlens array to spatially separate RGB colors

274 SINGLE AND DUAL PANEL LC PROJECTION SYSTEMS

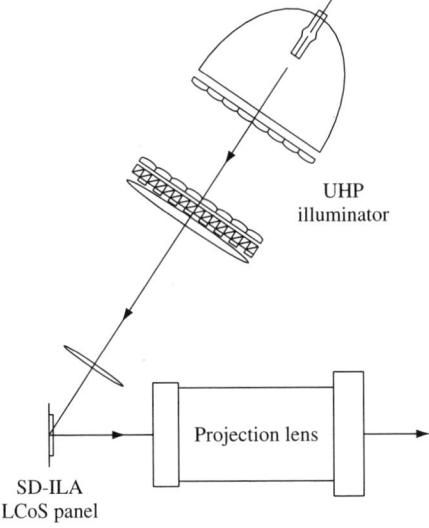

Figure 11.18 Conceptual holographic micro-optic single panel LCOS system

complexity and associated yield issues coupled with the less than adequate red throughput and color saturation ultimately caused the withdrawal of the product. The system, shown conceptually in Figure 11.18, is relatively simple and could yield low-cost projectors, should panel manufacturing and holographic technologies advance in the future.

11.5.3 Flat-panel LCD Projection

Based on the optics of the overhead projector (OHP) this system images a large, high-resolution, direct-view panel onto a projection screen. It is of intrinsically higher cost due to the panel size, though it is attractive in niche markets. It is currently commercialized by Clarity Inc. for large-area control panel applications. The system concept is shown in Figure 11.19.

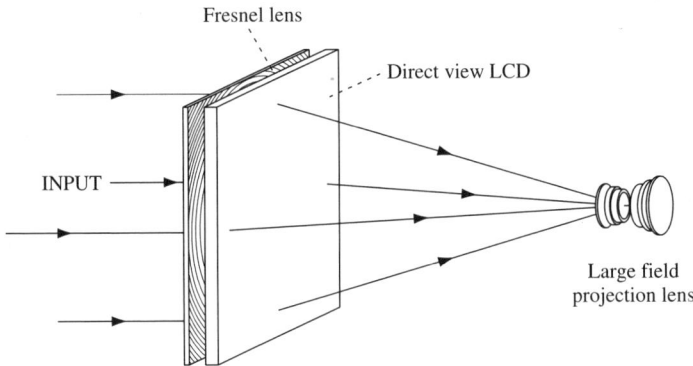

Figure 11.19 Large LCD single panel projection system

References

[Anderson J., 2002] J. E. Anderson, J. Gandhi, and J. Erdmann, Vertically aligned field-sequential microdisplay, SID'02 Digest, pp.958–961, 2002.

[Chen J., 2001] J. Chen, M. G. Robinson, and G. Sharp, Two-panel architecture for reflective LCD projector, SID'01 Digest, pp.1084–1087, 2001.

[Dewald D., 2001] D. S. Dewald, S. M. Penn, and M. Davis, Sequential color recapture and dynamic filtering: a method of scrolling color, SID'01 Digest, pp.1076–1089, 2001.

[Hamada H., 1994] H. Hamada *et al.*, A new bright single panel LC-projection system without a mosaic color filter, IDRC'94, pp.422–423, 1994.

[Hamada H., 1996] H. Hamada, Optical systems for high-luminance LC rear projection, SID'96, pp.911–914, 1996.

[Hoffman J., 1995] J. Hoffman and K. Kanti, High-efficiency, energy-recycling exposure system, US Patent 5,473,408, 1995.

[Jepsen M., 2001] M. L. Jepsen, M. J. Ammer, M. Bolotski, J. J. Drolet, A. Gupta, Y. Lai, D. Huffman, H. Shi, and C. Vieri, 0.9″ SXGA liquid crystal on silicon panel with 450 Hz. field rate, Microdisplay, paper 4.11, p.106, 2001.

[Kurita T., 2001] T. Kurita and T. Kondo, Effect of motion compensation on color breakup reduction and consideration of its use for cinema displays, Asia Display/IDW'01, p.1629, 2001.

[Ling L., 2001] L. C. Ling, Y. I. Yen, and F. C. Ho, An efficient illumination system for single-panel LCD projector, SID'01 Digest, p.1184, 2001.

[Pentico C., 2003] C. Pentico, M. Newell, and M. Greenberg, Ultra high contrast color management system for projection displays, SID'03 Digest, pp.130–133, 2003.

[Pettitt G., 2001] Greg Pettitt and Brad Walker, Cinema™ technology: color management and signal processing, Proceedings of the IS&T/SID Ninth Color Imaging Conference, pp.348–354, 2001.

[Poradish F., 1999] F. Poradish and J. M. Florence, Full-color projection display system using two light modulators, US Patent 5,905,545, 1999.

[Post D., 1997] D. L. Post, P. Monnier, and C. S. Calhoun, Predicting color breakup on field-sequential displays, *Proc. SPIE*, 3058, pp.57–65, 1997.

[Post D., 1998] D. L. Post, A. L. Nagy, P. Monnier, and C. S. Calhoun, Predicting color breakup on field-sequential displays: Part 2, SID'98 Digest, pp.1037–1040, 1998.

[Shimizu J., 1999] J. A. Shimizu, Single panel reflective LCD projector, *Proc. SPIE*, 3634, pp.197–206, 1999.

[Shimizu J., 2001] J. A. Shimizu, Scrolling color, LCOS for HDTV rear projection, SID'01 Digest, pp.1072–1075, 2001.

[Sterling R., 2000] R. D. Sterling and W. P. Bleha, D-ILA™ technology for electronic cinema, SID'00 Digest, pp.310–313, 2000.

[Van Dijk R., 2004] R. Van Dijk and J. A. Shimizu, System and method for motion compensation of image planes in color sequential displays, US Patent 6,831,948, 2004.

Appendix A

Static LC Director Calculation in LC Cell by Relaxation Technique [Khoo I. C. 1991; Li J. 1996]

In order to understand the EO performance of LCDs, the calculation of the LC director profile under an applied electric field is a necessary first step. In some circumstances, 2D and even 3D LC director calculations are needed to fully understand and optimize the EO effect of an LCD system due to the complex structure of a real AMLCD system. Nevertheless, 1D calculations are sufficient to illustrate the EO effect in LCDs in a majority of cases. To model the LC profile, the following assumptions are typically made:

1. The LC is a perfect insulator (i.e., no screening effect from ions).
2. The applied voltage is either DC or low-frequency AC.
3. There is very strong anchoring at the alignment layer boundary.
4. The problem can be considered 1D.

Throughout the following mathematical treatment the coordinates relating to the LC profile adhere to those shown schematically in Figure A.1.

Let V be the voltage across the sample. From Maxwell's equation $\nabla \times E = 0$, we have:

$$\nabla \times E = -i\frac{dE_y(z)}{dz} + j\frac{dE_x(z)}{dz} = 0 \tag{A.1}$$

APPENDIX A

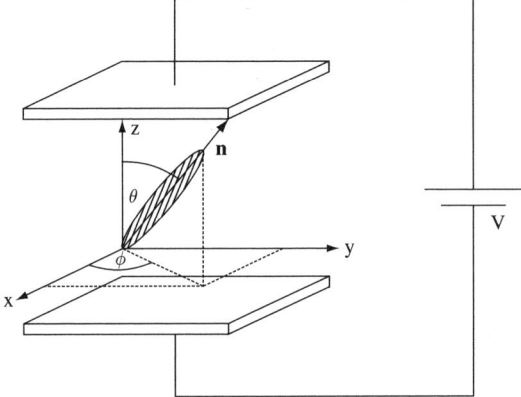

Figure A.1 Coordinate convention used to model LC profiles under the influence of an applied electric field

which directly implies that $E_x = E_y = 0$ and the electric field is therefore along the z-direction. Since there is no bulk charge inside the LC cell, $\nabla \cdot \mathbf{D} = 0$, and $dD_z(z)/dz = 0$. As a result, we get:

$$D_z = \varepsilon_{zz} E_z = \text{constant} \tag{A.2}$$

The dielectric tensor can be expressed as:

$$\varepsilon_{ij} = \varepsilon_\perp \delta_{ij} + (\varepsilon_\parallel - \varepsilon_\perp) n_i n_j \tag{A.3}$$

where:

$$n_x = \sin\theta \cos\phi \qquad n_y = \sin\theta \sin\phi \qquad n_z = \cos\theta$$

From equation (A.2), we get:

$$\int_0^d E_z dz = \int_0^d \frac{D_z}{\varepsilon_\perp + \Delta\varepsilon \cos^2\theta} dz = V$$

$$D_z = V \Big/ \int_0^d \frac{dz}{\varepsilon_\perp + \Delta\varepsilon \cos^2\theta} \tag{A.4}$$

At the equilibrium state, the following Euler–Lagrange equations must be satisfied:

$$F_T = F_B + F_{EM} = F_B - \tfrac{1}{2}\mathbf{D}\cdot\mathbf{E} = F_B - \tfrac{1}{2}D_z E_z = F_B - \tfrac{1}{2}D_z^2/\varepsilon_{zz}$$

$$= \frac{1}{2}\{K_1 (\sin^2\theta)(\theta')^2 + K_2[(\sin^4\theta)(\phi')^2 + 2q_0(\sin^2\theta)\phi' + q_0^2]$$

$$+ K_3[(\cos^2\theta)(\theta')^2 + (\sin^2\theta \cos^2\theta)(\phi')^2]\} - \frac{1}{2}D_z^2/(\varepsilon_\perp + \Delta\varepsilon \cos^2\theta) \tag{A.5}$$

and:

$$\frac{\partial F_T}{\partial \theta} - \frac{d}{dz}\frac{\partial F_T}{\partial \theta'} = 0$$

$$\frac{\partial F_T}{\partial \phi} - \frac{d}{dz}\frac{\partial F_T}{\partial \phi'} = 0 \quad \text{(A.6)}$$

where θ' and ϕ' are $d\theta/dz$ and $d\phi/dz$ respectively.

Here F_B is the Frank elastic free energy of the LC from equation (5.5), and F_{EM} is the Gibbs free energy from the electromagnetic field from equation (5.7). In principle equations (A.4) and (A.6) with the relevant boundary conditions can uniquely determine the 1D LC director profile under the voltage V. In practice, since there is no analytical solution for these equations, it is necessary to use numerical methods such as a relaxation technique. An initial guess for the LC profile is made which would not in general satisfy the equilibrium condition. Dynamic equations are then used to determine the migration of the LC orientation in terms of θ and ϕ until the equilibrium state is reached. The relevant dynamic equations come from balancing the friction torque with the elastic and electric torques. Friction torque is derived from the following nematic LC dissipation function:

$$Dis_{LC} = \frac{1}{2}\gamma \sum_{x,y,z}\left(\frac{dn_i}{dt}\right)^2 = \frac{1}{2}\gamma(\dot{\theta}^2 + \sin^2\theta\,\dot{\phi}^2) \quad \text{(A.7)}$$

where γ is the rotation viscosity of the LC, and $\dot{\theta}$, and $\dot{\phi}$ are $d\theta/dt$ and $d\phi/dt$ respectively. The balanced dynamic equations are then:

$$\frac{\partial Dis_{LC}}{\partial \dot{\theta}} = \gamma \frac{d\theta}{dt} = -\left(\frac{\partial F_T}{\partial \theta} - \frac{d}{dz}\frac{\partial F_T}{\partial \theta'}\right)$$

$$\frac{\partial Dis_{LC}}{\partial \dot{\phi}} = \gamma \sin^2\theta \frac{d\phi}{dt} = -\left(\frac{\partial F_T}{\partial \phi} - \frac{d}{dz}\frac{\partial F_T}{\partial \phi'}\right) \quad \text{(A.8)}$$

In summary the following steps are taken to determine the LC profile:

Step 1: Stratify the LC into a finite number of layers (normally about 40) using alignment boundary conditions to define the initial LC profile through the cell.

Step 2: Using equation (A.4) D_z is calculated for an applied voltage V across the LC director profile obtained in step 1.

Step 3: Equation (A.8) is then used to obtain the derivatives of θ and ϕ.

Step 4: A new LC director is then calculated using a finite time interval and an assumed viscosity.

Step 5: Steps 2 to 4 are then iterated until they converge on stable values of $\theta(z)$ and $\phi(z)$.

Figure A.2 shows one example of an LC director profile calculated for a TN cell under various applied voltages. The typical LC material MLC4431 is assumed aligned with a pretilt angle of 5°. The physical properties of this LC are:

- $n_e = 1.669$, $n_o = 1.505$ (at 589 nm and $T = 20°C$)
- clearing point 100°C

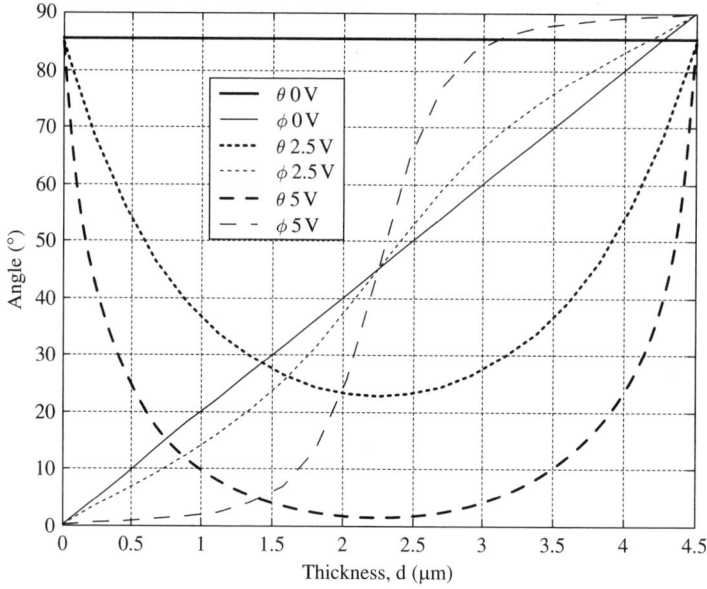

Figure A.2 Example of calculated director profile for a typical 90° TN display with 0, 2.5, and 5 V applied

- dielectric anisotropy $\Delta\varepsilon = 7.1$
- $d/p = 0.033$
- $K_1 = 13.3\,\text{pN}$, $K_2 = 6.6\,\text{pN}$, and $K_3 = 19.4\,\text{pN}$.

References

[Khoo I. C., 1991] I. C. Khoo and F. Simoni, *Physics of Liquid Crystalline Materials*, chapter 9, Gordon and Breach Science Publishers, Philadelphia, 1991.

[Li J., 1996] J. Li, J. E. Anderson, C. D. Hoke, T. Nose, and P. J. Bos, The high-field cured polymer networks in nematic liquid crystals, *Jpn. J. Appl. Phys.*, 35, L1342–1344, 1966.

Index

1-panel system, *see* projection system, single-panel
2-panel system, *see* projection system, two-panel
1 and 2-panel systems, 269
3M reflective PBS, *see* polarizing beam splitter (PBS)
3-panel system, *see* projection system, three-panel shared PBS
3 x PBS/X cube, 96
45° reflective TN mode, 120, 154
63.6° mixed TN (MTN) mode, 121, 154
90° MTN mode, 123
90° TN, 117, 154

A

absolute phase, 50
absorption, 102
absorptive polarizer, 17, 83
achromatic, 78
achromatic circular polarizer, 147
achromatic half-wave retarder, 147
active matrix liquid crystal display (AMLCD), 1
actuated mirror array (AMA), 16
additive primary color, 22
addressing electronics, 10
alignment, *see* LC alignment

alignment tolerance, 174
ambient discrimination, 33
amorphous silicon active matrix LCD (AMLCD), 10
amplitude, 48, 49
analog, 8
anchoring direction, 117
angle encoding, 214–215, 273
ANSI contrast, 28, 30, 182, 186, 244, 272
ANSI lumens, 30
anti-reflection (AR) coating, 17, 61, 88, 223
aperture ratio, 9, 10, 224
a-plate, 78, 160, 178
arc, 40, 42
artifact, *see* color, break-up
astigmatism, 242
auto-iris, 41
AV-61S901, 273

B

back focal length (bfl), 35, 203, 218, 239, 243–244
balanced white point, 17, 201, 222, 260, 261
barrel distortion, 29
bend LC mode, 110, 115
Berreman 4 x 4, 68, 155, 172
biaxial, 62, 78, 81

birefringence, 62, 82, 85, 108
  achromatic, 78
  achromatic circular polarizer, 147
  achromatic half-wave retarder, 147
  a-plate, 78, 160, 178
  biaxial, 62, 78, 81
  calcite, 79
  c-plate, 78, 84, 100, 159, 178, 227
  dielectric tensor, 62
  dispersion, 78
  field of view, 81
  form, 79
  half-wave, 167, 170
  mica, 79
  negative, 78, 80, 84, 159
  network, 137
  NRZ factor, 81
  oblique (o) plate, 78, 169, 171, 174
  off-axis, 66
  o-plate, 169, 178, 230, 232
  positive, 78
  quarter wave, 88, 89, 167
  quartz, 79, 171
  residual, 183
  retarder, 77
  sapphire, 84, 227
  stress induced, 101
  uniaxial, 62, 78, 81, 85
birefringent network, 137
BK7, 101
black body curve, *see* Planckian locus
black matrix, 11
bragg reflection, 114
Brewster's angle, 58, 94
brightness, 22
  see also system efficiency
brightness enhancement film (DBEF), 86

C

c-plate, 78, 84, 100, 159, 178, 227
calcite, 79
canon, 129
cathode-ray tube (CRT), 1, 3–4
cholesteric, 106, 210
cholesteric-LC film, 34
chromaticity, 3
CIE 1931 (x, y), 24
CIE 1976 (u', v'), 24
cinema film, 2
circular polarization, 50, 54, 146

clarity, 274
CMOS, 10
color
  additive primary, 22
  balance, 134
  balanced white, 17, 201, 222, 260, 261
  break-up, 10, 265–267, 269–270
  chromaticity, 3
  CIE 1931 (x, y), 24
  CIE 1976 (u', v'), 24
  coordinates, 24
  display metrics, 22
  distinction, 28
  EBU gamut, 26, 198
  fidelity, 153
  gamut, 6–7, 9
  just noticeable difference (JNDs), 28
  matching functions, 24
  Planckian locus, 27, 28, 222
  primary, 3, 11, 222
  sensitivity, 22
  SMPTE gamut, 26
  temperature, 134
color corner, 247
color enhancement filter, 135
color filter transition values, 199
color management, 37, 197–215
  angle encoding, 214–215, 273
  color balance, 200–201
  color wheel, 209, 210
  ColorSwitch™, 210–213, 258–259
  filter transition values, 199
  primary band determination, 198–201
  RSF based, *see* shared PBS
  scrolling color wheel, 214, 258–259, 263–264
  scrolling prism, 213–214, 267–268
  sensitivity to filter transition values, 200
  shared PBS, 204–208
  spatial color recombination, 203–204
  spatial color separation, 202–203
  temporal, 209–214
  temporal color balance, 259–261, 269
  temporal color mixing (cross-talk), 262–263, 264
color matching functions, 24
color sensitivity, 22
color temperature, 134
color wheel, 10, 209, 210, 214
ColorQuad™, 14, 248
ColorSwitch™, 210–213, 258–259

compensation, 122, 126, 153
  C curve, 178
  compensator, 183
  general LCOS mode, 178
  geometrical PBS compensation, 166
  head-on, 153
  homeotropic LCs, 159, 175, 225, 227
  HTPS, 227–234
  in-plane, 154
  MacNeille PBS/sheet polarizer
      combination, 169
  off-axis, 153, 159
  PBS/LCOS, 174
  PBS/o-pate/mirror, 174
  PBS/QWP/mirror, 174
  self, 97, 160, 178
  sheet polarizers, 164
  VA LCOS, 176
complex number representation of
    polarization, 50
contrast, 9, 12, 16, 153–195
  ANSI, 28, 30, 182, 186, 244, 272
  3 x PBS/X-cube, 240–242, 244–245, 246
  on-axis, 154
  sensitivity, 24
  sequential, 28, 182
  shared PBS, 254–255
  single-panel, 261–262
  system, 194–195
  transmissive HTPS system, 227–234
  two-panel, 271–272
convergent back reflection, 36
converted band, 143
cosine series, 143
CP5600, 129
CQ3 3-panel LCOS system, 248–255
critical angle, 58
cross-talk, 30
crossed cylindrical lens arrays, 34
CRT-based projection system, 3–5

D

DBEF, see brightness enhancement film
    (DBEF)
degree of polarization, 53
depolarization, 153
design of retarder stack filters, 137
dichroic beam splitter, 13, 92, 270–271
  see also dichroic mirror

dichroic filter, 11, 89–93, 203
  angular dependence, 91
dichroic mirror, 92, 202, 215, 220–221
dichroic X-cube, 11, 204, 208, 220, 222
dielectric anisotropy, 108, 109, 280
dielectric properties, 108
dielectric tensor, 62, 278
diffraction efficiency, 5
diffractive light valve, 5
diffuser, 32
diffusive optics, 226
digital, 8
digital micromirror device (DMD), 7, 8
DILA, see light amplifiers
direct-view display, 1
disclination, 6
discotic, 227
dispersion, 78
display
  active matrix liquid crystal (AMLCD), 1
  cathode-ray tube (CRT), 1, 3–4
  direct view, 1
  electronic, 1
  organic light emitting diode, 1
  plasma, 1, 2
  virtual, 1
display metrics, 22
dissipation function, 279
distortion
  barrel, 29
  pin-cushion, 29
  sag, 29
distortion, 5, 29, 35
DLA-SX21SU, DLA-HX1U,
    DLA-HD2K-SYS, 129
DLP, 8, 257–258
double notch RSF, 135
dye polarizer, 83
dynamic aperture stop, see auto-iris
dynamic equations, 279

E

EBU color gamut, 26, 198
eigenvector, 63
elastic constant, 111
elastic energy, 227
  see also Frank free energy
electric displacement D, 47
electrically controlled birefringence
    (ECB), 114, 116

electro-optical effect, 114
electrode, 7
electromagnetic field, 110
electromagnetic wave, 47
electronics display, 1
elliptical polarization, 51
elliptical reflector, 40, 43
ellipticity, 49, 100
étendue, 39, 41–44
e-type polarizer, 85, 86, 165, 166, 192
Euler–lagrange equations, 278
evaporated surface LC alignment, 111
expansion coefficient, 102
extended Jones' matrix, 65
extraordinary (**e**)-mode, 63
eye's sensitivity curve, 23

F

field, 35
field of view (FOV), 81, 124, 159, 226, 227
field strength E, 47
film rotation drum, 1
finite impulse response (FIR), 138
flat panel projection, 274
flicker, 30
f-number ($f_{/\#}$), 35–36
folding mirror, 30, 34, 220, 224
form birefringence, 79
Fourier transform, 137
Frank free energy, 110, 279
frequency transfer function, 138
Fresnel lens, 32
Fresnel reflection, 58, 91, 220, 223
Fresnel reflection coefficients, 5
fringe field, 11, 226
front projection, 1
Fuji Film, 227–230

G

gain, *see* screen gain
geometric plane, 96
Gibbs free energy, 279
glass, 102
    absorption, 102
    expansion coefficient, 102
    photoelastic constant, 102
    Young's modulus, 102
graded dichroics, 203

H

half-wave, 167, 170
handedness, 49
Harris, Amman and Chang (HAC) algorithm, 140
HD-52Z575, HD-61Z575, 129
head-on compensation, 153
high-angle capture, 35
high definition TV, 1
high-temperature polycrystalline silicon (HTPS), 10, 117
Hitachi, 129
hologram, 215, 273–274
homeotropic, 111, 159, 227
homogeneous, 111
homogenization, 37
human visual system, 10, 22
hybrid LC mode, 115

I

image sticking, 112
impulse response, 137
index ellipsoid, 70
Indium Tin Oxide (ITO), 11, 223, 226
in-plane compensation, 154
in-plane switching, *see* IPS
integrated circuit (IC), 7
integrator rod, *see* light-pipe
interference, 131
interference filter, *see* dichroic filter
interferometer, 129, 130
iodine polarizer, 83
IPS, 226
irradiance, 23

J

Jones' matrix formulation, 64
Jones' vector representation, 51
just noticeable difference (JNDs), 28
JVC, 129, 215, 238, 248, 273

L

lateral color, 35
lateral group, 107
LC
    alignment, *see* LC alignment
    bend, 110
    cholesteric, 106

dielectric anisotropy, 108, 109, 280
dielectric properties, 108
disk-like, 105
dissipation function, 279
dynamic equations, 279
elastic constant, 111
elastic energy, 227
    see also Frank free energy
electro-optical effect, 114
field of view (FOV), 124, 159, 226, 227
Frank free energy, 110, 279
Gibbs free energy, 279
image sticking, 112
lateral group, 107
LC polymer (LCP), 78, 80, 172, 233
linking group, 107
lyotropic, 105
magnetic susceptibility, 109
melting point, 108
mesophase, 105
mixture, 108
nematic, 106
odd–even effect, 108
order parameter, 106
orientational order, 105
pitch, 106
positional order, 105
response time, 16, 113, 263
rod-like, 105
smectic, 106, 107
splay, 110
terminal group, 107
thermotropic, 105
twist, 110
viscosity, 108, 113
voltage holding ratio, 112
LC alignment, 111
    anchoring direction, 117
    evaporated surface, 111
    homeotropic, 111, 159, 227
    homogeneous, 111
    polyimide (PI), 12, 111
    pretilt angle, 111
    rubbing, 111
    vertically aligned (VA), 108
LC diffractive light valve, 6
LC polymer (LCP), 78, 80, 172, 233
LCD modes
    45° reflective TN, 120, 154
    63.6° mixed TN (MTN), 121, 154
    90° MTN, 123
    90° TN, 117, 154
    bend, 115
    Bragg reflection, 114
    electrically controlled birefringence
        (ECB), 114, 116
    hybrid, 115
    LC diffractive light valve, 6
    left handed TN, 6
    light scattering, 114
    non-90° TN LC, 178
    normal black, 114
    normally white, 114
    polarization conversion efficiency
        (PCE), 116, 120, 121, 122, 123
    reflective, 6
    right handed TN, 6
    splay, 115
    super twisted nematic (STN), 112
    transmissive, 11, 157
    twisted nematic (TN), 6, 112
    VA 90° TN, 117, 154
    vertical aligned ECB, 116
LCOS, 13, 88, 129, 136, 153–195, 238–255,
    257–272
left handed TN, 6
lens array, 11, 34
light amplifiers, 17
light-pipe, 37, 100, 258
light-pipe polarization conversion system, 40
light scattering, 114
light source, 40–41
    arc, 40, 42
    auto-iris, 41
    elliptical reflector, 40, 43
    parabolic reflector, 40, 43
    ultra-high-pressure (UHP), 38, 40–41, 219
light valve, 1
linear PBS array conversion system, 39, 218, 258
linear polarization, 50
linear polarized polymer, see LPP
linking group, 107
liquid crystal, see LC
liquid crystal on silicon, see LCOS
low-temperature polycrystalline silicon
    (LTPS), 10
LPP, 233
lumen, 12
luminance, 1
lux (lm/m^2), 30

Lyot and Ohmann filter, 132
lyotropic, 105

M

Mach–Zehnder interferometer, 130
MacNeille PBS, 14, 94, 157, 171, 175, 238
magnetic field H, 47
magnetic flux density B, 47
magnetic susceptibility, 109
magnification, 35
Maxwell's field equation, 47, 277
MD, 226
melting point, 108
mesophase, 105
mica, 79
micro-optical elements, 32
microdisplay-based projection systems, 7
microlens array, 12, 34, 215, 224, 273
micromechanical diffractive grating light valves, 16
mirror, 100
misalignment, 175
modeling technique, 17, 72
modulation transfer function (MTF), 35
Moiré fringes, 34
multidomain, see MD
multilayer birefringent cube (MBC) PBS, 97, 245
multilayer birefringent polarizers, 85

N

N-BK7, 102
negative birefringence, 78, 80, 84, 159
nematic, 106
nits (cd/m^2, lm/sr/m^2), 24
non-90° TN LC modes, 178
normally black mode, 114
normally white mode, 114
NRZ factor, 81

O

oblique (o) plate, see o-plate
odd–even effect, 108
off-axis, 4, 13, 14, 115, 158, 192–193, 246
off-axis compensation, 153, 159
off-axis effects
  geometrical PBS, 166
  homeotropic LCs, 159
  MacNeille/sheet polarizer, 169
  paired MacNeille PBS, 166
  PBS/LCOS, 175
  sheet polarizers, 160
off-telecentric, see telecentricity
on-axis contrast, 154
on-axis LCOS system, 115
o-plate, 78, 169, 171, 174, 178, 230, 232
order parameter, 106
ordinary (**o**)-mode, 63
organic light emitting diode display, 1
orientational order, 105
orthogonal polarization states, 52
o-type polarizer, 85, 86, 97, 165, 166, 192, 240
overall system contrast, 194
overhead projector (OHP), 274

P

panel, 1
parabolic reflector, 40, 43
parasitic capacitance, 11
partial polarization, 53
PBH56, 102
PBS, see polarizing beam splitter (PBS)
p-component, 58, 91
perception, 22
perturbation, 153
phase, 48
phase compensator, 226
Philips color prism, 13, 208
photoelastic constant, 102
photometry
  ANSI lumens, 30
  brightness, 22
  eye's sensitivity curve, 23
  irradiance, 23
  lumen, lm, 12
  luminance, 1
  lux (lm/m^2), 30
  nits (cd/m^2, lm/sr/m^2), 24
  radiance, 23
pi cell (π cell), 213
  see also LCD modes; bend LC mode
pile of plates polarizer, 59
pin-cushion distortion, 29
pitch, 106
pixel, 1
Planckian locus, 27, 28, 222
plasma display, 1, 2
Poincaré sphere, 54, 118, 154

polarization
  absolute phase, 50
  amplitude, 48, 49
  azimuthal, 49
  Berreman 4 x 4, 68, 155, 172
  Brewster's angle, 58, 94
  circular, 50, 54, 146
  complex number representation, 50
  critical angle, 58
  degree of, 53
  depolarization, 153
  eigenvector, 63
  ellipse, 49
  elliptical, 51
  ellipticity, 49, 100
  ellipticity angle, 49, 54
  extended Jones matrix, 65
  extraordinary (**e**)-mode, 63
  Fresnel reflection coefficients, 5
  handedness, 49
  index ellipsoid, 70
  Jones' matrix formulation, 64
  Jones' vector representation, 51
  left-handed, 49
  linear, 50
  matrix formulation, 60, 61
  modulating, 1, 5
  modulation, 5
  ordinary (**o**)-mode, 63
  orthogonal polarization states, 52
  partially polarized, 53
  p-component, 58, 91
  phase, 48
  Poincaré sphere, 54, 118, 154
  polarization pupil mapping, 167, 170
  quasi-monochromatic light, 53
  reflection, 57, 182
  refraction, 57
  right-handed, 49
  s-component, 58, 91
  small birefringent approximation, 63
  state of (SOP), 50
  Stokes' parameters, 53
  total internal reflection (TIR), 58, 101
  unmodulating, 5
  unpolarized, 53
polarization conversion efficiency, *see* LCD
  modes
polarization conversion system (PCS), 39, 44
  light-pipe, 40
  linear PBS array, 39, 218, 258

polarization pupil mapping, 167, 170
polarizer, 83–88
  absorptive, 17, 83
  clean-up, 134, 137
  dye, 83
  e-type, 85, 86, 165, 166, 192
  iodine, 83
  multilayer birefringent, 85
  o-type, 85, 86, 97, 165, 166, 192, 240
  post, 157
  pre-, 87, 88, 157
  PVA (polyvinyl alcohol), 80, 83
  reflective, 17, 85
  sheet, 158
  TAC (triacetyl cellulose), 83, 227
  wire grid, 86
polarizing beam splitter (PBS), 13, 94, 114
  3M reflective, *see* multilayer birefringent
    cube (MBC) PBS
  dichroic cube (MacNeille), 14, 94–96, 157,
    171, 175, 238
  geometric plane, 96
  multilayer birefringent cube (MBC) PBS, 14,
    97, 245
  orthogonal coated surface, 188
  parallel coated surface, 188
  p-reflectivity R_p, 95, 99, 157, 194, 241
  p-transmittance T_p, 99, 157, 194
  rotation of polarization, 96
  skew ray, 94, 166, 187
  s-transmittance T_s, 99, 158, 194, 240
  wire grid plate (WGP), 98, 175, 191,
    242, 270
polarizing film, 34
polycarbonate, 80, 129
polycrystalline silicon (PS), 10
  high-temperature (HTPS), 10, 117
  low-temperature (LTPS), 10
  parasitic capacitance, 11
polyimide (PI) LC alignment, 12, 111
polymer, 78, 86
polymer-dispersed LC (PDLC) device, 16
positional order, 105
positive birefringence, 78
pretilt angle, 111
primary color, 3, 11, 222
primary color band determination, 198–201
projection lens, 4, 5, 14, 35–36, 224
  back focal length, 35
  convergent back reflection, 36
  distortion, 35

projection lens (*Continued*)
  efficiency, 224
  $f_{/\#}$, 36
  field, 35
  high-angle capture, 35
  lateral color, 35
  magnification, 35
  resolution, 35
  telecentricity, 35
  throw, 35
projection system
  CRT, 4–5
  DLP, 7–10, 257–258
  MacNeille 3 x PBS/X-cube, 14, 15
  off-axis, 13, 14, 192–193, 246
  PBS/X-cube, 208
  Philips prism, 13, 14, 208
  Schlieren, 5–6
  single-panel, 257–270, 273–274
  three-panel shared PBS, 15, 136–137, 204–208, 247–255
  three-panel wire grid polarizer (WGP), 88
  transmissive HTPS, 10–13, 217–234
  two-panel, 270–272
  Vikuiti, *see* multilayer birefringent cube (MBC) PBS
projection system performance
  artifacts, *see* color, break-up
  brightness, *see* throughput
  color break-up, 10, 265–267, 269–270
  color saturation, 10
  contrast, 3, 153, 194–195, 225–234, 240–242, 244–245, 246, 254–255, 261–262
  $f_{/\#}$, 167
  gray scale, 153
  magnification, 35
  rainbow effect, 10
  resolution, 3, 22, 239
  size, 22, 239
  throughput, 3, 12, 24, 219–225, 240, 254, 255
  trapezoidal (keystone distortion), 5
  uniformity, 22, 24, 102, 239
PVA (polyvinyl alcohol), 80, 83

Q

quarter wave, 88, 89, 167
quarter-wave stack, 89
quartz, 79, 171
quasi-monochromatic light, 53
QWP, 182

R

radiance, 23
rear projection, 1
rear projection screen, 31
rear projection TV, *see* RPTV
recirculation, 263–264
reflection, 57, 182
reflection invariance, 188
reflective mode, 5
reflective polarizer, 17, 85
reflectivity, 58, 85
refraction, 57
refractive index, 85
reliability, 10
resolution, 35
response time, 16, 113, 263
retarder, 77
retarder, *see* birefringence
retarder stack, 14, 17, 129–151
retarder stack filter (RSF), 129
  color enhancement filter, 135
  color splitters/combiners, 136
  converted band, 143
  cosine series, 143
  design, 137
  diagonal, 146
  double notch, 135
  even symmetry, 144
  filter spectrum, 138
  finite impulse response (FIR), 138
  Fourier transform, 137
  frequency transfer function, 138
  Harris, Amman and Chang (HAC) algorithm, 140
  impulse response, 137
  Lyot and Ohmann filter, 132
  off-diagonal, 146
  optical filter, 134
  reflection invariance, 188
  reverse angle symmetry, 144
  reverse order reflection/rotation, 150
  reverse order rotation, 151
  reverse order symmetry, 144, 145
  reverse order with reflection, 148
  rotational invariance, 143, 188
  RSF angular profile mapping, 140
  sine series, 143
  skew ray compensated, 187
  solc filter, 132
  symmetric design, 143

three-panel projection, 247–255
two-panel projection, 272
unconverted band, 143
unitary Jones' matrix, 143
yellow notch filter, 134
retinal irradiance, 22
right handed TN, 6
rod-like, 105
rotating prism, 213
rotational invariance, 143, 188
RPTV, 10, 24, 31, 33
rubbing LC alignment, 111

S

sag, 29
sapphire, 84, 227
schlieren projection system, 5–6
s-component, 58, 91
screen, 31–34
screen gain, 33
SE 5500c, 220–225
Seiko-Epson, 217
self-compensation, 97, 160, 178
sensitivity to filter transition values, 200
sequential, 9, 122
sequential contrast, 28, 182
SF1, 102
SF2, 102
SF57HHT, 102
SF6, 102
shared PBS, 204, 205, 207, 245, 247, 248, 254, 270, 272
sharp, 273
sheet polarizer, 158
signal-dependent non-uniform, 30
single-panel system, *see* projection system
skew ray, 94, 166, 187
smectic, 106, 107
SMPTE color gamut, 26
Snell's law, 58
solc filter, 132
Sony, 10, 217
Sony Grand WEGA, 10
source, *see* ultra-high-pressure (UHP) mercury lamp
spatial color recombination, 203–204
spatial color separation, 202–203
spatial frequency, 24
spectrum, 22
splay, 110, 226

angle, 233
splay LC mode, 115
state of polarization (SOP), 50
static random access memory (SRAM), 7
Stokes' parameters, 53
stress induced birefringence, 101
stretch, 80, 129
substrate, 101
super twisted nematic (STN), 112
surface-stabilized ferroelectric LCs (SSFLCs), 16
SX21, 129, 248
SX50, 129
symmetric RSF design, 143
sync shutter, 1
system efficiency
    3 x PBS/X-cube, 240, 244, 245, 255
    collection, 41–44, 219
    color correction, 222–223, 259–260
    color management, 220, 259
    illumination, 219, 258
    imaging system, 260
    modulation system, 223–224, 260
    panel overfill, 220
    RSF systems, *see* shared PBS
    shared PBS, 249, 254, 255
    single-panel, 258–261, 262
    transmissive HTPS system, 224–225
    two-panel, 269

T

TAC (triacetyl cellulose), 83, 227
telecentricity, 35, 37, 42, 126, 178, 191–192, 203, 246
temporal color balance, 259–261, 269
temporal color mixing (cross-talk), 262–263, 264
terminal group, 107
thermal expansion, 244
thermotropic, 105
three-panel system, *see* projection system
throughput, *see* system efficiency
throw, 35
time multiplexing, 9
total internal reflection (TIR), 58, 101
total internal reflection (TIR) prism, 8, 59, 258
transmissive mode, 5
transmissivity, 58
trim filters, 203
tristimulus curves, *see* color, matching functions
TV cabinet, 34
twist, 110

twisted nematic (TN), *see* LCD modes
two-panel system, *see* projection system

U

ultra-high-pressure (UHP) mercury lamp, 38, 40–41, 219
    lifetime, 219
    output, 219
    spectrum, 41
ultrex, *see* 1 and 2-panel systems; wire grid polarizer
uniaxial, 62, 78, 81, 85
unitary Jones' matrix, 143
unpolarized, 53
UV stability, 108
UV/IR filter, 220

V

VA 90° TN, 117, 154
vertical aligned ECB, 116
vertically aligned LC (VA), 108
virtual display, 1
viscosity, 108, 113
voltage holding ratio, 112

W

white frame, 210, 260
wire grid plate (WGP) PBS, 98, 175, 191, 242
wire grid polarizer, 86
WV compensating film, *see* Fuji Film

X

X-cube, *see* dichroic X-cube

Y

yellow notch filter, 134
Young's modulus, 102